# Recent Advances in Water and Wastewater Treatment with Emphasis in Membrane Treatment Operations

# Recent Advances in Water and Wastewater Treatment with Emphasis in Membrane Treatment Operations

Special Issue Editors

**Anastasios I. Zouboulis**
**Ioannis Katsoyiannis**

MDPI • Basel • Beijing • Wuhan • Barcelona • Belgrade

**MDPI**

*Special Issue Editors*
Anastasios I. Zouboulis
Aristotle University of Thessaloniki
Greece

Ioannis Katsoyiannis
Aristotle University of Thessaloniki
Greece

*Editorial Office*
MDPI
St. Alban-Anlage 66
4052 Basel, Switzerland

This is a reprint of articles from the Special Issue published online in the open access journal *Water* (ISSN 2073-4441) from 2017 to 2019 (available at: https://www.mdpi.com/journal/water/special_issues/membrane_treatment_operations)

For citation purposes, cite each article independently as indicated on the article page online and as indicated below:

LastName, A.A.; LastName, B.B.; LastName, C.C. Article Title. *Journal Name* **Year**, *Article Number*, Page Range.

**ISBN 978-3-03897-624-0 (Pbk)**
**ISBN 978-3-03897-625-7 (PDF)**

# Contents

# About the Special Issue Editors

**Anastasios I. Zouboulis** is (Full) Professor of Chemical & Environmental Technology, Department of Chemistry, Aristotle University, Thessaloniki (GR) since 2007. He is performing research in the fields of water and wastewater treatment technologies, including Environmental Biotechnology, wastewater management, reclamation and reuse, as well as in industrial toxic solid wastes stabilization and/or recovery/reuse options. He is author/co-author of more than 250 papers, published in refereed scientific journals, and of more than 150 papers published in the Proceedings of National & International conferences, attracting more than 12,000 citations, according to Google Scholar (current Scopus H-factor 49, Google Scholar 61). He has supervised 11 Ph.D. (defended), 25 M.Sc. Theses and more than 70 Diploma theses. He has participated in more than 75 national and international research and demonstration projects, funded by E.U. as well as from local sources (ministries, industries, companies etc.), while he was Scientific Responsible in more than 60 of them, relevant to his specific scientific interests. His international reputation was recognised by his election as (foreign) member in the Russian Academy of Sciences (since 2003). He has received several prizes and scholarships from national and international organizations. He is also an active consultant for several local industries in the field of environmental protection, a (previous) national expert for the horizontal activities of FP6 and FP7 (EU), and he is acting often as a reviewer/evaluator for scientific publications, as well as for several EU or national research submitted proposals/projects. He is/was also member of the Editorial Board and/or Associate Editor of several scientific publications/journals, such as Separation Science & Technology, Water Research, Journal of Hazardous Materials, Chemosphere etc., as well as member of the Organizing and Scientific Committees of several national and international conferences (over 40) and he was often invited as plenary speaker. Editor of several special issues (7) in different scientific journals, as well as of Conference Proceedings (7), which have been organized with his direct supervision and co-author of student textbooks and of several Book chapters, published by International Publishing Houses.

**Ioannis Katsoyiannis** is a Greek environmental chemist, currently Assistant Professor at the Department of Chemistry at Aristotle University of Thessaloniki. He is recipient of several prestigious fellowships, such as from the Greek State Scholarship Foundation, from the Alexander von Humboldt foundation, the German Academic Exchange Service (DAAD) and from the Swiss national Science Foundation. He has been also recipient of a Marie Curie Individual fellowship. He has authored or co-authored 60 papers in peer reviewed international journals, which have attracted 3300 citations and has a current h-index of 24. He has participated in several national and international research projects and has been the coordinator in 5 research projects. Before joining the Aristotle University in 2014, he has spent research years abroad firstly, as a post doc (from 2004–2009) at the Technical University of Berlin and then at the water research institute of ETH Zurich. Next, from 2009 to 2014 he worked as supervisor chemical engineer for the giant engineering companies Alstom Power and Hitachi Power Europe, in the commissioning of some of the bigger thermal power plants of Europe, in Italy, Germany, UK and the Netherlands. Since March 2018, he is the chairperson of the Division of Chemistry and Environment of the European Chemical Society and serves as Editor for the journal of Environmental Science and Pollution Research and as a member of the editorial board of the journal Sustainability (Switzerland). Since 2018, he is the member of the selection committee for the historical landmarks of Chemistry, a project organized by EuChemS. He was the chair of the organizing committee of the 2016 Panhellenic Chemistry Conference, which was held in Thessaloniki and attracted more than 1000 participants. He is the conference chair of the upcoming International Conference of Chemistry and the Environment (ICCE), in 2019, in Thessaloniki, which is the official environmental chemistry conference of EuChemS.]

# Preface to "Recent Advances in Water and Wastewater Treatment with Emphasis in Membrane Treatment Operations"

Worldwide, an estimated 800 million people remain without access to an improved source of drinking water and in parallel global water demand for manufacturing is expected to increase by 400% between 2000 and 2050. Although between 2001 and 2010 more than one billion Euros were invested by the European Commission to tackle this problem, it is evident that the Millennium Development Goals targets have not been fully met by 2015. In September 2015, the United Nations adopted the sustainable development goals for 2030 and one of them is to ensure availability and sustainable management of water and sanitation for all. Given the time frame of only 15 years, the achievement of such goals requires immediate action and translation of knowledge into practice. Therefore, the papers in this special issue are dealing with this problematic situation, providing novel solutions which can have a broad impact in the development of novel and energy efficient water treatment technologies.

**Anastasios I. Zouboulis, Ioannis Katsoyiannis**

*Special Issue Editors*

water

MDPI

*Editorial*

# Recent Advances in Water and Wastewater Treatment with Emphasis in Membrane Treatment Operations

**Anastasios I. Zouboulis * and Ioannis A. Katsoyiannis**

Department of Chemistry, Laboratory of Chemical and Environmental Technology,
Aristotle University of Thessaloniki, 54124 Thessaloniki, Greece; katsogia@chem.auth.gr
* Correspondence: zoubouli@chem.auth.gr

Received: 6 December 2018; Accepted: 22 December 2018; Published: 27 December 2018

**Abstract:** The present Special Issue brought together recent research findings from renowned scientists in this field and assembled contributions on advanced technologies that have been applied to the treatment of wastewater and drinking water, with an emphasis on novel membrane treatment technologies. The 12 research contributions highlight various processes and technologies that can achieve the effective treatment and purification of wastewater and drinking water, aiming (occasionally) for water reuse. The published papers can be classified into three major categories. (a) First, there are those that investigate the application of membrane treatment processes, either directly or in hybrid processes. The role of organic matter presence and fouling control is the main aim of the research in some of these studies. (b) Second, there are studies that investigate the application of adsorptive processes for the removal of contaminants from waters, such as arsenic, antimony, or chromate, with the aim of the efficient removal of the toxic contaminants from water or wastewater. (c) Lastly, there are studies that include novel aspects of oxidative treatment such as bubbleless ozonation.

**Keywords:** membranes; adsorption; natural organic matter; arsenic; chromate; antimony; ozonation

---

Worldwide, an estimated 800 million people remain without access to an appropriate source of drinking water, and in parallel, the global water demand for manufacturing is expected to increase by 400% between 2000–2050 [1]). Although between 2001–2010 more than one billion Euros were invested by the European Commission to tackle this problem, it is evident that the targets of the Millennium Development Goals have not been fully met by 2015 [2]. In September 2015, the United Nations adopted the Sustainable Development Goals for 2030, one of which was to ensure the availability and sustainable management of water and sanitation for all. Given the time frame of less than 15 remaining years to accomplish this, the achievement of these important goals requires immediate action and the translation of knowledge into practice. This special issue was intended to bring together recent research findings from renowned scientists in this field, presenting certain recent advanced technologies as applied to the treatment of wastewater and of drinking water, with a specific emphasis given to novel membrane treatment operations. Twelve research contributions have highlighted various processes and technologies that can achieve the effective treatment and purification of wastewater and of drinking water, aiming (occasionally) for water reuse.

In particular, the work of Keucken et al. [3], entitled "Combined Coagulation and Ultrafiltration Process to Counteract Increasing NOM in Brown Surface Water", describes and evaluates a 30-month UF pilot (coagulation-coupled ultrafiltration) trial, treating the surface water of Lake Neden (Sweden) and providing drinking water to 60,000 residents. With an optimal aluminum coagulation dosing $(0.5–0.7\ \text{mg L}^{-1})$, efficient NOM removal was achieved. UV absorbance, the freshness index, and the liquid chromatography-organic carbon detection (LC-OCD) measurements were used to optimize the respective treatment process.

The next study reviewed one of the most pressing problems regarding membrane operation, i.e., the fouling caused by organic matter. The study entitled "Membrane Fouling for Produced Water Treatment: A Review Study from a Process Control Perspective", by Jepsen et al. [4] reviewed and analyzed the fouling detection, removal, prevention, and the dynamical and static modeling approaches, with an emphasis on how the membrane process can be manipulated from a process control perspective to overcome this problem. They showed that the majority of the respective models rely on static descriptions that are limited to a narrow range of operating conditions, which subsequently limits their usability. This work concluded that although the membrane filtration has been successfully applied and matured in several industrial areas, important challenges regarding the cost-effective mitigation of fouling, especially for the offshore de-oiling applications, still exist.

In a relevant study [5] it was examined the "Membrane Fouling Patterns in Biofilm Ceramic Membrane Bioreactor". This study dealt with the determination of the fouling propensity of filtered biomass in a pilot-scale biofilm membrane bioreactor to enable the prediction of fouling intensity. The system was designed to treat domestic wastewater with the application of ceramic microfiltration membranes. Partial least squares regression analysis of the data, which was obtained during the long-term operation of the used biofilm membrane bioreactor (MBR) (BF-MBR) system, demonstrated that the Mixed Liquor Suspended Solids (MLSS), the Diluted Sludge Volume Index (DSVI), the Chemical Oxygen Demand (COD), and their slopes were detected as the most significant parameters for the estimation and prediction of fouling intensity, whereas the normalized permeability and its slope were found to be the most reliable fouling indicators. Three models were derived, depending on the applied operating conditions, which enabled an accurate prediction of the fouling intensities in the system. These results can help prevent severe membrane fouling through the appropriate modification of operating conditions, aiming to prolong the effective operative lifetime of membrane modules, and saving energy and resources for the efficient maintenance of this treatment system.

Further studies regarding membrane fouling were carried out by Sun et al. [6], performing "Quantitative Analysis of Membrane Fouling Mechanisms", which are involved in the "Microfiltration of Humic Acid–Protein Mixtures at Different Solution Conditions". This paper argues that a systematical quantitative understanding of different mechanisms, although of fundamental importance for the better control of fouling, is still unavailable for the microfiltration (MF) of humic acid (HA) and protein mixtures. Based on extended Derjaguin–Landau–Verwey–Overbeek (xDLVO) theory, the major fouling mechanisms, i.e., Lifshitz–van der Waals (LW), electrostatic (EL), and acid–base (AB) interactions were for the first time quantitatively analyzed, considering model HA–bovine serum albumin (BSA) mixtures at different experimental conditions. The obtained results indicated that the pH, ionic strength, and calcium ion concentration of the solution can significantly affect the physicochemical properties and the interaction energy between the poly(ether sulfone) (PES) membrane and the HA–BSA mixtures. The free energy of cohesion of the HA–BSA mixtures was minimized at pH 3.0, ionic strength 100 mM, and $[Ca^{2+}]$ 1.0 mM. The AB interaction energy was a key contributor to the total interaction energy when the separation distance between the membrane surface and the HA–BSA mixtures was less than three nm, while the influence of EL interaction energy was found to be of lower importance, considering the total interaction energy. The attractive interaction energies of membrane–foulant and foulant–foulant were increased at lower pH values and at higher ionic strength and calcium ion concentrations, thus aggravating the membrane fouling problem, which was also supported by the respective experimental results. The obtained findings would provide valuable insights for the quantitative understanding of membrane-fouling mechanisms, which were caused by the treatment of mixed organics during the MF process.

The next study by Arahman et al. [7] investigated the functionalization of membranes to increase efficiency by studying the "Effect of Ca and Mg Ions on the Filtration Profile of Sodium Alginate Solution in a Poly(ether sulfone)-2-(methacryloyloxy) Ethyl Phosphorylchloline Membrane". This work explained the filtration performance of a hollow fiber membrane that was fabricated from poly(ether sulfone)-2-(methacryloyloxy) ethyl phosphorylchloline, while using a sodium alginate (SA) feed solution. The filtration process was applied in a pressure-driven cross-flow module, using a single-piece hollow fiber membrane in a flow of outside–inside operating mode. The effect of the presence of Ca and Mg ions in the SA solution on the relative permeability, membrane resistance, cake resistance, and cake formation on the membrane surface was examined. Furthermore, the performance of membrane filtration was predicted by using mathematical models, which were developed based on Darcy's law. The results show that the presence of Ca ions in the SA solution has the most prominent effect on the formation of a cake layer, which showed a significant effect for lowering the relative permeability.

Further to these studies, Lintzos et al. [8] examined the influence of "Backwash Cleaning Water Temperature on the Membrane Performance in a Pilot SMBR Unit". In this work, different backwash (BW) schemes were applied on identical hollow fiber (HF) membranes in a membrane bioreactor (MBR), treating municipal wastewater. The effect of BW duration (one minute, three minutes, and eight minutes) and of water temperatures (8 °C, 18 °C, 28 °C, and 38 °C) on membrane fouling was investigated. Specifically, the transmembrane pressure (TMP) drop and the membrane permeability increase, due to the BW, were investigated. Furthermore, the time required for the membrane to return to the initial stage, just before each BW experiment, was also examined. It was found that the membranes presented better operating performance as the BW temperature and the backwash duration were increased, which also improved the membrane permeability. By using higher BW water temperatures, more hours were required to return the membranes to the initial condition (just before cleaning), noting also that the examined BW water temperatures did not adversely affect the permeate quality.

The following work by Karanasiou et al. [9] examined the efficiency of a modified membrane process by using "Vacuum Membrane Distillation and Employing Hollow-Fiber Modules". Vacuum membrane distillation (VMD) is an attractive variant of the novel membrane distillation process, which is promising for various separations, including water desalination and bioethanol recovery (through the fermentation of agro-industrial by-products). This publication is part of an effort to develop a capillary membrane module for various applications, as well as a model that would facilitate the process design of VMD. Experiments were conducted in a pilot-scale VMD unit, comprising polypropylene capillary membrane modules. Performance data that was collected at modest temperatures (37–65 °C) with deionized and brackish waters confirmed the improved system productivity with the increase of feed water temperature; simultaneously, excellent salt rejection was obtained. The recovery of ethanol from ethanol–water mixtures—as well as from fermented winery by-products—was also studied, in continuous, semi-continuous, and batch operating modes. At a low feed solution temperature (27–47 °C), ethanol solution was concentrated 4–6.5 times in continuous operation and two to three times in the semi-continuous mode. Taking advantage of the small variation of properties in the module axial-flow direction, a simple VMD process model was developed, satisfactorily describing the obtained experimental data. This VMD model appears to be promising for several practical applications and warrants further research and development (R&D) work.

Apart from the studies dealing with membrane treatment applications for water and wastewaters, in this special issue, some interesting research, examining other processes, such as ozonation, coagulation, and adsorption, were also included. In particular, in the work of Zoumpouli et al. [10], entitled "A Single Tube Contactor for Testing Membrane Ozonation", a membrane ozonation contactor was built to investigate ozonation by using appropriate tubular membranes. Non-porous tubular poly(dimethylsiloxane) (PDMS) modified membranes of 1.0–3.2 mm inner diameter was tested at ozone gas concentrations of 110–200 g/m$^3$ and liquid side velocities of 0.002–0.226 m/s. In this case, the application of modeling could sufficiently predict the final ozone concentrations. Model contaminant degradation experiments (evaluated by UV light absorption measurements) of ozonated water samples were also used to generate information on the reactivity of ozone with different water matrices. The combination of simple membrane contactors with modeling approaches has allowed the prediction of ozonation performance under a variety of experimental conditions, leading to an improvement in the efficiency of bubbleless ozone systems, especially applied for water treatment.

The study titled the "Removal of Antimony Species, Sb(III)/Sb(V), from Water by Using Iron Coagulants", by Mitrakas et al. [11] systematically investigated the removal of the most commonly found antimony species in water, namely, Sb(III) and Sb(V), by the addition of iron-based coagulants. The applied coagulants were Fe(II), Fe(III), and the equimolar mixed Fe(II)/Fe(III) salts, and the experiments were performed using realistic (low) antimony concentrations in the range 10–100 µg/L, by examining artificially polluted tap water solutions. Sb(III) removal by Fe(III) provided better adsorption capacity at a residual concentration that was equal to the drinking water regulation limit (five µg/L), that is, $Q_5 = 4.7$ µg Sb(III)/mg Fe(III) at pH 7, which was much higher than the respective value achieved by the addition of Fe(II) salts, i.e., $Q_5 = 0.45$ µg Sb(III)/mg Fe(II), at the same pH value. Similarly, Sb(V) was removed more efficiently by Fe(III) addition, than by the other examined coagulant agents. However, the Fe(III) uptake capacity for Sb(V) was found to be significantly lower, i.e., $Q_5 = 1.82$ µg Sb(V)/mg Fe(III), than the corresponding value for Sb(III). The obtained results can give a realistic overview of the efficiency of conventionally used iron-based coagulants and their mixture for achieving Sb concentrations below the respective drinking water regulation limit; therefore, they can be subsequently applied for the design of real-scale water treatment units.

The next study, which was by Smoczynski et al. [12], analyzed certain important aspects of coagulation efficiency. The authors studied the sludge particles that were formed during the coagulation of synthetic and municipal wastewaters for increasing the sludge dewatering efficiency. In this work, municipal wastewater sludge was produced by the chemical coagulation of synthetic wastewaters (SWW), which was based on Synthene Scarlet P3GL disperse dye, as well as on real municipal wastewater (MWW), which was coagulated by the addition of commercial coagulants PAX (i.e., preliminarily hydrolyzed aluminum coagulant) and PIX (i.e., a ferric pre-polymerized coagulant based on $Fe_2(SO_4)_3$). It was found that the presence of phosphate ions in the system facilitates the removal efficiency of the examined dye due to the interaction between the dye molecules and $H_2PO_4^-$ ions. These results suggested that flocs composed of spherical {Al(OH)$_3$} units possessed more internal space for water than the respective aggregates, consisting of rod-shaped {Fe(OH)$_3$} units. The obtained results showed that smaller-sized particles are dominating in SWW sludge, whereas larger-sized particles are prevalent in MWW sludge. The parameters studied were the size distribution and the specific surface area of the particles.

The following work by Usman et al. [13] examined the "Efficiency of Small-Sized Powdered Ferric Hydroxide as Arsenic Adsorbent". In this study, batch adsorption experiments were carried out to remove arsenic species from water. The dust ferric hydroxide (DFH) was characterized in terms of zero point charge, zeta potential, surface charge density, particle size, and moisture content. Batch adsorption isotherm experiments indicated that the Freundlich model described the isothermal adsorption behavior of arsenic species notably well. The results indicated that the adsorption capacity of DFH in deionized ultrapure water, when targeting a residual equilibrium concentration of 10 µg/L at the equilibrium pH value of $7.9 \pm 0.1$ and with contact time of 24 h (i.e., $Q_{10}$), was 6.9 µg/mg and 3.5 µg/mg for As(V) and As(III), respectively, whereas the measured adsorption capacity of the conventionally used granular ferric hydroxide (GFH), under similar conditions, was found to be lower, i.e., 2.1 µg/mg and 1.4 µg/mg for As(V) and As(III), respectively. Furthermore, the adsorption of arsenic species onto DFH in the Hamburg tap water matrix, as well as in the National Science Foundation (NSF) challenge water matrix, was found to be significantly lower. The lowest recorded adsorption capacity at the same equilibrium concentration was 3.2 µg As(V)/mg and 1.1 µg As(III)/mg for the NSF water. Batch adsorption kinetics experiments were also conducted to study the impact of different water matrixes on the behavior of removal kinetics for the As(V) and As(III) species by the addition of DFH, and the respective data were best fitted to the second-order kinetic model. The outcomes of this study confirmed that the small-sized iron oxide-based material, being a by-product of the production process of GFH adsorbent, has significant potential to be used for the adsorptive removal of arsenic species from water, especially when this material can be combined with the subsequent application of low-pressure membrane filtration/separation in a hybrid water treatment process, as has been previously demonstrated by the use of Fe(II) and microfiltration, and provided excellent overall results [14].

The last work, which was by Tatoulis et al. [15], investigated the possibilities for the "Simultaneous Treatment of Agro-Industrial and Industrial Wastewaters: Case Studies of Cr(VI)/Second Cheese Whey and Cr(VI)/Winery Effluents". Hexavalent chromium (Cr(VI)) was co-treated either with second cheese whey (SCW) or with winery effluents (WE) by using pilot-scale biological trickling filters in series under different operating conditions. Two pilot-scale filters in series, using plastic support media, were used for each case. The first filter (i.e., Cr-SCW-filter or Cr-WE-filter) aimed at Cr(VI) reduction and the partial removal of dissolved Chemical Oxygen Demand (d-COD) from SCW or WE, and it was inoculated with indigenous microorganisms, originating from industrial sludge. The second filter in the series (i.e., SCW-filter or WE-filter) aimed for the further removal of d-COD, and it was inoculated with indigenous microorganisms that were isolated from the raw SCW or WE wastewaters. Various Cr(VI) concentrations (5–100 mg $L^{-1}$) and SCW or WE wastewaters (having d-COD, 1000–25,000 mg $L^{-1}$) were tested as feed concentrations. Based on the experimental results, the sequencing batch reactor operating mode with a recirculation of 0.5 L $min^{-1}$ proved very efficient, since it led to the complete Cr(VI) reduction in the first filter in series, and achieved overall high Cr(VI) reduction rates (up to 36 mg $L^{-1}d^{-1}$ and 43 mg $L^{-1}d^{-1}$ for the SCW and WW cases, respectively). The percentage of d-COD removal for the SCW and WE wastewaters in the first filter was rather low, ranging 14–42.5% and 4–29% for the cases of the Cr-SCW-filter or Cr-WE-filter, respectively. However, the addition of the second filter in the series enhanced the total d-COD removal to above 97% and 90.5% for the SCW and WE cases, respectively. These results indicated that agro-industrial wastewater could be used as a carbon source for the efficient Cr(VI) reduction, while the use of two trickling filters in the series could effectively treat both industrial and agro-industrial wastewaters with relatively low installation and operational costs. This technology uses indigenous microorganisms, and therefore can be classified as a biological treatment method.

In conclusion, this special issue contains 12 studies with important results, covering several aspects of water and wastewater treatment, by the application mainly of membranes, but also examining the application of other important technologies, such as ozonation, adsorption, and coagulation. Some of these studies refer specifically to the pollution problems of the Mediterranean region, covering important issues, because of the severe problems of water scarcity that this area is particularly facing and is expected to face more severely in the near future.

**Author Contributions:** A.I.Z. and I.A.K. have both conceived the idea and wrote the manuscript.

**Acknowledgments:** We thank all the authors of the papers for the great support to the realization of the present special issue.

**Conflicts of Interest:** The authors declare no conflict of interest.

## References

1. Mekonnen, M.M.; Hoekstra, A.Y. Sustainability: Four billion people facing severe water scarcity. *Sci. Adv.* **2016**, *2*, e1500323. [CrossRef] [PubMed]
2. Hering, J.G.; Maag, S.; Schnoor, J.L. A call for synthesis of water research to achieve the sustainable development goals by 2030. *Environ. Sci. Technol.* **2016**, *50*, 6122–6123. [CrossRef] [PubMed]
3. Keucken, A.; Heinicke, G.; Persson, K.M.; Köhler, S.J. Combined coagulation and ultrafilltration process to counteract increasing NOM in brown surface water. *Water* **2017**, *9*, 697. [CrossRef]
4. Jespen, K.L.; Bram, M.V.; Pedersen, S.; Yang, Z. Membrane fouling for produced water treatment: A review study from a process control perspective. *Water* **2018**, *10*, 847.
5. Kulesha, O.; Maletskyi, Z.; Ratnaweera, H. Multivariate chemometric analysis of membrane fouling patters in biofilm ceramic membrane bioreactor. *Water* **2018**, *10*, 982. [CrossRef]

6. Sun, C.; Zhang, N.; Li, F.; Ke, G.; Song, L.; Liu, X.; Liang, S. Quantitative anallysis of membrane fouling mechanisms involved in microfiltration of humic acid-protein mixtures at different solution conditions. *Water* **2018**, *10*, 1306. [CrossRef]
7. Arahman, N.; Satria, S.; Razi, F.; Bilad, R.M. The effect of Ca and Mg ions on the filtration profile of sodium alginate solution in a polyethersulfone 2-(methacryl-oxy)-ethyl-phosphorychloline membrane. *Water* **2018**, *10*, 1207. [CrossRef]
8. Lintzos, L.; Chatzikonstantinou, K.; Tzamtzis, N.; Malamis, S. Influence of the backwash cleaning water temperature on the membrane performance in a pilot SMBR unit. *Water* **2018**, *10*, 238. [CrossRef]
9. Karanasiou, A.; Kostoglou, M.; Karabelas, A. An experimental and theoretical study on separation by vacuum membrane distillation employing hollow fiber modules. *Water* **2018**, *10*, 947. [CrossRef]
10. Zoumpouli, G.A.; Baker, R.; Taylor, C.M.; Chippendale, M.J.; Smithers, C.; Ho, S.S.X.; Mattia, D.; Chew, Y.M.J.; Wenk, J. A single tube contactor for testing membrane ozonation. *Water* **2018**, *10*, 1416. [CrossRef]
11. Mitrakas, M.; Mantha, Z.; Tzollas, N.; Stylianou, S.; Katsoyiannis, I.; Zouboulis, A. Removal of antimony species, Sb(III)/Sb(V) from water by using iron coagulants. *Water* **2018**, *10*, 1328.
12. Smoczynski, L.; Kalinowski, S.; Cretescu, I.; Smoczynski, M.; Ratnaweera, H.; Trifescu, M.; Kosobucka, M. Study of sludge particles formed during coagulation of synthetic and municipal wastewater for increasing the sludge dewatering efficiency. *Water* **2018**, in press.
13. Usman, M.; Katsoyiannis, I.; Mitrakas, M.; Zouboulis, A.; Ernst, M. Performance evaluation of small sized powdered ferric hydroxide as arsenic adsorbent. *Water* **2018**, *10*, 957. [CrossRef]
14. Katsoyiannis, I.A.; Zouboulis, A.I.; Mitrakas, M.; Althoff, H.-W.; Bartel, H. A hybrid system incorporating a pipe reactor and microfiltration for biological iron, manganese and arsenic removal from anaerobic groundwater. *Fresenius Environ. Bull.* **2013**, *22*, 3848–3853.
15. Tatoulis, T.I.; Michailides, M.K.; Tekerlekopoulou, A.G.; Akratos, Ch.S.; Pavlou, S.; Vayenas, D.V. Simultaneous Treatment of Agro-Industrial and Industrial Wastewaters: Case Studies of Cr(VI)/Second Cheese Whey and Cr(VI)/Winery Effluents. *Water* **2018**, *10*, 382. [CrossRef]

*water*

MDPI

*Article*

# Combined Coagulation and Ultrafiltration Process to Counteract Increasing NOM in Brown Surface Water

Alexander Keucken [1,2,*], Gerald Heinicke [3], Kenneth M. Persson [2,4] and Stephan J. Köhler [5]

1    Vatten & Miljö i Väst AB (VIVAB), 311 22 Falkenberg, Sweden
2    Water Resources Engineering, Faculty of Engineering, Lund Technical University, 221 00 Lund, Sweden;
     Kenneth.Persson@sydvatten.se
3    DHI, Agern Allé 5, 2970 Hørsholm, Denmark; ghe@dhigroup.com
4    Sweden Water Research AB, Ideon Science Park, 223 60 Lund, Sweden
5    Department of Aquatic Sciences and Assessment, Swedish University of Agriculture Sciences,
     750 07 Uppsala, Sweden; Stephan.Kohler@slu.se
*    Correspondence: Alexander.Keucken@vivab.info; Tel.: +46-70-598-99-62

Received: 14 August 2017; Accepted: 7 September 2017; Published: 13 September 2017

**Abstract:** Membrane hybrid processes—coagulation coupled with ultrafiltration (UF)—have become a common method to comply with the legal, chemical, and microbiological requirements for drinking water. The main advantages of integrating coagulation with membrane filtration are the enhanced removal of natural organic matter (NOM) and reduced membrane fouling. With in-line coagulation, coagulants are patched into the feed stream directly prior to the membrane process, without removing the coagulated solids. Compared with conventional coagulation/sedimentation, in-line coagulation/membrane reduces the treatment time and footprint. Coagulant dosing could be challenging in raw water of varying quality; however, with relatively stable specific ultraviolet absorbance (SUVA), dosing can be controlled. Recent studies indicate that *UV* absorbance correlates well with humic substances (HS), the major fraction to be removed during coagulation. This paper describes and evaluates a 30-month UF pilot trial on the surface water of Lake Neden (Sweden), providing drinking water to 60,000 residents. In this study, automatic coagulant dosing based on online measurement was successfully applied. Online sensor data were used to identify the current optimal aluminium coagulation conditions (0.5–0.7 mg $L^{-1}$) and the potential boundaries (0.9–1.2 mg $L^{-1}$) for efficient future (2040) NOM removal. The potential increase in NOM could affect the Al dose and drinking water quality significantly within 20 years, should the current trends in dissolved organic carbon (DOC) prevail. *UV* absorbance, the freshness index, and liquid chromatography-organic carbon detection (LC-OCD) measurements were used to optimise the process. Careful cross-calibration of raw and filtered samples is recommended when using online sensor data for process optimisation, even in low-turbidity water (formazin nephelometric unit (FNU) < 5).

**Keywords:** ultrafiltration; hollow fibre; natural organic matter (NOM); coagulation; optical sensors

## 1. Introduction

In the late 1980s, an increase of natural organic matter (NOM) concentration was first reported in Swedish surface waters as a link between increased amount of humic substances (HS) and the darkening of Swedish lakes [1]. Over the last few decades, several other reports have confirmed that the occurrence of NOM in water (browning of surface waters) was a worldwide phenomenon [2–4]. Changes in the climate (temperature, quality, and amount of precipitation) [5] and the decline in acid deposition are reasonable explanations for the increasing NOM concentrations [6]. NOM is a complex mixture of organic compounds present in all fresh water, particularly surface water [7].

The presence of NOM could have severe effects on drinking water quality and its treatment processes. These problems include (i) negative effects on water quality relevant to colour, taste, and odour; (ii) increased disinfectant dose requirements, which in turn result in potential harmful disinfection by-product (DBP) production [8]; (iii) promoted biological growth in the distribution system; and (iv) increased levels of complex heavy metals and adsorbed organic pollutants [9].

Among the available technologies to remove NOM, the most common and economically feasible method is coagulation and flocculation, followed by sedimentation/flotation and filtration. Other treatment options for NOM removal include the magnetic ion exchange resin (MIEX®) technique, activated carbon filtration, advanced oxidation processes, and membrane filtration [10–15].

Early studies of filtration processes showed that membranes were effective in removing dissolved organic matter (DOM)—including precursors of trihalomethane (THM)—from surface- and groundwater sources [16]. Over the last decade, the combination of membrane processes with other unit processes has become a common way to achieve the removal of NOM and function as a barrier against microorganisms.

Recently, several studies have focused on evaluating NOM removal by capillary nanofiltration (NF) in Swedish surface water sources. These studies indicate that a process combining coagulation and NF could remove more than 90% of the dissolved organic carbon (DOC), and 96% of the *UV* absorbance at 254 nm from lake water [17]. Using direct NF resulted in 93% removal of *UV*-absorbance ($UV_{abs}$), and 88% total organic carbon (TOC) [18].

Membrane processes need pre-treatment for enhanced NOM removal and decreased membrane fouling. Hybrid processes may therefore be superior to the individual processes. The integration of coagulation with membrane filtration has two main advantages: enhanced removal of NOM molecules and reduction of membrane fouling. The most recent mode of combining coagulant with microfiltration (MF) or ultrafiltration (UF) is to add coagulant into the feed stream immediately prior to the membrane process, without removal of the coagulated solids (in-line coagulation). The advantages of in-line coagulation are the reduced footprint and lower coagulant dose, as settleable flocs are not needed [19]. Coagulant selection and dosing can be optimised specifically for NOM removal, as particle removal is assured by the membrane [20]. Careful dosing is required to produce large enough flocs to avoid pore blocking, while avoiding increased fouling [21].

Coagulant dosing can be challenging in raw water, with large variations in water quality. In raw water sources with relatively stable specific ultraviolet absorbance (SUVA), dosing can be controlled with the *UV* signal in the incoming raw water. *UV* has been shown to correlate well with the presence of HS [17], which compose the major fraction removed during coagulation. In this instance, the use of optical sensors for dose control could be a viable method of process control. In this study, a combination of sensor-based in-line coagulation was investigated for the removal of NOM by UF. A 30-month pilot test was carried out on surface water from Lake Neden, a drinking water source for more than 60,000 residents on the west coast of Sweden.

This study aims to evaluate a combined coagulation/UF process, with respect to

- Its sensitivity and the limits of the coagulant dose for NOM removal
- The use of optical sensors for online dosing of coagulants
- Its vulnerability to a further decrease in raw water quality.

The merit of this research is to advance our operational understanding of the effect of membrane hybrid processes, combining UF and in-line coagulation for efficient NOM removal. Particular focus was on dosing control by optical sensors.

## 2. Material and Methods

### 2.1. Raw Water Source Quality

The raw water source used in this study was a mixture (20%/80%) of water from a nearby alkaline groundwater well (pH 8, TOC = 0.6 mg $L^{-1}$, $\sigma$ = 380 $\mu S$ $cm^{-1}$) and a slightly acidic clear-water lake (pH = 6.7, TOC = 3.4 mg $L^{-1}$, and $\sigma$ = 60 $\mu S$ $cm^{-1}$). The surface water source was an oligotrophic lake, surrounded by mixed woodland. The average feed water quality of the various pilot trials is described in Table A1. Lake Neden was heavily contaminated by acid rain during the 1980s and 1990s, and was subsequently treated with lime, as were most of the lakes in that area of southwestern Sweden. As a result, the organic matter concentration was suppressed temporarily but is currently recovering to its natural values, similar to many other lakes in the area [22]. In this area, it was observed that the water colour more than doubled during 2007–2012 [23]. The time series of colour (Abs_420) and TOC indicated an increase in colour and carbon content during 1995–2010 in several lakes in the area (Figures A1 and A2). In addition, it is assumed that prolonged vegetation periods will cause higher concentrations of organic matter in the future [24]. Compared with the other lakes in the area, the water of Lake Neden is clear, low in TOC (3.4 $\pm$ 0.4 mg $L^{-1}$), and has a comparatively low SUVA value (3.2 $\pm$ 0.4). The removal of organic matter by flocculation is limited by the amount of HS in the water. This characteristic can be determined by using either the SUVA value or by using more advanced DOC characterisation techniques [7]. In all of our experiments, conventional NOM analyses (TOC, DOC, and *UV* absorbance) and fluorescence excitation emission matrices (EEMs) [7] were combined with liquid chromatography-organic carbon detection (LC-OCD) analysis for feed water, concentrate, and permeate. This was done to elucidate the retention of specific NOM fractions as a function of varying operating conditions.

Consistent with the continuing browning of lakes and rivers in large parts of Scandinavia, a rising trend in colour and chemical oxygen demand (COD) has been observed in the surface water abstracted by the Kvarnagården water treatment plant (WTP). No significant reduction in HS was achieved with the old full-scale treatment process, consisting of rapid sand filtration, pH-adjustment, and *UV* irradiation.

### 2.2. UF Full-Scale Design and Pilot Studies

#### 2.2.1. Retrofit of Full-Scale Plant and Pilot Trials

Preliminary pilot trials with one-stage UF (UF-HF-P1) and hollow-fibre NF (NF-HF-P) were performed from June 2010 to May 2012 [25,26]. More extensive field testing with a two-stage UF pilot plant has been carried out since January 2015. An overview of the pilot studies is given in Table 1.

**Table 1.** Summary of different pilot studies at Kvarnagården water treatment plant (WTP). HF: hollow fibre; NF: nanofiltration; PES: polyethersulfone; UF: ultrafiltration.

| Pilot Plant Type (Module Type) | Code | Scale | Start | End | Membrane Type |
|---|---|---|---|---|---|
| UF HF one-stage (KOCH, HF 10-48 35) | UF-HF-P1 | Pilot | 1 June 2010 | 15 August 2011 | PES |
| UF-HF two-stage (Pentair, XIGA/AquaFlex) | UF-HF-P2 | Pilot | 1 January 2015 | Running | PES |
| NF (Pentair, HFW 1000) | NF-HF-P | Pilot | 2 November 2011 | 4 May 2012 | PES |
| UF-HF two-stage HF (Pentair, XIGA/AquaFlex) | UF-HF-F | Full | 15 February 2017 | Running | PES |

In November 2016, the WTP was upgraded with a UF facility (capacity of 1080 $m^3$ $h^{-1}$ net permeate flow rate). In brief, the full-scale plant consists of a two-stage UF membrane filtration process, with in-line coagulation of a primary UF membrane stage that provides NOM retention and a barrier function against microorganisms (Figure 1). Because of coagulant residues in the backwash water and the limited sewer capacity of the site, a second-stage UF membrane system was installed to increase the recovery of the plant to >99%.

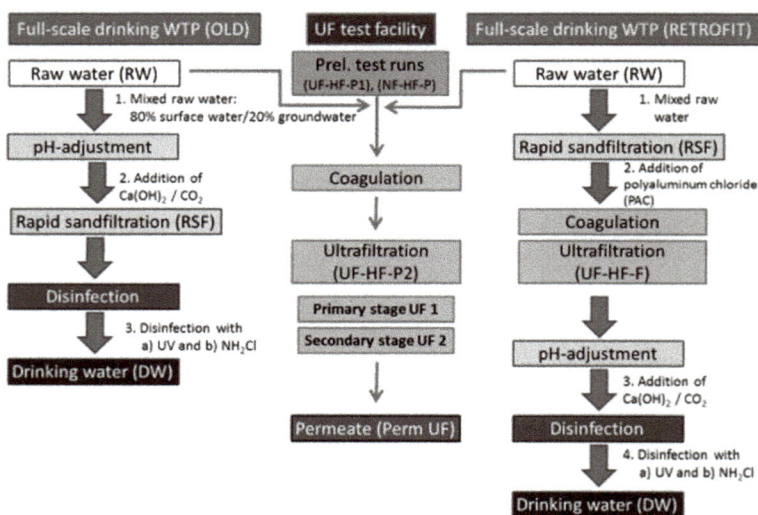

**Figure 1.** Treatment train for full-scale process and test facility at Kvarnagården WTP.

### 2.2.2. Control Philosophy of Pin-Floc Coagulation for Full-Scale UF Plant

For the full-scale plant, the coagulation philosophy was to create "pin flocs" (i.e., flocs of limited size) for the operation of the UF hybrid process. On the one hand, these pin flocs are of sufficient size to be retained by the UF membranes and to create a relatively open cake structure on the membrane surface. On the other hand, the floc size required could be limited because the removal of solids is determined by the size difference between the flocs and the membrane pores, and does not depend on gravitational separation. The critical parameters for optimum pin-floc coagulation are proper distribution of the coagulant into the UF feed stream, a sufficiently high mixing energy at the coagulant dosing point during a minimum contact time, and an optimum pH depending on the coagulant selected. The specific coagulation dosing conditions for the UF full-scale plant in the present study are shown in Figure 2. A similar but more simplified set-up was applied for the UF pilot studies.

Based on extended UF-trials [25] (year 2011), the effective coagulant concentration required to improve UF operation was found to be in the range 0.4–1.5 mg Al L$^{-1}$. As the efficiency of pin-floc coagulation strongly depends on the absolute number of collisions at the dosing point, the proper distribution of these relatively low coagulant concentrations and UF feed water is vital. For this reason, a maximum dilution factor of 500 was applied in the present study. This means that direct dosing of the coagulant stock concentration into the UF feed stream was avoided. Instead, a small separate dispersion pump was installed to create a carrier water supply, which obtained water from the UF feed line. The coagulant stock concentration was dosed into the suction line of the dispersion pump. For operational flexibility, the coagulant dosing pump was frequency controlled to enable flow-ratio dosing control based on the actual UF feed flow.

At the coagulant dosing point, the mixing energy should be sufficiently high to maximize the absolute number of collisions and consequent pin flocs. This was achieved by installing a static in-line mixer, providing a plug-flow contacting environment. A residence time of 10 s between the coagulant dosing point and the first UF unit was sufficient to allow for a limited flocculation period.

If pH correction of the UF feed stream was required, a frequency-controlled dosing pump for acid/caustic chemicals would be necessary to achieve stable pH control, based on pH and flow fluctuations of the UF feed stream. Since coagulant dosing influenced the pH of the UF feed, pH was measured downstream of the static mixer. For proper mixing of the acid/caustic chemicals into the

UF feed stream, a low-pressure drop static in-line mixer was installed upstream of the coagulant dosing point.

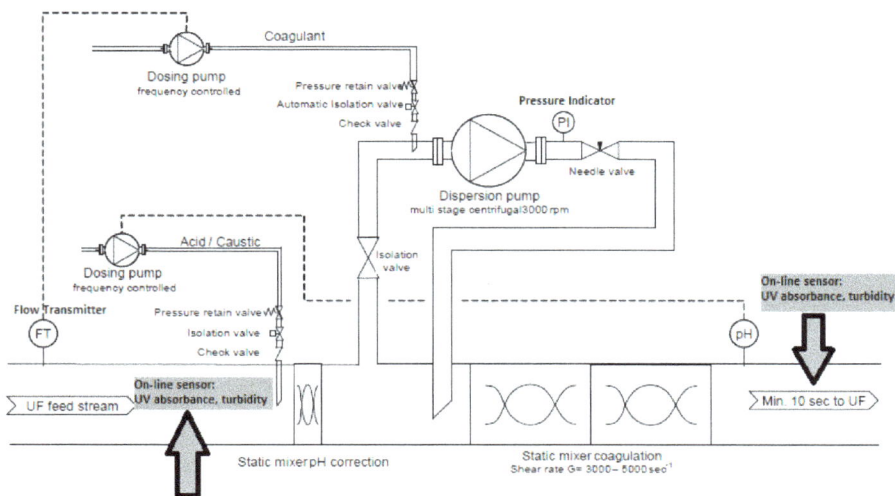

**Figure 2.** Design for in-line coagulation of full-scale process at Kvarnagården WTP.

### 2.2.3. Two-Stage UF Test Facility (UF-HF-P2)

During the construction period of the membrane plant, an extensive test facility was established for various long-term trials to verify the membrane performance of the full-scale design (Figure 3).

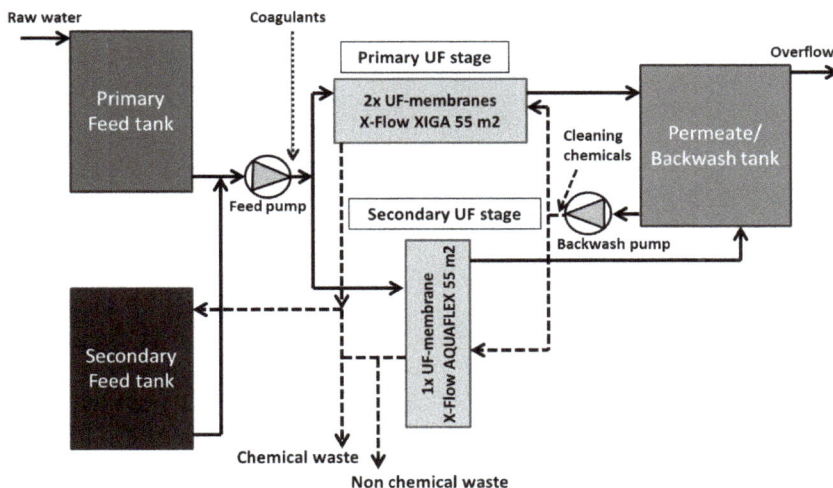

**Figure 3.** Schematic overview of the pilot plant process for the two-stage UF test facility (UF-HF-P2).

A 40 foot (12 m)-long container pilot plant (designed as a stand-alone unit to mimic the full-scale UF plant) has been in operation since January 2015, with a treatment capacity of 170 m$^3$/day (150 m$^3$/day permeate production). The pilot plant consists of the following main sections:

- Feed section, including dosing equipment for coagulant and chemicals for pH correction,
- Membrane system (two-stage UF), including air integrity testing,
- Permeate and backwash section, including chemical dosing for membrane cleaning.

The pilot plant was equipped with a computer for systems control and automatic operation. All the necessary process parameters were logged and trended on the computer. The plant allowed remote access, providing the same functionality as local access.

The two-stage membrane system comprised (a) a primary UF unit (horizontal dead-end filtration), with two membrane modules (Pentair X-Flow XIGA, 55 m$^2$); and (b) a secondary UF unit (vertical dead-end filtration), with one membrane module (Pentair X-Flow AQUAFLEX, 55 m$^2$). The raw water was supplied from the intake of the WTP to the primary feed tank by a pressurised line. The feed pump obtained water from the primary feed tank during the filtration of the primary UF unit. If pH correction of feed water was required, both H$_2$SO$_4$ and NaOH were added in front of the feed pump to obtain an optimal pH window for coagulants. During filtration, the UF permeate was directed to the permeate tank and discharged by overflow to the recipient. Permeate water from the permeate tank was used during backwashing of the UF units. The backwash waste from the primary UF unit was directed to the secondary feed tank. Chemical waste from the chemically-enhanced backwashing (CEB) programme at the primary UF unit was directed to the chemical waste discharge. During the backwash of the primary UF unit, the secondary feed tank was filled. When the primary UF unit finished a filtration cycle and the secondary feed tank reached a specific level, the primary UF unit stopped after a final backwash. Subsequently, the secondary UF unit was activated and started production. During filtration of the secondary UF unit, the feed pump obtained water from the secondary feed tank. The secondary permeate was directed to the permeate tank. During the hydraulic cleaning (feed water plus permeate) of the secondary UF unit, the waste water was directed to the non-chemical waste discharge facility. The waste water of the CEB for the secondary unit was directed to the chemical waste discharge facility. One feed pump was available for both the primary and the secondary units. In addition, one backwash pump was available for both the primary and secondary UF units. For the CEBs, dosing systems for H$_2$SO$_4$, NaOH, and NaOCl were available. The dosing points were placed in the backwash inlet line, and were used by both the primary and secondary units.

In addition to the hydraulic cleaning of the membranes with a combined backwash and forward flush, several automatic cleaning sequences were pre-programmed for specific cleaning protocols. In general, cleaning took place on an elapsed-time interval. A cleaning cycle consisted of flushing with clean water (permeate), followed by soaking with a maximum of two cleaning agents (acidic and caustic). The average feed water quality, cleaning protocols, and operational conditions during long-term test runs are summarised in Tables A2 and A3, whereas the membrane key performance parameters and manufacturer-reported properties of the hollow fibre membranes are listed in Table A4.

### 2.2.4. Coagulant Dosing System for UF Test Facility

The online measurements of turbidity and $UV$ absorbance (254 nm) in feed water ($UV_{Raw}$) were used to control the coagulation dosing rate in the feed line in order to meet the target values for NOM removal and permeate quality. The dosing rate of coagulants could be adjusted depending on current feed water quality and flow rates, according to Equation (1). The dependence was derived based on empirical evaluations of laboratory experiments and pilot studies, as part of the current study and previous studies by the authors [25].

$$Dos_{Coag} = A + B*^{(TURB)} +C*^{(UV\ abs)} \tag{1}$$

where

- $Dos_{Coag}$ is the coagulant dosing concentration (mg Metal L$^{-1}$)
- $B$ is the conversion factor for turbidity [-] (was set to zero during the pilot trials)

- $C$ is the conversion factor for $UV$ absorbance [-] (range: 0.005–0.035)
- $TURB$ is the feed water turbidity (FTU)
- $UV\ abs$ is the feed water $UV$ absorbance ($m^{-1}$)
- $A$ is a set point for the base coagulant dosage (range: 0.2–2.0).

2.2.5. Evaluation of Coagulation Efficiency

The variation of and differences in $UV$ signals between the raw ($UV_{Raw}$) and permeate ($UV_{Perm}$) is the most efficient way to evaluate the removal efficiency of the UF flocculation process. The removal efficiency of $UV$ absorbance at different coagulant dosing situations was evaluated by comparing the absolute change in $UV$ absorbance (Equations (2)–(4)).

$$\Delta\,UV = UV_{Raw} - UV_{Perm}. \tag{2}$$

The observed values for $UV$ were subsequently plotted as a function of the aluminium dose and fitted to a power relationship of the form:

$$\Delta\,UV = a + b*Al_{DOS} + c*Al_{DOS}{}^2 \tag{3}$$

With $a$, $b$, and $c$ used as empirical fitting factors to describe the observed curvature. In addition, a change in the $UV$ removal efficiency that is normalised to the target coagulant dose ($Al_{DOS*}$) under regular operational conditions of 0.6 mg $L^{-1}$ was calculated as:

$$UV_{norm} = \Delta\frac{UV}{Al_{DOS(t)}}*Al_{DOS*} \tag{4}$$

After initial separate treatment of the data from experiments UF-HF-P1 and UF-HF-P2 (Table 1), the datasets were combined, as the resulting curvature from the experiments was not different.

*2.3. Characterisation of Organic Fractions in Feed Water and Treated Water*

2.3.1. Determination of UV, TOC, and DOC

The following three measurements were done by a commercial laboratory. The $UV$ absorbance at 254 nm was measured with a 5-cm cuvette ($UV_{Lab2\ unfilt}$) using a Hach DR 5000 $UV$-Vis spectrophotometer (Loveland, CO, USA). The TOC and dissolved organic carbon (DOC) were determined using an Elementar Vario TOC Cube analyser (Langenselbold, Germany), with precision of 0.2 mg $L^{-1}$.

2.3.2. Evaluation of NOM Retention by LC-OCD

The composition of the organics that were present in the water samples of this study were characterised by using $UV$ absorbance at 254 nm and DOC-LABOR liquid chromatography-organic carbon detection (LC-OCD). The LC-OCD technique is based on a polymethacrylate size-exclusion column (Toso, Japan), coupled with three detectors (organic carbon, organic nitrogen, and $UV$-absorbance). This technique facilitates the subdivision of organic matter into six major subfractions: biopolymers, HS, building blocks, low-molecular-weight acids, low-molecular-weight neutrals, and hydrophobic organic carbon. Detailed information on the LC-OCD technique is available from the following studies [17,27,28].

2.3.3. Absorbance and Fluorescence Characterisation and Additional DOC and TOC

The presence of organic carbon was determined on unfiltered (TOC) and filtered samples (DOC) by using pre-combusted (4 h at 450 °C) GF/F filters (effective pore size of 0.7 µm) and acidified to pH 2 by using 37% HCl on a Shimadzu TOC-VCPH (Kyoto, Japan). The TOC and DOC were within the analytical precision parameters of the measurements (0.3 mg L$^{-1}$). Fluorescence excitation emission matrices (EEMs) were collected by using an Aqualog (Horiba, Edison, NJ, USA) spectrofluorometer [7]. Previously established indices were calculated from the corrected EEMs—namely humification index (HIX), fluorescence index (FI), and freshness index ($\beta{:}\alpha$) according to Cory and McKnight [29]; Ohno [30]; and Parlanti et al. [31]. The freshness index ($\beta{:}\alpha$) has been shown to be particularly valuable for the characterisation of coagulation [7,17].

### 2.4. Optical Sensors for Online Process Control and Dosing

Two online instruments (i::scan™; s::can Messtechnik GmbH, Vienna, Austria) were installed to detect the changes in the *UV* absorbance, colour, and turbidity in the feed water after the groundwater and lake water had been mixed (UF$_{Feed}$), and alternatively, in the permeate from both UF stages (UF$_{Perm}$). In addition to the *UV* absorbance probes, pH-sensors, pressure transmitters, flow meters, and temperature sensors were used for online monitoring of membrane performance and water quality. The absorbance spectra in the wavelength range 230–350 nm were acquired, with the online sensor using a flow-cell with a path length of 35 mm. The empirical relationships from particle-rich waters were used to calibrate the absorbance measurements against both TOC and turbidity, with algorithms developed by the probe manufacturer (so-called global calibration).

The correctness of the absorbance values of the online sensors were monitored by using three other independent *UV* absorbance measurements. On a biweekly basis, the *UV* was measured directly in unfiltered water samples by the operators at the WTP, using a 4-cm cuvette (*UV*$_{Lab1\ unfilt}$) and at a commercial laboratory (*UV*$_{Lab2\ unfilt}$; see Section 2.3.1). Furthermore, the filtered water samples were sent to an external research laboratory and measured by using a 1-cm flow-through cuvette (*UV*$_{Lab3\ filt}$) in combination with a high-precision combined fluorescence/absorbance spectrophotometer (Aqualog Horiba Jobin Yvon). During the second year of the study, an internal standard (60 ppm K-phthalate, with approximately A = 0.7 @ 254 nm) was added to all the sample runs to determine whether *UV* lamp drift had occurred.

The correctness of the calculated turbidity values of the online sensors were monitored with regular laboratory measurements of turbidity, using a HACH Model 2100N IS® Turbidimeter (Loveland, CO, USA), designed for turbidity measurement in accordance with ISO 7027.

The presence of particles probably led to deposition on the sensor, particularly on the raw water side. To prevent the degradation of the online signals, both probes were cleaned at regular intervals by following the procedures suggested by the manufacturer. For automatic cleaning, a rotating brush was mounted inside the flow cell in such way that the brush fibres reached the measuring windows on both sides of the measurement path of the spectrometer probes. The optimal autobrush cleaning settings were defined to 10 brush rotations every 20 min for the permeate and 10 brush rotations every 5 min for the feed water. Manual cleaning of the probes was carried out preventively once a month by using a mild alkaline cleaning agent provided by the manufacture and cleaning tissue. In the event of persistent fouling, pure alcohol (ethanol) and 3% hydrochloric acid (to prevent a mineral film/residue forming on the measuring windows) were used as cleaning liquids. The observed changes in *UV* absorbance and the calculated turbidity before and after manual cleaning typically ranged between 0.2 and 0.3 m$^{-1}$ and 0.1 and 0.2 FTU, respectively.

## 3. Results and Discussion

### 3.1. Preliminary Membrane Trials Using NF (NF-HF-P) and UF (UF-HF-P1)

Preliminary feasibility tests were carried out between 2010 and 2012 with several membrane processes to reduce the NOM concentration in the drinking water. Ultrafiltration with and without coagulant dosage was compared with hollow-fibre NF. The results from these test runs are summarised in Table 2.

Whereas UF alone did not remove NOM sufficiently, UF with in-line coagulation and hollow-fibre NF produced permeate that complied with the regulatory requirements for colour and COD.

Three operational modes of the UF pilot plant (UF-HF-P1) were investigated during the feasibility tests, namely: (1) dead-end filtration; (2) cross-flow with continuous bleed; and (3) cross-flow with intermittent flush. Dead-end filtration with raw water resulted in rapid increase of trans-membrane pressure (TMP). Cross-flow with continuous bleed produced approximately 20% of the feed flow as concentrate, requiring further treatment or having to be discarded. The operational mode that facilitated constant, high flux, and high recovery was cross-flow with intermittent flush. The UF was run with a polyaluminium coagulant dose of 1 mg/L Al per cubic metre of raw water [25].

UF with in-line coagulation could be run at a flux of approximately 60 L/m$^2$/h (LMH), which was four times the flux for capillary NF. This has major implications related to investment costs and the footprint of the plant. Accordingly, it was decided to continue the investigation focusing on a retrofit of the old full-scale plant with a UF hybrid process, with the aim of achieving NOM removal (as $UV_{abs}$) of at least 50% and feed water recovery of at least 99%.

**Table 2.** Comparison of water quality parameters in raw water, current drinking water, and after membrane treatment (median and standard deviations) during the feasibility tests year 2010–2012. The NF was run as one stage, with a recovery rate of 50%. During the early UF and NF trials, most turbidity values in permeate were below the detection limit. For the NF, most total organic carbon (TOC) and dissolved organic carbon (DOC) values were below the detection limit (affected values in italics). Values below the detection limit were reported as half the detection limit. COD: chemical oxygen demand; SUVA: specific ultraviolet absorbance.

| Parameter | Raw Water | Drinking Water (Full-Scale WTP OLD) | UF without Coagulant | UF with Coagulant | NF$_{50\%}$ | Target Value |
|---|---|---|---|---|---|---|
| Code | RAW | DW | UF-HF-P1 | UF-HF-P1 | NF-HF-P | |
| COD (mg/L) | $2.2 \pm 0.2$ | $2.1 \pm 0.2$ | $2.1 \pm 0.1$ | $1.3 \pm 0.2$ | $0.9 \pm 0.5$ | <4 limit |
| TOC (mg/L) | $2.6 \pm 0.2$ | $2.7 \pm 0.2$ | $2.5 \pm 0.2$ | $2.1 \pm 0.3$ | *1.0* | |
| DOC (mg/L) | $2.5 \pm 0.2$ | $2.4 \pm 0.2$ | $2.3 \pm 0.1$ | $2.0 \pm 0.2$ | *1.0* | |
| $UV_{254}$ (L/m) | $8.7 \pm 0.5$ | $8.1 \pm 0.5$ | $7.3 \pm 0.2$ | $3.7 \pm 0.5$ | $2.8 \pm 0.1$ | |
| SUVA (L/mg, m) | $3.6 \pm 0.4$ | $3.3 \pm 0.4$ | $3.3 \pm 0.1$ | $1.8 \pm 0.3$ | *2.8* | |
| Colour$_{405nm}$ (mg Pt/L) | $14.0 \pm 1.2$ | $13.0 \pm 1.3$ | $10.0 \pm 0.7$ | $2.5 \pm 1.3$ | $2.0 \pm 0.8$ | <15 limit <5 rec. |
| Turbidity (FNU) | $0.21 \pm 0.06$ | $0.22 \pm 0.05$ | 0.05 | 0.05 | 0.04 | <0.5 limit <0.1 rec. |
| pH | $7.6 \pm 0.1$ | $8.1 \pm 0.1$ | | | $7.4 \pm 0.1$ | 7.5–9 limit |

*3.2. Pilot Trials Using the Two-Stage UF Pilot Plant Test Facility (UF-HF-P2)*

The pilot trials with primary and secondary UF were used to identify the optimal operating conditions for the full-scale plant. At the design flux of 65 LMH and with a CEB interval after 20 filtration cycles, the primary UF stage showed stable operational conditions at permeability of approximately 400 LMH/bar over a period of 30 months. In this context, three episodes were analysed further, as they related to the robustness of the process. These are:

1. Impact of in-line coagulation on NOM removal and membrane performance
2. Effect of operation at high flux (maxflux and subsequent regaining of permeability)
3. Varying feed water quality (e.g., surface water only or variation in surface water NOM content).

3.2.1. Pilot Trials (UF-HF-P2): Episode 1—Effect of in-Line Coagulation

Initial trials with various coagulants (PAX XL 100 and PLUSPAC 1465) and varying doses resulted in long-term settings for coagulant dosing, Dos$_{coag}$ = 0.6 mg Al L$^{-1}$ (base coagulants dosage, A = 0.25; and a correction factor for $UV$ absorbance, C = 0.035).

Over a period of four days (7–11 December 2015), the pilot plant was operated without coagulant dosing, at a flux of 65 LMH. During that time, the permeability before the daily CEB quickly decreased from approximately 400 to below 250 L m$^{-2}$ h$^{-1}$ bar$^{-1}$ @ 20 °C. After two extended

CEBs, the permeability prior to the daily CEB started increasing slowly again. However, the initial permeability of 400 L m$^{-2}$ h$^{-1}$ bar$^{-1}$ @ 20 °C could not be restored completely within a week after the incident (Figure 4).

The simulated shutdown of coagulant dosing caused an instant decrease in the permeability of the primary UF, which could not be recovered fully by subsequent CEBs. The membrane obviously had lower "critical flux" without the coagulant. The critical flux of the membrane should be quantified for relevant feed water quality and coagulant dosing conditions in order to avoid irreversible loss of permeability. Investigation should be conducted to determine whether the permeability lost during such incidents could be recovered by a cleaning-in-place (CIP). Moreover, process strategies should be formulated on whether to reduce the flux automatically when a sudden loss of coagulant dosage occurs.

Once the full-scale plant is in stable operation, the pilot plant could be used to adjust the CEB and CIP conditions to ensure effective yet mild cleaning of the membranes. Furthermore, there should be a balance between the level of NOM removal and effective cleaning protocols, as well as the mechanical and chemical long-term stability of the membrane [26].

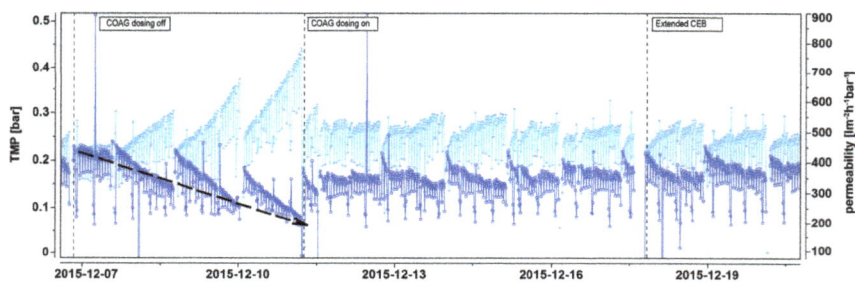

**Figure 4.** Hydraulic performance of the UF stage-1 container test modules at Kvarnagården WTP for the period 7–19 December 2015. Periods with an increase in transmembrane pressure (TMP; bar) (light blue on left y-axis) and a decrease in permeability (L m$^{-2}$ h$^{-1}$ bar$^{-1}$ @ 20 °C) (dark blue on right y-axis) are shown in relation to in-line coagulation (before, during, and after a stop in coagulant dosing, as displayed with vertical lines) and chemically-enhanced backwashing (CEB) cycles.

### 3.2.2. Pilot Trials (UF-HF-P2): Episode 2—High-Flux Testing

Over a period of six days (15–21 March 2016), the flux over UF stage-1 was increased from 65 to 70 LMH (i.e., by 7.7%), with a coagulation dosing concentration of 0.6 mg Al L$^{-1}$ (Figure 5). Before this specific episode, the permeability was stable at approximately 400 L m$^{-2}$ h$^{-1}$ bar$^{-1}$ @ 20 °C, with only minor reduction during a filter run, and 80 L m$^{-2}$ h$^{-1}$ bar$^{-1}$ @ 20 °C between daily CEBs. In addition, the permeability was restored by the CEB and was maintained in the long term. At 70 LMH, the permeability drop during a filter run was more rapid (40 L m$^{-2}$ h$^{-1}$ bar$^{-1}$ @ 20 °C), and there was a tendency toward decreasing permeability with every passing day (120 L m$^{-2}$ h$^{-1}$ bar$^{-1}$ @ 20 °C between daily CEBs). There was a strong indication that a certain maximum flux should not be exceeded in the long-term operation. These boundaries should be quantified by using the pilot plant for varying raw water qualities and temperatures. At the end of the high-flux testing period, a change in the raw water quality occurred (increase of $UV$ absorbance from 10 to 11.5 m$^{-1}$), which resulted in a further decrease in permeability (below 300 L m$^{-2}$ h$^{-1}$ bar$^{-1}$ @ 20 °C), although the coagulant dose was automatically adapted to 0.7 mg Al L$^{-1}$ in accordance with the current $UV$ correction factor and ordinary flux settings (65 LMH). The membrane system recovered first, after a further two days of normal operation, regaining levels of permeability and TMP comparable with the operational conditions prior to the testing period.

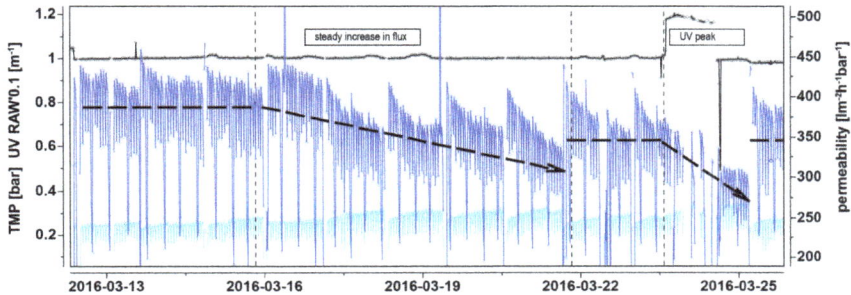

**Figure 5.** Hydraulic performance of the UF stage-1 container test modules at Kvarnagården WTP for the period 13 to 25 March 2016. The increase in flux was 65 to 70 L/m$^2$/h (LMH) during the period 15 to 21 March 2016. The TMP and permeability before, during, and after the increase in flux and *UV* absorbance are indicated with vertical lines. The TMP (bar) on the left-hand scale is indicated with light-blue lines, and permeability (L m$^{-2}$ h$^{-1}$ bar$^{-1}$ @ 20 °C) on the right-hand scale with dark-blue lines. *UV* absorbance (0.1*m$^{-1}$) on the left-hand scale is indicated with black lines.

### 3.2.3. Pilot Trials (UF-HF-P2): Episode 3—Varying Feed Water Quality

During a period of 21 h (23–24 March 2016), the feed water consisted of surface water only (no addition of alkaline groundwater). As shown in Figure 6, the change in the quality of the raw water resulted in an increase in TMP from 0.18 to 0.35 at a flux of 65 LMH. Despite recurring backwashing and one CEB during this period, the TMP could not be stabilised and the high levels could not be brought down, even after the addition of mixed raw water. The change in the feed water had no further effect on the removal efficiency of NOM, but the filtration behaviour indicated a tendency towards membrane fouling. Therefore, the cleaning protocols had to be adapted with regard to frequency and choice of cleaning chemicals.

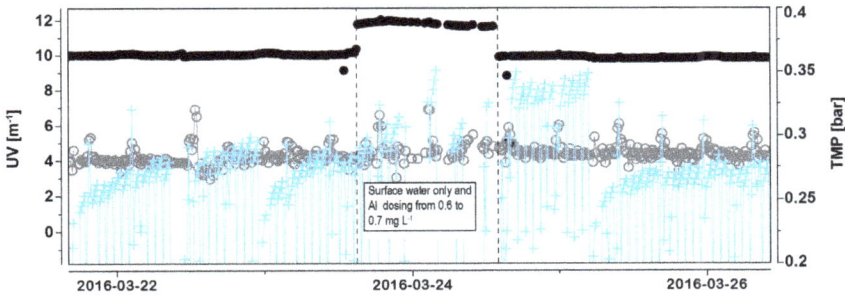

**Figure 6.** Membrane performance and natural organic matter (NOM) removal efficiency of the UF stage-1 container test modules at Kvarnagården WTP for the period 23 to 26 March 2016. A change occurred in the feed water quality for a period of 21 h. TMP (bar) is indicated on the right-hand scale (+), and *UV* absorbance (m$^{-1}$) on the left-hand scale: (●) *UV*$_{Raw}$, (○) *UV*$_{Perm}$.

### 3.3. Characterisation of Organic Matter in the Raw and Permeated Water

#### 3.3.1. Absorbance and Fluorescence Data Evaluation

The temporal changes in the organic matter concentration and character of the raw water source of Lake Neden were minor compared with other surface-water drinking-water plants in Sweden (e.g., Görväln WTP, Råberga WTP, and Ringsjö WTP) [18,32]. The turnover time of close to five years probably allowed for substantial removal of terrestrial-derived carbon, similar to the process observed in a larger Swedish lake, Lake Mälaren [33]. The comparatively low SUVA ($2.97 \pm 0.05$) and high freshness ($0.63 \pm 0.01$) in the raw water (Table 3) were indicative of mixing both with groundwater with low SUVA ($2.0 \pm 0.18$, Table 4) and with lake water containing internally produced carbon. This source water was more difficult to flocculate compared with many other boreal lakes, with shorter turnover times and a higher proportion of forest cover on the catchment. This notion is corroborated by the observation of a comparably high fraction (3%) of biopolymers in the lake water (Table 4).

The *UV* absorbance at 254 nm was measured with four different devices: online ($UV_{Sensor}$), in the laboratory of the WTP ($UV_{Lab1\ unfilt.}$), in a commercial laboratory ($UV_{Lab2\ unfilt.}$), and on filtered samples in a research laboratory ($UV_{Lab3\ filt.}$). Because of differences during the trials, the data were evaluated for the entire period and for each year (Table 3). The results are discussed in more detail in Section 3.4.

**Table 3.** Median and standard deviation for *UV* absorbance @254 nm ($m^{-1}$), DOC (mg $L^{-1}$), fluorescence index (FI), freshness index ($\beta$:$\alpha$), and SUVA (L/mg*m) for the period 2015–2016. Regarding the *UV* absorbance, four different measurements are available, namely those of three laboratories ($UV_{Lab1-3}$) and the sensor data ($UV_{Raw}$). The samples for laboratory 3 are all filtered (0.7 μm glass fibre filters, GFF) samples. Missing results are marked n.d., and DELTA (%) is the percentage of removal of *UV* calculated as $\Delta UV/UV$.

| Sample | $UV_{Raw}$ | $UV_{Lab1\ unfilt.}$ [$] | $UV_{Lab2\ unfilt.}$ [$] | $UV_{Lab3\ filt.}$ [#] | $DOC_{Lab3}$ | $FI_{Lab3}$ | $\beta/\alpha_{Lab3}$ | $SUVA_{Lab3}$ |
|---|---|---|---|---|---|---|---|---|
| Raw 2015 | $9.40 \pm 0.46$ | $9.23 \pm 0.44$ | $9.30 \pm 0.38$ | $8.57 \pm 0.36$ | n.d. | n.d. | n.d. | n.d. |
| Raw 2016 | $9.90 \pm 0.17$ | $9.30 \pm 0.37$ | $9.04 \pm 0.48$ | $8.60 \pm 0.39$ | n.d. | n.d. | n.d. | n.d. |
| Raw | $9.80 \pm 0.23$ | $9.40 \pm 0.45$ | $9.11 \pm 0.41$ | $8.59 \pm 0.37$ | $2.89 \pm 0.07$ | $1.47 \pm 0.02$ | $0.63 \pm 0.01$ | $2.97 \pm 0.05$ |
| Feed | n.d. | n.d. | n.d. | $5.93 \pm 1.48$ | $2.26 \pm 0.30$ | $1.58 \pm 0.05$ | $0.71 \pm 0.03$ | $2.51 \pm 0.32$ |
| Perm | $4.00 \pm 0.78$ | $4.00 \pm 1.00$ | $4.53 \pm 1.02$ | $4.41 \pm 1.05$ | $2.05 \pm 0.22$ | $1.61 \pm 0.05$ | $0.73 \pm 0.03$ | $2.12 \pm 0.25$ |
| DELTA | 59% | 57% | 50% | 48% | | | | |
| Perm 2015 | $4.10 \pm 0.74$ | $4.03 \pm 0.76$ | $4.50 \pm 1.09$ | $4.23 \pm 1.07$ | n.d. | n.d. | n.d. | n.d. |
| Perm 2016 | $3.80 \pm 0.34$ | $4.01 \pm 0.33$ | $4.62 \pm 0.95$ | $4.43 \pm 1.57$ | n.d. | n.d. | n.d. | n.d. |

Notes: [$] Measured in a 4-cm cuvette. [#] Measured in a 1-cm cuvette.

On average, less than 30% of the DOC and slightly more than 50% of the *UV* absorbance ($UV_{Raw}$) were removed during the coagulation UF process (Table 3). The SUVA decreased from approximately 3 to close to 2. Owing to a potential analytical error in the DOC ($\pm 0.2$ mg $L^{-1}$) and the filtered absorbance determined in a 1-cm cuvette ($\pm 0.5$ $m^{-1}$), the estimated error in SUVA for laboratory 3 was on the order of 0.3. The stability of the SUVA values over the entire period is shown in a DOC–SUVA plot in Figure A3. Of the three derived fluorescence indices, the freshness index ($\beta$:$\alpha$) was found to be an extremely valuable tool for evaluating removal as a function of the Al dose. The time series of freshness for both the raw water and the permeate indicated stable and reproducible time series (Figure A4). In addition, the excellent correlation of the Al dose and the freshness index facilitated the use of the freshness index as a superior indicator of the coagulation efficiency (Figure A5). This indicator is superior to SUVA, as it has higher precision and requires only one measurement compared with the two parameters for SUVA. This finding confirmed the results obtained by Köhler et al. [17] on the use of the freshness index.

### 3.3.2. LC-OCD Data Evaluation

Most of the LC-OCD measurements conducted over the last five years indicated quite stable conditions for both ground- and lake water (Table 4). On average, the raw water consisted of approximately 60% humic acids, only half of which could be removed by the current process.

**Table 4.** Average and standard deviation of liquid chromatography-organic carbon detection (LC-OCD) analysis for (n) samples of the groundwater well (Well), the lake (Neden), the feed (Feed, values in bold for comparison), and the UF and NF permeate (Perm), NF concentrate (Conc), and the drinking water from the retrofitted UF full-scale water treatment plant (DV). Missing results are marked as n.d. LMW: low molecular weight.

| Code | Sample | n | TOC (ppb-C) | DOC (ppb-C) | Biopolymers (ppb-C) | HS (ppb-C) | Building Blocks (ppb-C) | LMW$_{neutrals}$ (ppb-C) | LMW$_{acids}$ (ppb-C) | SUVA (L mg$^{-1}$ m$^{-1}$) |
|---|---|---|---|---|---|---|---|---|---|---|
| NF-HF-P | Lake | 3 | n.d. | 3235 ± 81 | 121 ± 8 | 2034 ± 54 | 568 ± 67 | 398 ± 33 | 9 ± 9 | 3.92 ± 0.03 |
| UF-HF-P1 | Lake | 4 | 3126 ± 202 | 3021 ± 204 | 100 ± 23 | 1919 ± 119 | 558 ± 49 | 401 ± 20 | 20 ± 9 | 3.82 ± 0.24 |
| UF-HF-P1 | GW | 2 | 600 ± 73 | 546 ± 213 | 5 ± 26 | 260 ± 85 | 0 ± 7 | 0 ± 2 | 0 ± 0 | 2.02 ± 0.18 |
| UF-HF-P1 | Feed | 4 | 2578 ± 402 | 2423 ± 343 | 72 ± 37 | 1603 ± 273 | 444 ± 57 | 344 ± 33 | 23 ± 11 | 3.65 ± 0.8 |
| UF-HF-P1 | Perm | 3 | 1655 ± 85 | 1714 ± 111 | 35 ± 10 | 855 ± 140 | 418 ± 28 | 315 ± 38 | 5 ± 7 | 2.62 ± 0.18 |
| UF-HF-F | Feed | 2 | 2541 ± 77 | 2337 ± 16 | 72 ± 10 | 1513 ± 10 | 409 ± 3 | 302 ± 5 | 42 ± 4 | 3.43 ± 0.22 |
| UF-HF-F | Perm | 2 | 1679 ± 34 | 1496 ± 24 | 27 ± 4 | 756 ± 53 | 424 ± 68 | 281 ± 1 | 17 ± 9 | 2.31 ± 0.26 |
| NF-HF-F | Feed | 3 | n.d. | 2593 ± 167 | 93 ± 2 | 1638 ± 55 | 508 ± 13 | 334 ± 19 | 0 ± 3 | 3.83 ± 0.04 |
| NF-HF-P | Perm | 3 | n.d. | 1158 ± 268 | 7 ± 3 | 592 ± 185 | 244 ± 56 | 210 ± 49 | 0 ± 3 | 3.09 ± 0.18 |
| NF-HF-P | Conc | 3 | n.d. | 4502 ± 1317 | 162 ± 99 | 2940 ± 980 | 804 ± 171 | 476 ± 123 | 1 ± 0 | 4.09 ± 0.16 |
| UF-HF-F | DV | 2 | 1742 ± 194 | 1473 ± 7 | 31 ± 1 | 754 ± 66 | 411 ± 70 | 272 ± 4 | 11 ± 5 | 1.88 ± 0.41 |

The DOC removal percentage and the stable conditions were in accordance with the minor changes in the inflowing raw water observed during the two-year trial period. The measured variation in DOC was in accordance with the two-year data time series presented above. The LC-OCD analysis indicated a lower DOC in the permeate (1.7 mg L$^{-1}$) compared with that of the classical DOC analysis (2.1 mg L$^{-1}$), but a higher DOC in the feed (2.5 mg L$^{-1}$) compared with that of the classical DOC analysis (2.3 mg L$^{-1}$). According to the LC-OCD, the combined coagulation/ultrafiltration processes removed approximately 30% of the DOC and 50% of the HS. However, all the other compounds (building blocks and low-molecular-weight (LMW) acids) remained unchanged compared with the feed and permeate composition. The removal of HS resulted in a decrease of SUVA from 3.4–3.7 to 2.3–2.6 compared with the LC-OCD, as well as a decrease from 3 to 2 compared with the classical DOC analysis. These differences fall just outside the analytical error of the classical method, and we are currently unable to explain why the LC-OCD-derived SUVA was higher throughout. In addition, the LC-OCD results indicated that both the biopolymers and the low molecular-weight neutrals (LMW$_{neutrals}$) originated from the lake water. Half of the biopolymers but none of the LMW acids were removed in the treatment process. These carbon fractions have the potential to cause regrowth in the distribution system, which in view of climate change, could probably have implications for the future adaptation of the process (see Section 3.6.2).

All the results from the pilot-scale trial (UF-HF-P1) conducted in 2011 agreed with the data obtained from the most recent (2017) full-scale process (UF-HF-F). This is a most satisfactory result, as it indicates that (a) the full-scale plant was successfully implemented and trimmed to resemble the pilot-scale plant closely and (b) the LC-OCD results were reproducible over time.

### 3.4. UV Sensor Data Evaluation

The removal of organic matter is controlled by the addition of aluminium salts, of which the dosage is a function of the variation of the incoming $UV$ absorbance in the raw water. The differences in the $UV$ absorbance in the raw water and the permeate can subsequently be used to assess the efficiency of the removal over time. As the online $UV$ sensor signal in the raw water ($UV_{Raw}$) could be affected by the presence of particles and changes in the optical behaviour owing to fouling, several control measurements were implemented to follow the $UV$ signal over time (Table 3). Two of the additional laboratory determinations (Lab 2 and Lab 3, also displayed in Table 3) of the $UV$ absorbance are compared with the online data in Figure 7.

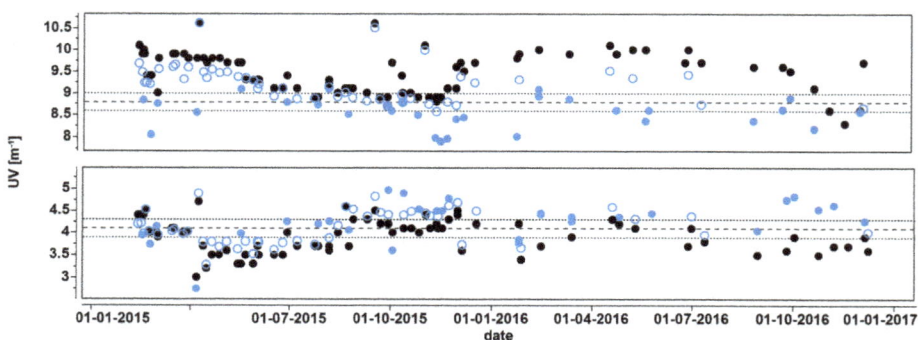

**Figure 7.** Time series of $UV$ (m$^{-1}$) measured over time using the $UV$ sensor ($UV_{Raw}$, ●), the unfiltered data of Lab 2 ($UV_{Lab2\ unfilt}$, ○), and the filtered data of Lab 3 ($UV_{Lab3\ filt}$, ●) for the raw water (above) and the permeate (below).

The four $UV$ measurement methods differed significantly—especially with respect to the raw water (Table 3). The $UV_{Lab2\ unfilt}$ (9.1) and $UV_{Lab3\ filt}$ (8.6) had lower values for the feed water compared with both the $UV_{\text{Lab1 unfilt}}$ (9.3) and the sensor $UV_{\text{Raw}}$ (9.9). From the above time series, it is concluded that a systematic deviation between the different signals was obtained, especially in the later period (Figure 7). All three laboratories found systematically lower $UV$ values in the raw water compared with the sensor ($UV_{\text{Raw}}$). Regarding the permeate, the differences in $UV$ absorbance were still systematic, with a lower $UV$ signal for the sensor compared with that of the laboratory analysis, but they were much less pronounced. The time series of measured differences between the sensor ($UV_{\text{Raw}}$) and $UV_{\text{Lab3 filt}}$ are shown in Figure A6, which confirm these systematic differences over time. These differences highlight the importance of cross-calibration. In particular, the significant difference in the measured values between the sensor ($UV_{\text{Raw}}$) and the filtered samples from Lab 3 ($UV_{\text{Lab3 filt}}$, 1.2 m$^{-1}$ in Table 3) in relation to the raw water was considered unusual. This is because the difference in the TOC and DOC was below 0.05 mg L$^{-1}$ on average (i.e., within the error of the method). Absolute and systematic differences in the different laboratory photometer signals should be excluded, as the filtered samples were not systematically lower for Lab 3 ($UV_{\text{Lab3 filt}}$ 4.4 > 4.0 m$^{-1}$ $UV_{\text{Raw}}$). As the data from Lab 3 were obtained with a 1-cm cuvette, error margins of 0.4 m$^{-1}$ could be avoided only with extreme precaution. In the instance of higher precision being required, a 5-cm cuvette had to be used in a separate measurement of the filtered samples. However, currently, no commercial 5-cm cuvette is available for coupled fluorescence–absorbance measurements similar to those performed in our study. The internal quality control in Lab 3 using K-phthalate standards revealed absolute differences in the monthly samples, which were below 3% on average (Figure A7). Furthermore, we could not exclude the possibility of some smaller fraction of the $UV$ absorbance being lost between the time of sampling and the time of analysis (within a few days for most of the samples). The values of both Lab 2 and Lab 3 were lower than were those of Lab 1 measured onsite.

The presence of turbidity in the raw water (on average 0.6 FTU), in addition to the fouling on the sensor ($UV_{\text{Lab2 unfilt}}$ < $UV_{\text{Raw}}$)—particularly later (from the spring of 2016, Figure 7)—could be the origin of the observed differences between the sensor, laboratory, and filtered data. Turbidity affected both the calculated $UV$ signal and the modelled colour (Figure A8 of the Appendix A). Based on the average measured turbidity in the raw water (0.6 FTU), it was possible to estimate a systematically higher $UV$ value of 0.3 m$^{-1}$ (0.6 × 0.523 = 0.34) in the unfiltered samples by using the established equation between turbidity and measured $UV$ (Figure A8). The remaining difference in the $UV$ absorbance between the raw water (0.9 m$^{-1}$) and the filtered sample could be ascribed to smaller but reasonably systematic differences (up to 0.4 m$^{-1}$). This could be related to fouling and the removal of some of the extremely dark hydrophobic fractions during filtration (TOC − DOC = 0.15 mg L$^{-1}$ in Table 4), requiring the removal of approximately 0.5 m$^{-1}$. This latter observation was corroborated by our own observations on the removal of $UV$ when no Al dosage was applied. The observed difference ($UV_{\text{Raw}}$ − $UV_{\text{Lab3 filt}}$) of 1.2 m$^{-1}$ was therefore split into three fractions, which were 0.3 m$^{-1}$ caused by turbidity, 0.4 m$^{-1}$ potentially caused by a systematic error between the sensors, and 0.5 m$^{-1}$ caused by filtration or fouling.

This evaluation revealed that the sensor data alone were not enough to track the performance of this UF process (DELTA in Table 3 varies between 48% and 59%) and that the use of both filtered and unfiltered control samples was required to support and calibrate the sensor data. At WTPs with higher turbidity and a higher fraction of particulate organic carbon, measurements that are even more precise would be required.

*3.5. DOC Removal Efficiency*

The sensor data and coagulant dosing were used to identify the optimal coagulation conditions over the entire pilot period. For this purpose, the variation in the observed $UV$ signals between the feed and permeate was plotted against the utilised coagulant dose. The observed relationship was nonlinear, with decreasing efficiency of $UV$ removal when the Al dose was increased. This behaviour could be mimicked with a second-degree curve, as shown in Figure 8.

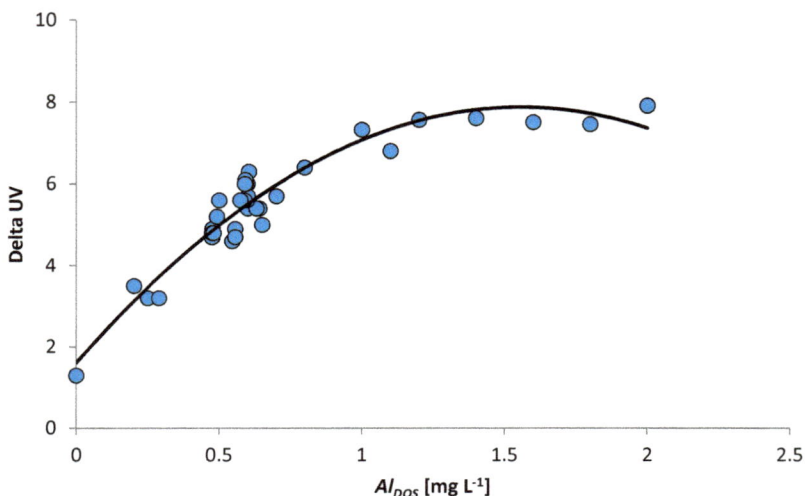

**Figure 8.** Sensor-based change in *UV* signal between the feed and permeate, as a function of added Al. $\Delta UV = 2.63 + 5.15 \times \text{Al dose} - 4.30 \times (\text{dose} - 0.529)^2$.

Based on this relationship, all the results related to dosing could be benchmarked. Occurrences of removal ($\Delta UV$) at least 25% below the optimum removal (according to the fitted equation) are coloured red, whereas those at least 10% lower (but within 25% of the optimal performance) are marked black (Figure A9). Most such data points occurred in the initial trimming phase of the pilot experiment, during the first three months. The increasing coagulation efficiency as a function of time is evident. During the first three months, a larger number of data points had a coagulation efficiency below 75%, whereas almost all the data points were above 90% after the nine-month trial.

The available sensor data and those obtained from the extra trials could be used to identify the potential limits of the coagulation dosing beyond 0.6 mg Al L$^{-1}$ currently applied. While there was a somewhat linear relationship between the dose and change in *UV* (Delta *UV*) in the range 0 to 0.7 mg Al L$^{-1}$, the relationship flattened out at higher doses. Dosing above 1.2–1.4 mg Al L$^{-1}$ did not bring about further reductions in *UV* (Figure 8). This limit was related to the raw water content in the HS and the low SUVA, as was described earlier (Table 4).

The evaluation of normalised coagulation efficiency (Equation 4) could help to determine the efficiency of increasing the dosage compared with the current dose of 0.6 mg Al L$^{-1}$. These data revealed the sharply decreasing efficiency of Al addition (Figure A10), and in accordance with Figure 8, the curve flattened out at higher doses. It could be calculated how much *UV* was removed with an increase in the dose ($\Delta UV/\Delta$Al). Dosing at 0.95 mg L$^{-1}$ (1.4 mg L$^{-1}$) achieved only 80% (respectively 66%) of the *UV* removal per dose achieved with the current dose (0.6 mg L$^{-1}$). Furthermore, higher doses produced more sludge, which from both an environmental and economical viewpoint was not a desired result. Based on the other limitations of the process (e.g., changes in TMP and flux), it was concluded that the current process with the current raw water probably had an optimal technical and economic dosing limit of close to 0.95 mg L$^{-1}$. This value was used in the scenario below for assessing the potential deterioration in water quality (see Section 3.6.2).

### 3.6. Adaptation and Resilience of the UF Process (UF-HF-P2)

3.6.1. Determination of Maximum Coagulation Dosage

The critical operational conditions were evaluated during a one-week stress testing period (25 June 2017 to 4 July 2017). This was done to define the maximum coagulation dosage for enhanced NOM removal, in consideration of the overall membrane performance (Table 5). Whereas the membrane performance at the primary UF stage was stable regardless of the incremental increase in the coagulant dosage, the secondary UF stage reached a critical level at 2.0 mg Al $L^{-1}$. This resulted in a distinct decrease in the permeability during a single filtration cycle (from 740 to 150 L $m^{-2}$ $h^{-1}$ $bar^{-1}$ @ 20 °C). Furthermore, doubling of the coagulant dosage (1 to 2 mg Al $L^{-1}$) resulted in a limited decrease in *UV* absorbance in the permeate with (0.55 $m^{-1}$), raising concerns about the economic and operational aspects of process strategies for higher NOM removal.

**Table 5.** The effect of incremental increase of the coagulant dose on *UV* absorbance and membrane performance. Data for TMP and permeability relate to an ordinary filtration cycle for UF stage-1 and UF stage-2 of the UF test facility (UF-HF-P2) at Kvarnagården WTP.

| Dose (mg Al $L^{-1}$) | 1.00 | | 1.40 | | 1.60 | | 1.80 | | 2.00 | |
|---|---|---|---|---|---|---|---|---|---|---|
| UF stage | UF1 | UF2 | UF1 | UF2 | UF1 | UF2 | UF1 | UF2 | UF1 | UF2 |
| Flux (LMH) | 49 | 33 | 49 | 33 | 49 | 33 | 49 | 33 | 49 | 33 |
| TMP (bar) | 0.16–0.19 | 0.05–0.14 | 0.17–0.19 | 0.06–0.17 | 0.16–0.18 | 0.06–0.22 | 0.16–0.18 | 0.06–0.24 | 0.16–0.19 | 0.06–0.25 |
| Permeability (L $m^{-2}$ $h^{-1}$ $bar^{-1}$ @ 20 °C) | 420–360 | 760–320 | 420–360 | 760–240 | 420–370 | 740–200 | 420–360 | 740–170 | 420–360 | 740–150 |
| Feed water, $UV_{254}$ ($m^{-1}$) | 10.02 | n.a. | 9.9 | n.a. | 9.8 | n.a. | 9.7 | n.a. | 10.05 | n.a. |
| Permeate, $UV_{254}$ ($m^{-1}$) | 2.70 | 4.1 | 2.3 | 3.2 | 2.3 | 2.8 | 2.25 | 2.7 | 2.15 | 2.8 |

3.6.2. Scenario Analysis and Evaluation of UF Performance during Constant Rise of DOC

The DOC concentrations have been rising in large parts of the boreal zone, with trends reported to be in the range 0.1 to 0.2 mg $L^{-1}$ per year, and significant changes in the water colour over time.

The following scenario was based on the *UV* data obtained from Lab 1 ($UV_{Lab1\ unfilt}$), assuming an initial *UV* value of 9.4 $m^{-1}$ and a removal level down to approximately 4 $m^{-1}$ using a dose of 0.6 mg $L^{-1}$. In this projection, it is assumed that the current character of DOC in the raw water sources from Lake Neden and the current groundwater well (i.e., SUVA) would not change over time, while a steady increase for DOC is defined with 0.05 mg $L^{-1}$ per year in the surface water from Lake Neden. This low annual increase, in comparison with the other sites in the area [22], was chosen to account for the potential breakdown of terrestrial TOC during the five-year turnover time. In addition to this change in DOC, the effect of increasing the fraction of groundwater to raw water (i.e., 20% and 25%, instead of the current 15% contribution) was studied. The exact assumptions for the values of DOC and SUVA are shown in Table A5. These three scenarios gave rise to a change in Al dosing that are shown below. A constant rise in DOC over time from the current DOC of 3 mg $L^{-1}$ (2015) to 4.25 (2040) would imply a sharp rise in the required Al dose over time, if the quality of the drinking water (UF permeate) was to remain the same (Figure 9). In 2031, as Al dosing would reach its maximum removal capacity (1.6 mg $L^{-1}$), the model predicted a decrease in the drinking water quality, as the rising DOC could not be removed any further. Increasing the fraction of GW from 15% to 20% would postpone this eventuality to 2040 and even beyond 2040, if 25% of the raw water could be obtained from the groundwater source. The fourth scenario assumed that the economic limit of Al dosing would be reached at a dose of approximately 0.95 mg $L^{-1}$. In this scenario, the Al dose would rise until the maximum dose was reached, leading to a deterioration of the *UV* absorbance in the drinking water from the current 4 $m^{-1}$ to approximately 6.3 $m^{-1}$ in 2040.

This analysis (Figure 9) clearly reveals that comparatively minor but reasonable changes in DOC over time owing to climate change or continuing recovery from acid rain would require adaptations to the process. Therefore, careful monitoring of changes in the raw water sources in future is recommended, similar to that currently being conducted by the regular Swedish lake-monitoring programme that is coordinated by the Swedish EPA.

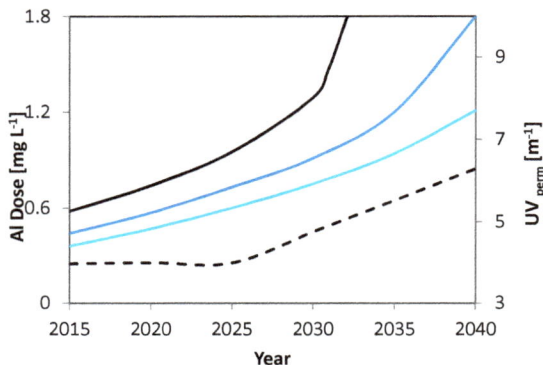

**Figure 9.** Predicted Al dose as a function of time for current mixing conditions (15% groundwater (GW), 85% surface water, black line), optional (20% GW and 80% surface water, dark-blue line), and increased (25% GW and 75% surface water, light-blue line), assuming constant SUVA in both sources over time, but an increase of 0.05 mg DOC $L^{-1}$ per year in the surface water (Lake Neden) over time. The black stippled line shows the predicted $UV$ absorbance in the permeate (UF perm) from the retrofitted UF full-scale water treatment plant, if current mixing conditions were maintained and the maximum economically feasible dosing was assumed to be 0.95 mg Al $L^{-1}$.

## 4. Conclusions

Long-term pilot studies are valuable to determine the optimal conditions for ultrafiltration of mixed raw water rich in organic material, with minor seasonal variations. The NOM removal based on in-line coagulation combined with UF membranes from modified polyethersulfone (PES) proved to be suitable for direct ultrafiltration of surface water from Lake Neden, with high removal of NOM and minimal membrane-fouling potential. The major findings of the study are:

- The in-line coagulation/UF process produced stable water quality and facilitated the calculation of a dose–response curve for optimal dosing conditions (0.5–0.7 mg Al $L^{-1}$) and potential boundaries (0.9–1.2 mg Al $L^{-1}$).
- The secondary UF stage reached a critical level at 2.0 mg Al $L^{-1}$, resulting in a distinct decrease in permeability during a single filtration cycle (from 740 to 150 L m$^{-2}$ h$^{-1}$ bar$^{-1}$ @ 20 °C).
- Doubling the coagulant dosage (1.0 to 2.0 mg Al $L^{-1}$) resulted in a limited decrease in $UV$ absorbance in the permeate (0.55 m$^{-1}$), raising concerns about the economic and operational aspects of process strategies to facilitate higher NOM removal.
- Systematic differences in the sensor and laboratory data must be taken into account for the different procedures to allow for the correct calculation of removal efficiency (quality control).
- The surface-water quality scenarios (up to the year 2040) indicated a potential increase in NOM, with significant effects on the coagulant dose and the quality of drinking water.

**Acknowledgments:** The financial support of the GenoMembran-project, funded by the Swedish Water and Wastewater Association (Swedish Water Development, SVU) is gratefully acknowledged. The financial support for S.J. Köhler for the duration of this study was assured through DRICKS (Framework programme for drinking water research at Chalmers University in Gothenburg, Sweden). Pentair X-Flow (The Netherlands) is acknowledged for the provision of various test modules. The authors would also like to acknowledge Jennie Lindgren for her contribution of reagents, materials, and analytical tools, as well as Erik Scharenborg for advice on data analysis.

**Author Contributions:** Alexander Keucken and Gerald Heinicke designed and planned the experiments; Alexander Keucken performed the experiments; Stephan J. Köhler, Kenneth M. Persson, and Alexander Keucken contributed the analysis tools. All the authors analysed the data and contributed to writing the paper.

**Conflicts of Interest:** The authors declare no conflict of interest. The funding sponsors had no role in the design of the study; in the collection, analyses, or interpretation of the data; in the writing of the manuscript; or in the decision to publish the results.

## Glossary

| Term | Definition |
|---|---|
| BB | Building Blocks |
| CEB | Chemical Enhanced Backwashing |
| CEEF | Chemical Enhanced Forward Flushing |
| CIP | Cleaning-in-Place |
| COD | Chemical Oxygen Demand |
| Da | Dalton |
| DBP | Disinfection By-Product |
| DOC | Dissolved Organic Carbon |
| FNU | Formazin Nephelometric Unit |
| GW | Groundwater |
| HS | Humic Substances |
| LC-OCD | Liquid Chromatography–Organic Carbon Detection |
| LMWneutrals | Low Molecular Weight Neutrals |
| MW | Molecular Weight |
| MWCO | Molecular Weight Cut-Off |
| NF | Nanofiltration |
| NOM | Natural Organic Matter |
| PES | Polyethersulfone |
| SUVA | Specific Ultraviolet Absorbance |
| TMP | Transmembrane Pressure |
| TOC | Total Organic Carbon |
| $UV_{abs}$ | Absorption of $UV$ light at 254 nm |
| UF | Ultrafiltration |
| WTP | Water Treatment Plant |

## Appendix A

**Table A1.** Feed water quality (median and standard deviations) of primary and secondary UF stages (UF-HF-P2).

| Parameters | Unit | Range (UF-Stage 1) | Range (UF-Stage 2) |
|---|---|---|---|
| Temperature | °C | $4.5 \pm 0.7$ | $4.6 \pm 0.5$ |
| pH | (-) | $7.4 \pm 0.1$ | $7.1 \pm 0.2$ |
| Turbidity | (FTU) | $0.6 \pm 0.2$ | $20.0 \pm 2.9$ |
| Hardness | °dH | $1.5 \pm 0.10$ | $1.6 \pm 0.15$ |
| Alkalinity | (mg/L $HCO_3$) | $19.0 \pm 2.1$ | $18.0 \pm 3.0$ |
| COD | (mg/L $O_2$) | $2.0 \pm 0.1$ | $25.0 \pm 7.5$ |
| TOC | (mg C/L) | $2.9 \pm 0.06$ | $27.0 \pm 3.2$ |
| DOC | (mg C/L) | $2.6 \pm 0.1$ | $4.4 \pm 0.45$ |
| $UV_{254}$ | (/5 cm) | $0.380 \pm 0.22$ | $3.955 \pm 3.3$ |
| Pt-Co | (mg Pt/L) | $13 \pm 1.0$ | $15.0 \pm 5.8$ |
| Conductivity | (µS/cm) | $110 \pm 6.9$ | $110 \pm 5.7$ |
| Iron | (mg/L Fe) | $0.026 \pm 0.02$ | $0.57 \pm 0.1$ |
| Manganese | (mg/L Mn) | $0.035 \pm 0.01$ | $0.044 \pm 0.01$ |

**Table A2.** Operating conditions and process parameters during long-term pilot trials (UF-HF-P2).

| Parameters | Unit | UF Primary | UF Secondary |
|---|---|---|---|
| Max. filtration time ($t_F$) | (min) | 90 | 60 |
| Max. filtration volume | ($m^3$) | 8.4 | 1.65 |
| Filtration flux ($J_F$) | ($L\,m^{-2}\,h^{-1}$) | 65–70 | 45 |
| $V_{CF}$ (cross flow velocity) | ($m\,s^{-1}$) | - | 0.5 |
| R (recovery during filtration) | (%) | 100 | 100 |
| $t_{BW}$ (backwash time) | (s) | 30 | 30 |
| $J_{BW}$ (backwash flux) | ($L\,m^{-2}\,h^{-1}$) | 250 | 250 |
| $t_{CEFF}$ (CEB interval) | (days) | 1.5 | 5 |
| CEB1 dosing solution (caustic) | (-) | 250–300 ppm NaOCl @ pH 12.2 with NaOH | 250–300 ppm NaOCl @ pH 12.2 with NaOH |
| CEB2 dosing solution (acidic) | (-) | 475 mg/L $H_2SO_4$ @ pH 2.4 | 475 mg/L $H_2SO_4$ @ pH 2.4 |
| $t_{SOAK}$ (Soak time CEB) | (min) | 10 | 10 |

**Table A3.** Membrane key performance parameters during pilot trials.

| Parameters | Unit | UF Primary | UF Secondary |
|---|---|---|---|
| Permeability | ($L\,m^{-2}\,h^{-1}\,bar^{-1}$ @ 20 °C) | 350–380 | 600–220 |
| Transmembrane pressure | (bar) | 0.18–0.28 | 0.12–0.25 |
| Total number of CEBs | (-) | 267 | 37 |
| Module age before replacement | (months) | 12 | 14 |
| Total amount of filtration volume (feed water) | $m^3$ | 57.150 [1] | 2.155 [2] |

Notes: [1] 150 $m^3$/day × 381 days, [2] 4.8 $m^3$/day × 449 days.

**Table A4.** Manufacturer-reported properties of the hollow fibre membranes.

| Parameter | Unit | Key Performance Values |
|---|---|---|
| **Membrane Material** | | **Sulfonated Polyethersulfone (PES)** |
| Max. backflush pressure | kPa | 300 |
| MWCO based on PEG [1] | kDa | 100 |
| Diameter (internal) | mm | 0.80 |
| Nominal pore size | nm | 20 |
| Membrane area | $m^2$ | 55 |
| Number of fibres | | ~15,000 |
| Reduction of bacteria | log | 6 (*Pseudomonas diminuta*) |
| Reduction of virus | log | 4 (MS2 coliphages) |
| Module hydraulic diameter | mm | 220.0 |
| Module length | mm | 1537.5 |

Notes: [1] PEG = Polyethylene glycol unit of molecular weight approximately 1000 dalton. MWCO: molecular weight cut-off.

**Table A5.** Scenario analysis (year 2040): Assumptions for source water values regarding DOC and SUVA (DELTA [DOC] 0.05 mg DOC/year).

| Year | GW DOC (mg L$^{-1}$) | GW SUVA (L/mg*m) | GW UV abs (m$^{-1}$) | Neden DOC (mg L$^{-1}$) | Neden SUVA (L mg$^{-1}$ m$^{-1}$) | Neden UV (m$^{-1}$) | GW Fraction (%) | UVabs MIX (m$^{-1}$) | DOC MIX (mg L$^{-1}$) | $AI_{DOS}$ S1 (mg L$^{-1}$) | $AI_{DOS}$ S2 (mg L$^{-1}$) | $AI_{DOS}$ S3 (mg L$^{-1}$) | Delta $UV^1$ | Delta $UV^1$ | Uvabs Permeate (m$^{-1}$) | AI DOS Max Dose (mg L$^{-1}$) |
|---|---|---|---|---|---|---|---|---|---|---|---|---|---|---|---|---|
| 2015 | 0.65 | 2 | 1.3 | 3 | 3.6 | 10.8 | 0.15 | 9.375 | 2.6475 | 0.58 | | | 5.40 | 5.40 | 3.97 | 0.58 |
| 2020 | 0.65 | 2 | 1.3 | 3.25 | 3.6 | 11.7 | 0.15 | 10.14 | 2.86 | 0.74 | | | 6.14 | 6.14 | 3.99 | 0.74 |
| 2025 | 0.65 | 2 | 1.3 | 3.5 | 3.6 | 12.6 | 0.15 | 10.905 | 3.0725 | 0.95 | | | 6.92 | 6.91 | 3.98 | 0.95 |
| 2030 | 0.65 | 2 | 1.3 | 3.75 | 3.6 | 13.5 | 0.15 | 11.67 | 3.285 | 1.29 | | | 6.92 | 7.68 | 3.98 | 0.95 |
| 2031 | 0.65 | 2 | 1.3 | 3.8 | 3.6 | 13.68 | 0.15 | 11.823 | 3.3275 | 1.48 | | | 6.92 | 7.84 | 3.97 | 0.95 |
| 2033 | 0.65 | 2 | 1.3 | 3.9 | 3.6 | 14.04 | 0.15 | 12.129 | 3.4125 | >2 | | | 6.92 | <8 | >4 | 0.95 |
| 2035 | 0.65 | 2 | 1.3 | 4 | 3.6 | 14.4 | 0.15 | 12.435 | 3.4975 | >2 | | | | <8 | >4 | 0.95 |
| 2040 | 0.65 | 2 | 1.3 | 4.25 | 3.6 | 15.3 | 0.15 | 13.2 | 3.71 | >2 | | | | <8 | >4 | 0.95 |
| 2015 | 0.65 | 2 | 1.3 | 2.9 | 3.6 | 10.44 | 0.2 | 8.612 | 2.45 | | 0.44 | | | 4.64 | 3.96 | |
| 2020 | 0.65 | 2 | 1.3 | 3.15 | 3.6 | 11.34 | 0.2 | 9.332 | 2.65 | | 0.57 | | | 5.35 | 3.98 | |
| 2025 | 0.65 | 2 | 1.3 | 3.4 | 3.6 | 12.24 | 0.2 | 10.052 | 2.85 | | 0.73 | | | 6.10 | 3.95 | |
| 2030 | 0.65 | 2 | 1.3 | 3.65 | 3.6 | 13.14 | 0.2 | 10.772 | 3.05 | | 0.91 | | | 6.78 | 3.98 | |
| 2035 | 0.65 | 2 | 1.3 | 3.9 | 3.6 | 14.04 | 0.2 | 11.492 | 3.25 | | 1.2 | | | 7.53 | 3.95 | |
| 2040 | 0.65 | 2 | 1.3 | 4.15 | 3.6 | 14.94 | 0.2 | 12.212 | 3.45 | | >2 | | | <8 | >4 | |
| 2015 | 0.65 | 2 | 1.3 | 2.9 | 3.6 | 10.44 | 0.25 | 8.155 | 2.3375 | | | 0.36 | | 4.16 | 3.98 | |
| 2020 | 0.65 | 2 | 1.3 | 3.15 | 3.6 | 11.34 | 0.25 | 8.83 | 2.525 | | | 0.47 | | 4.81 | 4.01 | |
| 2025 | 0.65 | 2 | 1.3 | 3.4 | 3.6 | 12.24 | 0.25 | 9.505 | 2.7125 | | | 0.6 | | 5.50 | 4.00 | |
| 2030 | 0.65 | 2 | 1.3 | 3.65 | 3.6 | 13.14 | 0.25 | 10.18 | 2.9 | | | 0.75 | | 6.18 | 3.99 | |
| 2035 | 0.65 | 2 | 1.3 | 3.9 | 3.6 | 14.04 | 0.25 | 10.855 | 3.0875 | | | 0.94 | | 6.88 | 3.97 | |
| 2040 | 0.65 | 2 | 1.3 | 4.15 | 3.6 | 14.94 | 0.25 | 11.53 | 3.275 | | | 1.21 | | 7.55 | 3.97 | |

Note: $^1$ Delta $UV = -2.5864*AI_{DOS}^2 + 8.0452* AI_{DOS} + 1.6065$.

**Figure A1.** Geographical location of lakes in the region for which long-term monitoring data from the Swedish monitoring programme are available.

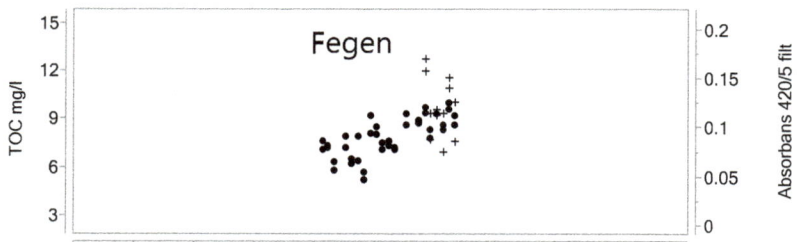

**Figure A2.** *Cont.*

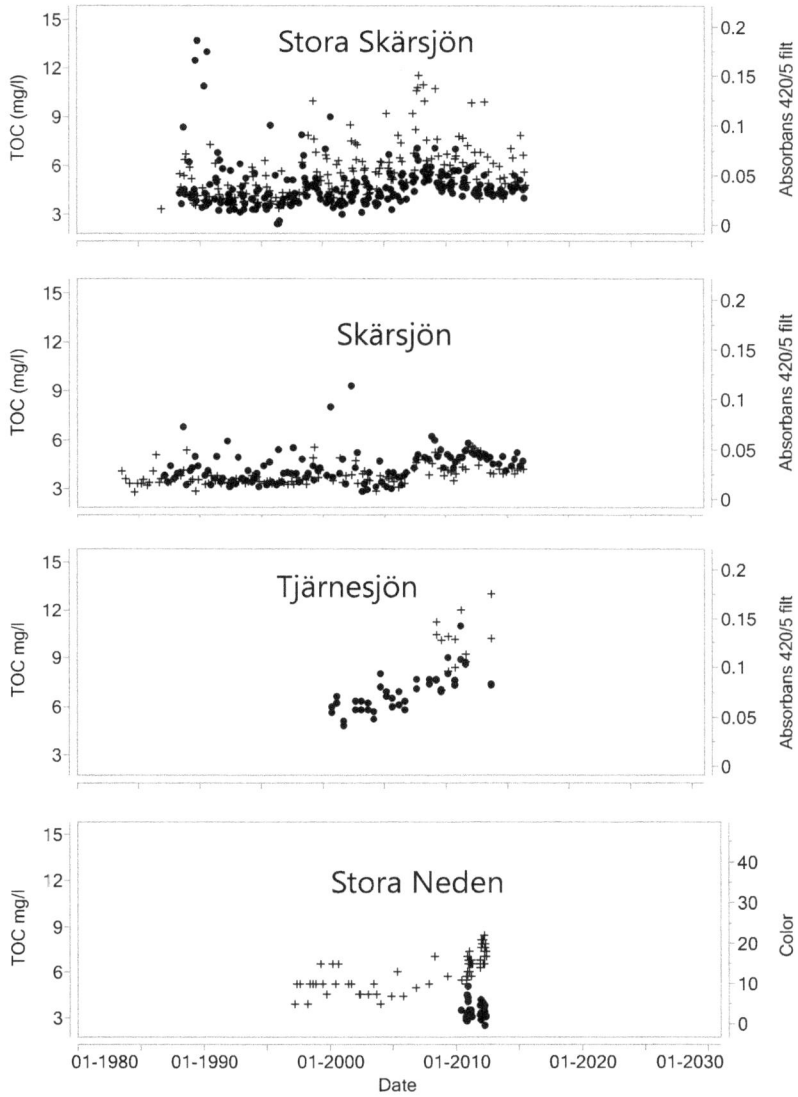

**Figure A2.** Time series of trends in TOC (+, left scale, mg L$^{-1}$) and colour measured as Abs_420 (●, right scale) or, as in the instance of Lake Stora Neden, in Pt colour that were acquired by the Swedish lake monitoring programme.

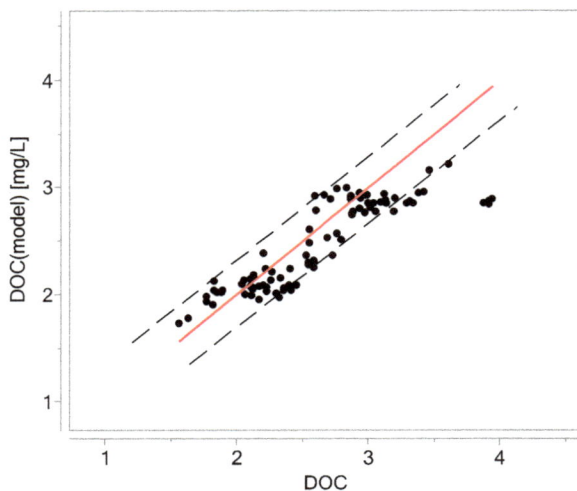

**Figure A3.** Measured DOC (mg L$^{-1}$) against predicted DOC (DOC (model) = a + b * $UV_{\text{Lab3 filt}}$). The close linear relationship is proof of extremely stable SUVA over time.

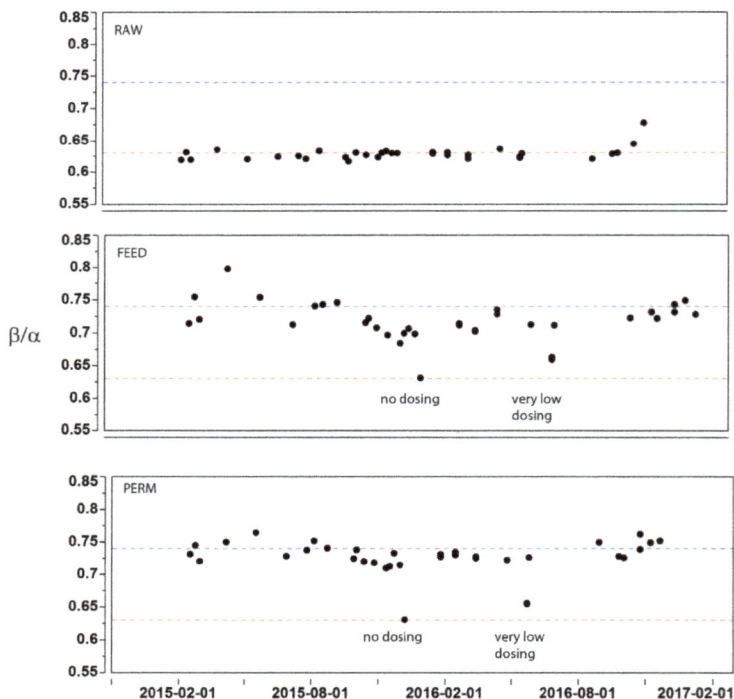

**Figure A4.** Time series of the freshness index of raw water (above), feed water (middle), and permeate (below) for 2015–2016.

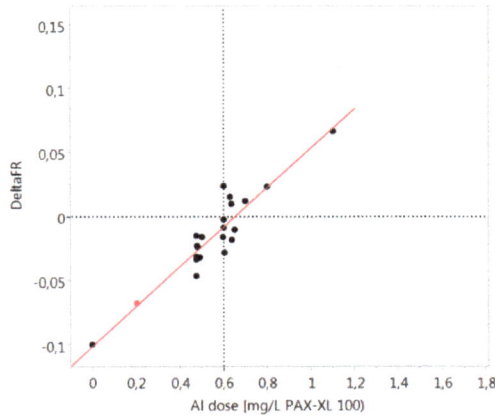

**Figure A5.** Change in the freshness index (DeltaFR) when comparing higher and lower doses than the current optimal dose of 0.6 mg L$^{-1}$. DeltaFR = $-0.100117 + 0.154*$Al DOS (mg L$^{-1}$ PAX-XL 100); R2 = 0.872, RMSE = 0.012.

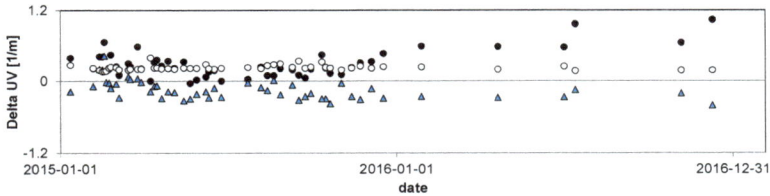

**Figure A6.** Comparison of *UV* signal from the sensor ($UV_{Raw}$, ●) Lab 2 ($UV_{Lab2\,unfilt}$, O), and from Lab 3 ($UV_{Lab3\,filt}$, ▲) over time.

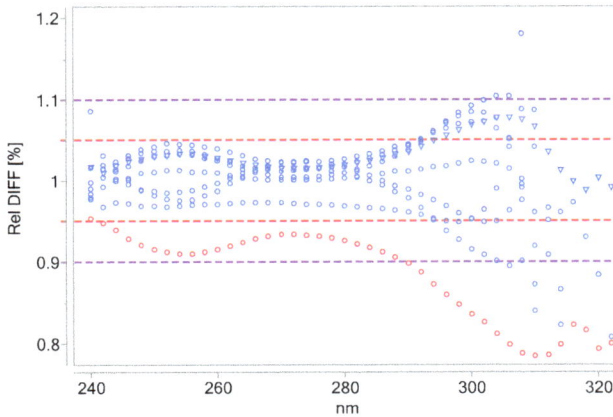

**Figure A7.** Differences in spectra normalised to average spectra of the K-phthalate internal standard solutions (60 ppm), as a function of wavelength that were measured monthly during 2016 for quality control purposes. On average, the relative deviation is below 3% in the *UV* range 240–290 nm. Different blue markers indicate the different sample months. One extreme sample is marked in red.

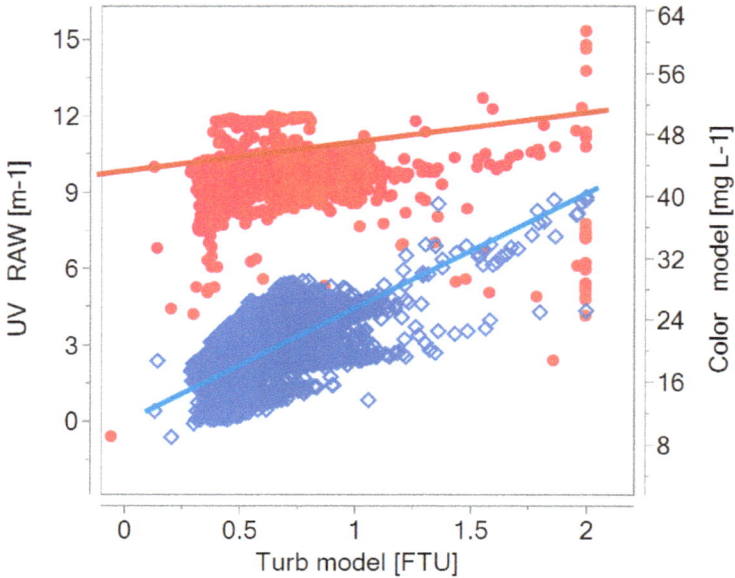

**Figure A8.** Estimation of effects of sensor-modelled turbidity (FTU) on sensor-modelled colour ($\diamond$), and sensor-measured *UV* (●). *UV* = 9.30 + 0.523 * Turb$_{model}$. Per one unit of modelled turbidity, approximately half a unit of *UV* is added, which is in accordance with unpublished stream data from the Fyris River (Uppsala, Sweden). At turbidity of 2, an additional *UV* signal of 1 is captured that is not related to the DOC. The offset 9.3 is extremely close to the average raw water *UV*. Regarding the colour, this effect is more pronounced, such that a turbidity of 2 could give rise to additional colour of 36 mg L$^{-1}$, therefore dominating the signal. Color = 11.6 + 18.4 * Turb$_{model}$.

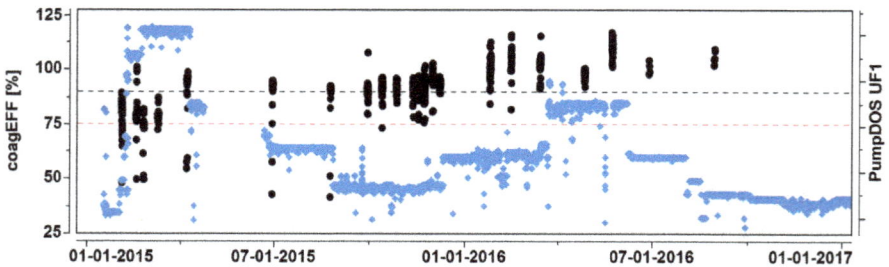

**Figure A9.** Change in coagulation efficiency (● coagEFF [%]) and dosing (◆ Al Dosing INT [L/h]) signal as a function of time for the two-year study period. The black and red horizontal lines highlight 90% and 75% coagEFF.

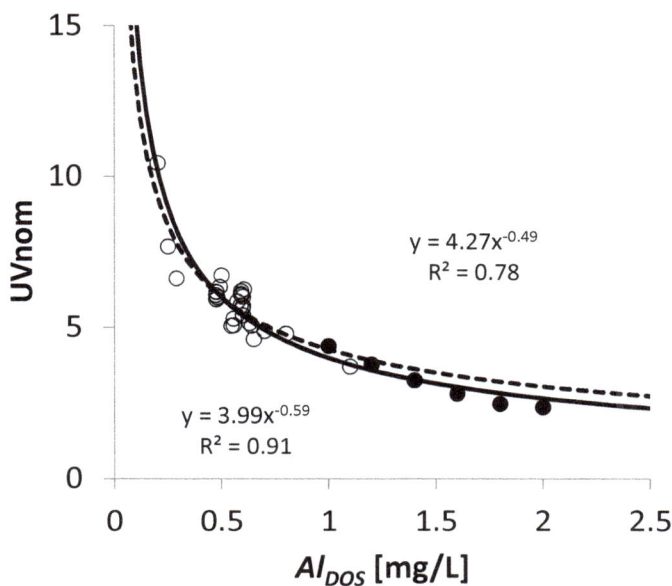

**Figure A10.** Estimation of normalised *UV* removal as a function of Al dose based on Equation (4). Black circles (●) are additional high-dose experiments, while white circles (O) are those from the optimization period. Two regression lines including (below) or excluding the additional points are displayed (above).

## References

1. Forsberg, C.; Petersen, R., Jr. A darkening of Swedish lakes due to increased humus inputs during the last 15 years. *Verhandlungen des Internationalen Verein Limnologie* **1990**, *24*, 289–292.
2. Eikebrokk, B.; Vogt, R.; Liltved, H. NOM increase in Northern European source waters: Discussion of possible causes and impacts on coagulation/contact filtration processes. *Water Sci. Technol. Water Supply* **2004**, *4*, 47–54.
3. Worrall, F.; Burt, T. Trends in DOC concentration in Great Britain. *J. Hydrol.* **2007**, *346*, 81–92. [CrossRef]
4. Evans, C.D.; Monteith, D.T.; Cooper, D.M. Long-term increases in surface water dissolved organic carbon: Observations, possible causes and environmental impacts. *Environ. Pollut.* **2005**, *137*, 55–71. [CrossRef] [PubMed]
5. Köhler, S.J.; Buffam, I.; Seibert, J.; Bishop, K.H.; Laudon, H. Dynamics of stream water TOC concentrations in a boreal headwater catchment: Controlling factors and implications for climate scenarios. *J. Hydrol.* **2009**, *373*, 44–56. [CrossRef]
6. Delpla, I.; Jung, A.-V.; Baures, E.; Clement, M.; Thomas, O. Impacts of climate change on surface water quality in relation to drinking water production. *Environ. Int.* **2009**, *35*, 1225–1233. [CrossRef] [PubMed]
7. Lavonen, E.E.; Kothawala, D.N.; Tranvik, L.J.; Gonsior, M.; Schmitt-Kopplin, P.; Köhler, S.J. Tracking changes in the optical properties and molecular composition of dissolved organic matter during drinking water production. *Water Res.* **2015**, *85*, 286–294. [CrossRef] [PubMed]
8. Lavonen, E.E.; Gonsior, M.; Tranvik, L.J.; Schmitt-Kopplin, P.; Köhler, S.J. Selective Chlorination of Natural Organic Matter: Identification of Previously Unknown Disinfection Byproducts. *Environ. Sci. Technol.* **2013**, *47*, 2264–2271. [CrossRef] [PubMed]
9. Jacangelo, J.; DeMarco, J.; Owen, D.; Randtke, S. Selected processes for removing NOM: An overview. *J. Am. Water Works Assoc.* **1995**, *87*, 64–77.
10. Jacangelo, J.G.; Rhodes Trussell, R.; Watson, M. Role of membrane technology in drinking water treatment in the United States. *Desalination* **1997**, *113*, 119–127. [CrossRef]

11. Singer, P.; Bilyk, K. Enhanced coagulation using a magnetic ion exchange resin. *Water Res.* **2002**, *36*, 4009–4022. [CrossRef]

12. Matilainen, A.; Iivari, P.; Sallanko, J.; Heiska, E.; Tuhkanen, T. The role of ozonation and activated carbon filtration in the natural organic matter removal from drinking water. *Environ. Technol.* **2006**, *27*, 1171–1180. [CrossRef] [PubMed]

13. Zularisam, A.; Ismail, A.; Salim, R. Behaviours of natural organic matter in membrane filtration for surface water treatment—A review. *Desalination* **2006**, *194*, 211–231. [CrossRef]

14. Toor, R.; Mohseni, M. UV-H2O2 based AOP and its integration with biological activated carbon treatment for DBP reduction in drinking water. *Chemosphere* **2007**, *66*, 2087–2095. [CrossRef] [PubMed]

15. Matilainen, A.; Sillanpää, M. Removal of organic matter from drinking water by advanced oxidation processes: A review. *Chemosphere* **2010**, *80*, 351–365. [CrossRef] [PubMed]

16. Amy, G.L.; Alleman, B.C.; Cluff, C.B. Removal of dissolved organic matter by nanofiltration. *J. Environ. Eng.* **1990**, *116*, 200–205. [CrossRef]

17. Köhler, S.J.; Lavonen, E.E.; Keucken, A.; Schmitt-Kopplin, P.; Spanjer, T.; Persson, K.M. Upgrading coagulation with hollow-fibre nanofiltration for improved organic matter removal during surface water treatment. *Water Res.* **2016**, *89*, 232–240. [CrossRef] [PubMed]

18. Lidén, A.; Persson, K.M. Comparison between ultrafiltration and nanofiltration hollow-fiber membranes for removal of natural organic matter—A pilot study. *J. Water Supply Res. Technol. AQUA* **2015**, *65*. [CrossRef]

19. Jung, C.-W.; Kang, L.-S. Application of combined coagulation-ultrafiltration membrane process for water treatment. *Korean J. Chem. Eng.* **2003**, *20*, 855–861. [CrossRef]

20. Vickers, J.; Thompson, M.; Kelkar, U. The use of membrane filtration in conjunction with coagulation processes for improved NOM removal. *Desalination* **1995**, *102*, 57–61. [CrossRef]

21. Tran, T.; Gray, S.; Naughton, R.; Bolto, B. Polysilicato-iron for improved NOM removal and membrane performance. *J. Membr. Sci.* **2006**, *280*, 560–571. [CrossRef]

22. Valinia, S.; Futter, M.; Fölster, J.; Cosby, B.; Rosén, P. Simple models to estimate historical and recent changes of total organic carbon concentrations in lakes. *Environ. Sci. Technol.* **2015**, *49*, 386–394. [CrossRef] [PubMed]

23. Vattenkvalitet i Hallands Sjöar 2012. *Resultat Från Omdrevsprogrammet 2007–2012 Länsstyrelsen i Hallands län Enheten för Naturvård & Miljöövervakning Meddelande*; Vattenkvalitet i Hallands Sjöar: Hallands Sjöar, Swedish, 2013; ISSN 1101-1084.

24. Finstad, A.; Blumentrath, S.; Tømmervik, H.; Andersen, T.; Larsen, S.; Tominaga, K.; Hessen, D.; De Wit, H. From greening to browning: Catchment vegetation development and reduced S-deposition promote organic carbon load on decadal time scales in Nordic lakes. *Sci. Rep.* **2016**, *6*, 31944. [CrossRef] [PubMed]

25. Keucken, A.; Heinicke, H. NOM characterization and removal by water treatment processes for drinking water and ultra pure process water. In Proceedings of the 4th IWA Speciality Conference on Natural Organic Matter, Costa Mesa, CA, USA, 27–29 July 2011.

26. Keucken, A.; Donose, B.-C.; Persson, K.-M. Membrane fouling revealed by advanced autopsy. In Proceedings of the 8th Nordic Drinking Water Conference, Stockholm, Sweden, 18–20 June 2012.

27. Keucken, A.; Wang, Y.; Tng, K.H.; Leslie, G.L.; Spanjer, T.; Köhler, S.J. Optimizing hollow fibre nanofiltration for organic matter rich lake water. *Water* **2016**, *8*, 430. [CrossRef]

28. Huber, S.A.; Balz, A.; Abert, M.; Pronk, W. Characterisation of aquatic humic and non-humic matter with size-exclusion chromatography–organic carbon detection–organic nitrogen detection (LC-OCD-OND). *Water Res.* **2011**, *45*, 879–885. [CrossRef] [PubMed]

29. Cory, R.; McKnight, D. Fluorescence spectroscopy reveals ubiquitous presence of oxidized and reduced quinones in dissolved organic matter. *Environ. Sci. Technol.* **2005**, *39*, 8142–8149. [CrossRef] [PubMed]

30. Ohno, T.; Bro, R. Dissolved organic matter characterization using multiway spectral decomposition of fluorescence landscapes. *Soil Sci. Soc. Am. J.* **2006**, *70*, 2028–2037. [CrossRef]

31. Parlanti, E.; Woerz, K.; Geoffroy, L.; Lamotte, M. Dissolved organic matter fluorescence spectroscopy as a tool to estimate biological activity in a coastal zone submitted to anthropogenic inputs. *Org. Geochem.* **2000**, *31*, 1765–1781. [CrossRef]

32. Lidén, A.; Keucken, A.; Persson, K.M. Uses of fluorescence excitation-emissions indices in predicting water efficiency. *J. Water Proc. Eng.* **2017**, *16*, 249–257. [CrossRef]
33. Köhler, S.J.; Kothawala, D.; Futter, M.N.; Liungman, O.; Tranvik, L. In-Lake Processes Offset Increased Terrestrial Inputs of Dissolved Organic Carbon and Color to Lakes. *PLoS ONE* **2013**, *8*, e70598. [CrossRef] [PubMed]

*water*

MDPI

*Review*

# Membrane Fouling for Produced Water Treatment: A Review Study From a Process Control Perspective

Kasper L. Jepsen *, Mads Valentin Bram, Simon Pedersen and Zhenyu Yang

Department of Energy Technology, Aalborg University, DK-6700 Esbjerg, Denmark;
mvb@et.aau.dk (M.V.B.); spe@et.aau.dk (S.P.); yang@et.aau.dk (Z.Y.)
* Correspondence: klj@et.aau.dk; Tel.: +45-6169-5368

Received: 9 May 2018; Accepted: 19 June 2018; Published: 26 June 2018

**Abstract:** The offshore oil and gas industry is experiencing increasing water cuts as the reservoirs mature. The increase in produced water stresses the currently deployed deoiling technologies, resulting in more oil in the discharged water. Deploying membrane filtration to reduce the hydrocarbon concentration inherits additional complications related to fouling of the membranes: A process where the accumulation of material within and on the membrane surface adds additional flow resistance. This paper reviews and analyses the fouling detection, removal, prevention, dynamical and static modeling, with emphasis on how the membrane process can be manipulated from a process control perspective. The majority of the models rely on static descriptions or are limited to a narrow range of operating conditions which limits the usability of the models. This paper concludes that although the membrane filtration has been successfully applied and matured in many other industrial areas, challenges regarding cost-effective mitigation of fouling in the offshore deoiling applications, still exist. Fouling-based modeling combined with online parameter identification could potentially expand the operating range of the models and facilitate advanced control design to address transient performance and scheduling of fouling removal methods, resulting in cost-effective operation of membrane filtration systems. With the benefits of membrane filtration, it is predicted that membrane technology will be incorporated in produced water treatment, if the zero-discharge policies are enforced globally.

**Keywords:** crossflow membrane filtration; produced water treatment; fouling; modeling; process control; separation; multiphase

## 1. Introduction

In offshore oil and gas production, an increasing environmental concern is the enormous amounts of produced water (PW) discharged into the oceans. Matured oil fields in the Danish North Sea produce three barrels of water for every barrel of oil [1]. The extraordinary amount of PW is considered the largest stream of contaminated water in the exploration and production of oil and gas [2]. The PW can be discharged to the sea if treated to comply with governmental regulations. The governmental regulations for discharge into the North Sea is a concentration of 30 mg/L oil-in-water (OiW) and a maximum of 202 tonnes of oil discharged in 2017 and 2018 [3]. In 2015 the Danish Environmental Protection Agency reported a total of 193 tonnes of dispersed oil discharged, which is remarkably close to the allowed amount [4], hence fundamental change is required to guarantee compliance with future governmental regulations.

The currently used technologies in the oil and gas sector, for water purification and oil removal, are mainly: Gas flotation, hydrocyclone, and gravity-based separator [5]. While these technologies provide sufficient oil and water separation to comply with the current regulation, a growing environmental concern may force regulation to become stricter. Common operational performance of the hydrocyclone reduces the OiW concentration to 20–80 mg/L [6], and the separation efficiency is highly depending on droplet size [7]. In general, larger oil droplets are easily separated compared to smaller droplets, especially for gravity-based methods [8,9].

Previous studies have investigated the characteristics and the available technologies for produced water treatment (PWT) and found membrane filtration to be a promising candidate for improving separation efficiency [10–14] examples of membrane filtration deployment is summarized in [11]. Membrane filtration is the process of using a semi-permeable material with very small pores to filter substances based on droplet and particle size. Especially ceramic membranes and their advantages, such as chemical, mechanical, and thermal stability and narrow pore size distribution, are well suited for PWT [10,15,16]. The pressure-driven membranes are commonly divided into four categories based on pore size; microfiltration (MF), ultrafiltration (UF), nanofiltration (NF) and reverse osmosis (RO), where UF where found to be superior for reducing OIW concentration [10,12–14,17,18]. NF and RO have also been deployed when silica, dissolved organic matter, and salt are to be removed from the produced water [11].

Studies in PWT using membrane filtration showed that fouling is a considerable problem [19–23]. Fouling, i.e., accumulation of contaminants inside the membrane and on the membrane surface, reduces the permeability, and thereby cost-effectiveness of the membrane. In short, fouling can be either reversible or irreversible, and appears as; scaling, silt, biofouling, and organic fouling [24]. The reduction in flux caused by fouling can be as high as 80%, even when antifouling measures such as backwash and crossflow (CF) are deployed [15].

The unavoidable fouling necessitates additional installation footprint (the space needed for the installation) to compensate for the reduced permeability caused by the fouling. In offshore cases, this leads to undesirable weight and space demands, which are crucial factors for cost-effective offshore installations [6,10,25].

Several recent review studies have been carried out for membrane filtration of produced water, see [10,15,26–31]. The studies address the complexity and composition of PW [15,26–28,30], chemical pretreatment [10,28,31], physical pretreatment [10], membrane materials and modification [27,30], membrane pore size and its effect on PWT [26–28,31], and steady state operating conditions and its effect on the filtration system [27]. However, none of the reviews address the membrane filtration system from a process control perspective, which is the main focus of this review. The compelling room for improvement in process control is confirmed in [32], regarding the following areas:

1. Scheduling of fouling removal measures.
2. Scheduling of fouling prevention measures.
3. Process optimization, to minimize operational and maintenance costs, where cost is a balance of fouling removal, process uptime, installation footprint, and process throughput.

It is these items that will be addressed in this study, whereas membrane material and chemicals can significantly improve membrane filtration performance [11,33,34], but the focus of this review remains control oriented. A series of models will be described and their potential application in process control design will be discussed. Fundamental hydrodynamic effects, interactions on a molecular level, and chemical effects are not considered in this work, as such effects would result in an unnecessary high model complexity which is not beneficial for control design. Figure 1 shows an overview of a CF membrane filtration system and the common terms associated with it.

The rest of this paper is organized as follows: Section 2 covers filtration of produced water; Section 3 introduces the critical flux concept; Section 4 presents methods deployed for fouling prevention and removal; Section 5 reviews the fouling models; Finally, the paper is concluded in Section 6.

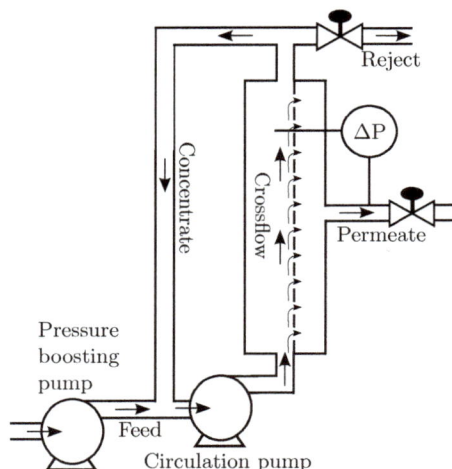

**Figure 1.** Membrane filtration overview.

## 2. Filtration of Produced Water

A detailed description of PW is covered in [26,30], but in general PW is an OiW emulsion, where oil is dissolved in water. The emulsion is stabilized by the naturally occurring surfactants from the reservoir. The properties and composition of the PW change according to the oil field, well, field maturity, and artificially added chemicals, such as corrosion inhibitors and biocides. In particular the immense variation between wells ensures that a standardized filtration solution is near impossible design, and therefore filtration systems are often designed for a specific well or oil field [35,36].

The essential part, from a control perspective, is that the PW properties are changing and with the large variance between different wells an unified control solution for the filtration unit must adapt to those conditions. Typical industrial applications of the membrane technology are food industry, pharmaceutical, biotechnological, and chemical sectors, all of which are very well designed controlled processes where the flow, pressure, and feed properties are predictable throughout the lifetime of the membranes. On the contrary, PWT conditions can significantly change with time, especially feed properties and flow rate can change with the maturity of the oil field.

The majority of the studies on membrane filtration deal with separation of liquid and non-deformable material. For the studies addressing OiW separation, the deformability of oil is often not considered, examples hereof [37–40]. For membrane filtration of PW, it is necessary to consider the ramifications that deformation of the oil droplets can have on the defined methods and models. Depending on the driving pressure and interfacial tension forces, oil droplets can be forced to deform and be pushed through the pores that are narrower than the droplet's diameter. In comparison, rigid solid particles either permeate or become rejected independently of pressure but determined by pore size [41], the deformation of an oil droplet is illustrated in Figure 2. To determine if an oil droplet permeates or is rejected by the membrane, a set of general parameters are defined in Equation (1), and a droplet is forced through the constricted channel if

$$P_d - P_u > \gamma(c_u - c_d) \tag{1}$$

is satisfied [42].

A model describing the TMP required for a droplet to deform and permeate the membrane (critical pressure) was proposed in [43], later corrected in [44] (Equation (2)), and validated in [41]. The critical pressure required is described as:

$$\Delta P^* = 2\gamma \frac{cos(\beta)}{r_p} \cdot \left[ 1 - \left( \frac{2 + 3cos(\beta) - cos(\beta)^3}{4(d/2r_p)^3 cos(\beta) - (2 - 3sin(\beta) + sin(\beta)^3)} \right) \right], \tag{2}$$

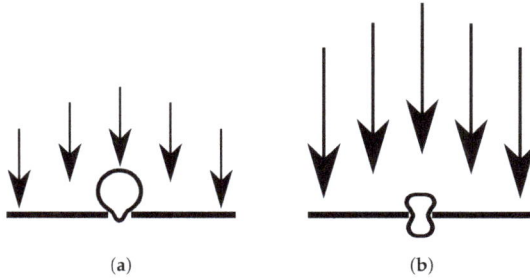

(a)                                    (b)

**Figure 2.** Oil droplet with sufficient and insufficient driving pressure to overcome interfacial tension. (a) Insufficient driving pressure; (b) Sufficient driving pressure.

Figure 3 is generated to show examples of $\Delta P^*$ as function of droplet size, where $\beta$ and $\gamma$ are defined in [41] to be 135° and 14 dyn/cm respectively. It should be noted that $\theta$ and $\gamma$ depend on oil composition and added chemical agents, which for PW are known to be varying over time [45].

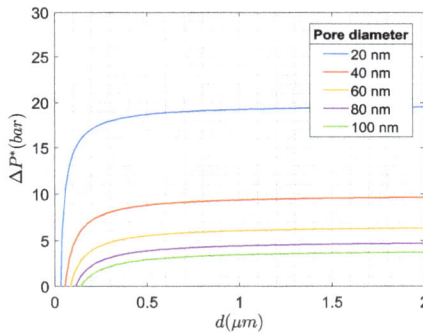

**Figure 3.** Required pressure to force a droplet with specific size through five different pore sizes.

The general trend of Figure 3 indicates that once above some droplet diameter, in this case, 0.5 μm, the $\Delta P^*$ only increases asymptotically to a constant value, whereas the $\Delta P^*$ is much more reliant on membrane pore size. This indicates that once above some droplet size, the oil droplets and solid particles act similarly regarding being fully blocked, similar conclusions were made in [41,45].

A comprehensive investigation into the deformation of droplets for CF MF of an OiW mixture (OiWM) has been made in [46]. According to simulations of CF filtration, at TMPs less than 1 bar and a pore size of 0.2 μm, droplets above 0.9 μm are rejected [41]. At higher CF velocity (CFV) the contact angle changes and higher pressures are required for oil droplets to permeate the membrane, thus increasing separation efficiency. However, oil droplet breakup occurs at too high CFV, reducing separation efficiency. Results, which have been replicated using computational fluid dynamics technique [47].

Analytic [18,41,48] and experimental [49–51] studies, summarized in Table 1, shows critical pressures found based on different experiments and models. Critical pressures for experimental studies are determined based on steady state removal efficiency and TMP, if the removal efficiency suddenly drops as pressure increases, it is assumed to be the critical pressure. In the experimental studies on oil removal using MF, the removal efficiency was observed to decrease once the TMP increases above 1.5 bar [50,51]. This verifies the results in [41,43], where the critical pressure was found to be within [1–2] bars. The analytic and experimental results, in Table 1, are within acceptable tolerance given the difference in pore and droplet size and model assumptions.

**Table 1.** Critical pressure for an OiWM found in different studies.

| Pore Size | Membrane Type | Critical Pressure | Method | Mean Droplet Size | Reference |
|---|---|---|---|---|---|
| 0.15 μm | Ceramic | 1 bar | Analytic | 0.9 μm | [41] |
| 0.5 μm | Ceramic | 2.8 bar | Analytic | 11 μm | [18] |
| 0.2 μm | Ceramic | 1.25 bar | Experiments | 3 μm | [49] |
| 0.2 μm | Ceramic | 1.55 bar | Experiments | Not reported | [50] |
| 0.05 μm | Ceramic | 2 bar | Experiments | Not reported | [51] |
| 0.2 μm | Inorganic Aluminum Oxide | 4 bar | Analytic | 1 μm | [48] |

The examined critical pressure models do not directly consider how fouling behavior of oil differs from solid particles, but still, some conclusions can be made:

- At steady state, an oil droplet larger than the pore size may permeate the membrane if the TMP is large enough.
- It is generally not considered how dynamic changes in TMP affect the oil droplets' ability to permeate the membrane.
- Ideally maintaining the TMP below the critical pressure causes an unrealistic low TMP given the droplet distribution.

The deformability of oil is not necessarily an entirely undesired effect, as applying high reverse pressure causes stuck oil droplets in the pores to deform and exit the membrane. Applying heat with the cleaning media lowers the viscosity and allows stuck oil droplets to easier deform and exit the membrane pores, a similar technique is exploited when extracting bitumen from the reservoir by injecting steam [11]. Furthermore, because oil droplets can deform, transient performance of especially the TMP is important, if the permeate quality is to be maintained.

## 3. Critical Flux Concept

The critical flux is a well defined and observed concept that is widely used within the field of membrane filtration. This section shortly introduces the critical flux concept, and highlights some studies where critical flux is observed. The critical flux hypothesis is defined as: There exists a critical flux, $j^*$, such that $j < j^*$ yields $\frac{dR_t}{dt} \approx 0$, or alternatively, while the flux is less than the critical flux, no or little fouling occurs [52–55]. The critical flux is frequently used as a measure of membrane performance and is dependent on solute density, particle diameter, particle form factor, porosity, hydrodynamics, and temperature [55]. The critical flux behavior is clearly observed inside and outside the laboratory environment [56,57], and its behavior for an OiWM was recently investigated in [58,59].

Operating membrane filtration at sub-critical flux leads to less fouling and resistance, hence reducing energy consumption. For offshore deployment, installation footprint is the critical factor to consider, especially as membrane filtration technology requires around 3 times larger footprint compared to the hydrocyclone technology [6]. To minimize the required footprint, it is essential to operate the membranes at supercritical flux (above the critical flux) with a manageable degree of fouling.

In some cases, fouling is observed even when the operational flux is below the estimated critical flux [60–62]. It can conceivably be caused by either an unnoticeable low fouling rate inside the membrane pores or that each droplet size has a specific critical flux resulting in a distribution of critical fluxes whereas the critical flux identified is often not the lowest critical flux [63]. The low fouling rate can cause flux to locally exceed the critical flux (illustrated in Figure 4), causing the fouling to suddenly accelerate [53,62]. From a control perspective, the problem could possibly be avoided by ensuring that the local flux remains below the critical flux. To do so, the control strategy must be able to control the local flux. As local flux is not a direct measurement it must be estimated based on available sensors, e.g., pressure and flow measurements.

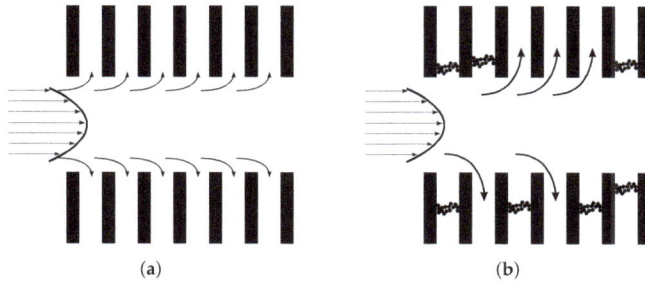

(a)                    (b)

**Figure 4.** Fouling under sub-critical flux operation. (**a**) Pre-blockage; (**b**) Post-blockage.

### 3.1. Critical Flux Identification

A widely used method for identifying the critical flux is flux-stepping. Flux-stepping is where the flux is increased in steps, and fouling accumulation is evaluated for each step [56,64–66]. At the flux where fouling begins to accumulate, the previous flux step is assumed to be the critical flux [52]. Alternative versions of flux-stepping exist, where relaxation (zero or nearly zero flux) of the membrane is applied between each increment in flux as illustrated in Figure 5. The CF in the relaxation phase removes some reversible fouling and provides an opportunity to estimate at which flux irreversible fouling occurs [53]. A slightly different procedure is proposed and investigated in [67,68].

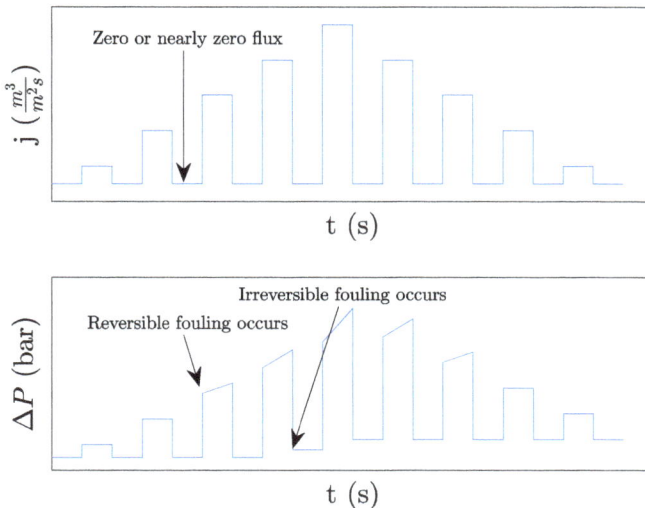

**Figure 5.** Concept illustration of the modified flux-stepping method.

A gray-box-model has be used as an alternative to experimentally identify the critical flux under different conditions, as seen in Equation (3) [69].

$$j^* = 0.37\lambda \left( \frac{C_w^2 (d/2)^4}{L} \right)^{\frac{1}{3}} ln \left( \frac{C_w}{C} \right) \tag{3}$$

The model was validated with satisfactory results based on CF filtration of an OiWM. Coefficients such as concentration at the membrane wall are difficult to identify and both a view cell and camera were used to visually identify model parameters [58]. Because of the instrumentation required for the identification process, it is neither practical nor cheap to implement on large-scale systems, especially as the variation between oil fields would require reidentification.

### 3.2. Fouling Detection

For identifying the critical flux, fouling detection is a necessity, a commonly used method for fouling detection is based on TMP and flux measurement to estimate the internal resistance of the membrane. The estimated changing rate of the internal resistance is then used for fouling detection. Under constant flux operation the fouling detection criteria, can be simplified to threshold detection of $\frac{d\Delta P}{dt}$, as shown in Equation (4) [53].

$$\frac{d\Delta P}{dt} > F \Rightarrow \text{Fouling occurrences} \tag{4}$$

Two alternative methods for fouling detection are proposed in [52,56]. The first alternative is to use concentration-based measurement in a closed loop system, and as the concentration is reduced, fouling is concluded to have settled. The second alternative is to use non-invasive microscopic-based observation method (DOTM), where each particle could be observed while settling on the membrane. Both methods require an additional and expensive instrument to be installed, and the concentration method requires a closed loop system, both factors are problematic for an industrial PWT plant.

### 3.3. Critical Flux Discussion

Even though fouling can occur below the critical flux, critical flux still provides a method to judge membrane performances across different operating conditions. However, the critical flux identification and the fouling detection procedure must be standardized to ensure comparability between studies. Of the three methods for fouling detection the (DOTM) and concentration-based methods result in far lower $j^*$ values, than the TMP-gradient-based method, conceivable because particles are depositing in blind membrane pores where permeability is not affected, as a result, the TMP-gradient-based $j^*$ describes the membrane's performance more accurately.

## 4. Fouling Prevention and Removal

Multiple methods for reducing fouling have been investigated. The methods can be branched into categories: Prefilters, surface shearing, chemical agents, operational conditions, and control thereof. This review will be limited to: Prefilters, surface shearing, and operational conditions and how the operating conditions of those methods affect both membrane and fouling prevention performance. Fouling removal techniques are extensively deployed to remove any reversible fouling, examples hereof are: Chemical cleaning, backwashing, and ultrasonic cleaning [70–73], whereas this review addresses backwashing and ultrasonic cleaning.

### 4.1. Pretreatment

Less expensive prefilters are often installed as pretreatment and protection for the relatively costly membranes [74]. Prefilters have a larger pore size which remove particles in the feed, undesirable to the membranes. Like membranes, prefilters suffers from fouling, and cleaning actions must be taken to maintain filter porosity [75], and thus less care is taken with respect to cleaning as they can be replaced for a low cost [76]. For produced water, the gravity-based separators and hydrocyclones are often deployed as pretreatment, as they are more efficient than prefilters for oil removal. Typically the OiW concentration is reduced by the gravity-based separator to 2000–5000 mg/L [77], and the hydrocyclones are capable of reducing the concentration further to 20 mg/L [78]. The hydrocyclones have several advantages compared to prefilters, as they inexpensive, requires less maintains, and are more compact. However, they are very sensitive to changes in flow rate, and control must be carefully designed [79].

### 4.2. Surface Shearing

Surface shearing is extensively used to prevent fouling accumulation. The methods deployed to generate shearing on the surface are: CF [49,80,81], vibrating/rotating filters [23,82–84], and membrane channel modifications [85,86]. The review covers only the surface shearing introduced by the often deployed CF technique.

CF filtration adds shearing to the membrane surface and reduces the probability of particles accumulating as fouling. CF UF of an OiWM, compared to dead-end filtration, can in some cases increase flux by over 100% depending on CFV [80]. However CFV can negatively affect the permeate quality for MF of an OiWM. In [49,87] it was observed that CFV decreases total organic content (TOC) removal efficiency, in the range of 0.75 m/s to 4 m/s. This observation is described with the model developed for an OIWM in [41], where higher CFV (shear rates) would cause droplet break-up and reduced removal efficiency. Furthermore, the model predicted that an optimal shear rate exists, where above and below removal efficiencies are reduced, an effect which has not been experimentally observed.

Different conclusions are drawn with respect to the effect of CFV on steady state flux. Some studies show increased CFV reduces the permeate flow, subject to constant TMP control [81,88,89], while in other studies (on OiWM) the opposite response is observed [18,49,90]. In [89] particle in the range of 2.7 µm to 27.5 µm where tested with different CFV. It is observed that CFV does cause higher specific cake resistance and less cake thickness, and it is the ratio between those two that can cause CFV to reduce the overall flux [81]. As for the cases with an OiWM, the tendency is consistent across the literature, that increased CFV results in higher permeate flow rates [49,90].

In [88] a filtration system was constructed, where the cross-section of the membrane could be monitored while under operation. It was observed that constant TMP and low CFV would cause particles to settle in a less packed pattern, compared to high CFV where a more structured packing was observed. Even though the thickness of the cake was reduced by a factor of 2.5, the resulting permeate flux was reduced by a factor of 2, hence CFV affects the cake porosity. The relationship between CFV and porosity can explain results where CFV reduces permeate flux, as in [89]. For an OiWM the effect is not observed, because the packing behavior of oil differs significantly from solids.

### 4.3. Operating Conditions

Operating conditions, such as CFV, temperature, TMP, and flux can be adjusted to reduce fouling and thereby effectively prevent fouling, this has been extensively studied in [91], where an OIWM was treated. Optimizing these operating conditions to reduce both fouling and running cost of membrane filtration systems subject to fulfill the filtration requirement, has been extensively investigated. Firstly, operating conditions for an OiWM filtration system was optimized by deploying the full factorial design methodology, where the different operating conditions was analyzed with respect to permeate flux, fouling resistance, and TOC rejection [92]. Secondly, optimal operating conditions for MF and UF systems treating an OiWM where found in [93]. The Taguchi method was deployed to find the optimal conditions among temperature, TMP, CFV, and back pulse time. However, only three different levels for each parameter was investigated, whereas the optimal point was found to be the highest level in temperature, TMP, and CFV, indicating that the considered levels poorly selected and higher flux could be achieved by extending the considered range. Other studies have shown that significant savings are also achieved by setpoint optimization in [94–97].

Commonly, a membrane filtration system is operated in one of two modes; (i) constant TMP, or (ii) constant permeate flux. Constant flux is often necessary to meet demands from the up- or down-stream processes [20]. Unfortunately, there are different claims to which mode is preferred and favorable with respect to fouling minimization [32,98]. Results in [98] showed that constant flux operation developed less fouling during filtration of a fixed permeate volume. On the contrary, for surface water treatment it was observed that in certain temperature ranges constant TMP operation resulted in less fouling [32]. Which control method is best suit for an OIWM have not been considered in the literature, but two factors must be accounted for; fouling and process requirements. For PWT the feed flow can be very irregular which must be accounted for by the controller, as such, either constant flow nor pressure are suitable, and it can be necessary to allow large oscillations to be directed to the reject.

The preferred controlled variable for membrane filtration is widely discussed in [32,98–100], while transient behavior and control structure of the system are often not addressed. Given the critical flux hypothesis, it is crucial to ensure the system is kept at the defined operating point even though process disturbances are present, which are likely for PWT [101]. Greater emphasis on the control structure and the design, in order to address transient system behavior and reference tracking, could improve membrane filtration effectiveness and efficiency. In [102] it is claimed that the MPC technique can improve the transient response, but the direct effect on fouling compared to the traditional deployed PID controller is not presented. The critical flux concept indicates that increasing permeate flow rate can cause increased resistance that is not reduced together with permeate flow rate, a fact that is not considered together with process disturbances and transient system behavior. Documenting how fouling is affected by process disturbances in permeate flow rate would highlight if filtration improvement could be made by deploying disturbance rejection control.

### 4.4. Backwashing

A common approach to fouling removal is backwashing; a process where the permeate flow direction is reversed. The reverse flow cause most of the fouling to be released back into the CF channel [49,73,103]. From the CF channel, the released fouling is commonly directed to a discharge, where it can be either stored or directed to a previous separation step for reprocessing [104].

For backwashing, the duration, frequency, pressure, temperature, and backwash media can be adjusted to achieve improved recovery. Throughout literature, different backwashing conditions have been tested, Table 2 summarises the conditions applied. The table shows that even in the field of OiWM filtration, there is a significant difference between chosen duration and frequency. The variance indicates that either the optimal configuration is not found in each case, or more likely, the optimal configuration is case specific.

**Table 2.** Backwashing configurations for ceramic membranes.

| Filtration Duration | Backwash Duration | TMP | Membrane/Feed-Type | Reference |
|---|---|---|---|---|
| 20 min | 60 s | −3 bar | UF/surface water | [105] |
| 240 min | 20 min | −2.4 bar | UF/reservoir water | [106] |
| 60 s | 0.7 s | −6 bar | UF/OiWM | [18] |
| 280 s | 15 s | −2 bar | MF/OiWM | [49] |
| 30 min | 60 s | −2 bar | UF/OiWM | [107] |

The choice of backwash frequency (time between backwash), duration (the time backwashing is applied), and pressure/flow is not explained in [18,49,108], and potentially higher permeate production could be achieved by finding the optimal interval and duration. This is confirmed in [109] which recently studied backwash optimization with respect to permeate production, where the interval, duration, and pressure is chosen based on experiment data, where 10 coefficients are identified based on 24 experiments at a duration of 6 h. The method requires time-consuming experiments at less then optimal operation, especially if the process is to be repeated as significant process change occurs. To extend this method to PW, the parameters should not be found based on a single set of experiments, but rather an online adaptive method where the system behavior is continuously monitored and backwash parameters are adjusted to compensate for the changes that occur in the PW properties and as irreversible fouling occurs.

An experimental study on an OiWM claims that up to around 95% of the original flux can be recovered and the average flux can be increased by 100% when deploying backwash [49]. The continuous recovery of 95% of the original flux indicates that no further irreversible fouling occurs which is unlikely and in direct conflict with the results in [18,107] where irreversible fouling continuously occurs. The definition of original flux used in [49] is likely to be interpreted as the flux recovered from previous backwash iteration, and this definition of original flux should be avoided, as it can be confused with the initial clean membrane flux.

A limitation of backwashing is that while the TMP is reversed no permeate is produced. If a typical backwash sequence, with 20 min backwash and 4 h normal operation is considered, the total downtime from backwashing is 7.7%. The significant downtime from applying backwashing gives considerable room for optimizing the backwash sequence. Additionally, the backwashing media commonly used is permeate produced from the membrane system itself, and therefore the overall average flux must be considered, as suggested in [110]. The overall average flux can be calculated as shown in Equation (5).

$$j_{avg} = \frac{\int_0^{t_f} j_f \, dt - \int_{t_f}^{t_f+t_r} j_r \, dt}{t_f + t_r} \tag{5}$$

For an OiWM, it was observed that for short-term operation backwashing provided a higher ratio between flux recovery and required downtime, compared to chemical cleaning [111]. Nonetheless, backwashing is only suited for short-term flux recovery, and chemical cleaning is necessary for long-term operation [49]. The chemical cleaning process requires considerable downtime and chemical agents [72,74,112]. Thus, alternative techniques where no or less downtime is required would be advantageous.

*4.5. Ultrasonic*

Ultrasonic cavitation can be used for both removal and prevention, ultrasonic cleaning is a technique where electrical energy is utilized to create ultrasonic cavitation. The method is attractive, as it can be used while the membranes are in operation and thus requiring no downtime [70,71]. Multiple parameters, such as frequency and intensity, can be adjusted to obtain the optimal energy efficiency.

Multiple frequencies and intensity have been investigated, and the studies are summarized in Table 3. The observation from these studies is that lower frequency and higher intensity result in the highest permeate flux. It is theorized that lower frequencies provide better efficiency as larger air droplets are created [113], which correlates well with the studies in Table 3. Based on results in [114], the effectiveness (flux gained per W/cm$^2$ spent) does decrease with intensity, which implies that a balanced point does exist. However, the results are somewhat questionable, as the presented figures in [114] indicate an operational flux that is significantly higher than the clean water flux. This is probably caused by incorrect unit conversion, as the general tendency seems reasonable.

For an OiWM, ultrasonic cleaning is rarely deployed, in [115] a single frequency of 38 kHz at an unknown intensity was applied. The ultrasonic cavitation reduced the permeate flow resistance by 19%. and the study concluded that the reduction in flow resistance would naturally lead to a reduced energy usage. However, the energy usage of the ultrasonic transducers is not addressed and compared to applying higher TMP or CFV. Without any comparison, it is difficult to postulate that applying ultrasonic cleaning is more efficient.

Table 3. Ultrasonic case studies for membranes filtration.

| Frequencies | Intensity | Application | Results | Reference |
|---|---|---|---|---|
| 28, 45, and 100 kHz | 23 W/cm$^2$ | Filtration of peptone and milk aqueous solutions | 28 kHz increased permeate flux by 100% | [116] |
| 20 kHz | 5 W/cm$^2$ | Membrane distillation | Permeate flux increased by 300% | [117] |
| 70–620 kHz | 0–2.2 W/cm$^2$ | Sulfate polystyrene latex particles | 2.2 W/cm$^2$ and 70 kHz provide the highest flux recovery | [113] |
| 28 kHz | 0–1.7 W/cm$^2$ | Filtration of dry baker's yeast | 1.7 W/cm$^2$ produced the highest permeate flux | [114] |
| 38 kHz | Non-specified | Filtration of an OiWM | 19% reduction in flow resistance | [115] |

Comparing ultrasonic results in the literature can be problematic, as the methods for determining the intensity are rarely debated nor explained. In [116] two methods are independently used: Intensity estimation based on changes in temperature of the liquid and measured with a pulse receiver. The two methods measure different intensities, the temperature-based method estimates the power dissipated as heat whereas the pulse receiver measures the ultrasonic power reaching the receiver. Either way, deploying different methods for measuring ultrasonic power results in incomparability across the literature.

*4.6. Fouling Prevention and Removal Discussion*

For backwashing, ultrasonic, and membrane operating conditions, there are several parameters that can be adjusted to improve the effectiveness. From a control perspective, these parameters should be carefully adjusted to either increase flux or the overall energy efficiency, depending on the requirements. Some studies have already investigated flux recovery and net permeate production optimization of the cleaning methods [18,95,109]. However, none have addressed the specific problem for PW, where feed properties and irreversible fouling state change with time. One way to account for this is to let the scheduling algorithm estimate system behavior online and adapt backwash intensity, duration, and forward filtration time to maintain the desired optimal, whether it is a balance between energy efficiency and permeate production or just permeate production.

In general, ultrasonic cleaning is an effective cleaning solution, without any necessity for downtime nor chemicals. However, significantly more energy is demanded by ultrasonic cleaning, compared to backwashing. Below are two comparable examples of energy consumption by usage of either ultrasonic cleaning or backwashing, respectively:

- Ultrasonic cleaning: With an intensity of 1 W/cm$^2$ to a filtration area of 1 m$^2$ nets 10 kW of power usage.
- Backwashing: At a relatively low TMP of 1 bar, with a membrane area of 1 m$^2$, the flux will be approximately $480 \frac{L}{h \cdot m^2 \cdot bar}$ [111]. In comparison a typical GRUNDFOS CRE 5-5 pump can provide 5 m$^2$/h at 4 bar, and consumes 1.1 kW while in operation [118].

Even though many studies investigate ultrasonic cleaning for fouling removal and find the method to be effective, the huge power requirement is rarely addressed [116,119]. It is not uncommon to conclude ultrasonic to be more effective then backwashing, purely based on an observed increase in flux, but clearly, an increase in flux does not necessarily lead to higher energy efficiency [115]. The unaddressed efficiency of ultrasonic cleaning, compared to backwashing, complicates the selection of the energy efficiency method. Additional studies to address energy efficiency of the two methods would benefit the field.

The ultrasonic cleaning method scales unfavorably with installation size compared to backwashing, as the ultrasonic transmitters must be placed with each membrane unit, whereas backwashing pressure can be supplied from a single unit. Especially, as the system scales to meet the huge amount of PW, where membrane area is measured in thousands of square meters, the required amount of installed ultrasonic transmitters is enormous.

## 5. Fouling Models

The models covered in this section are investigated for two purposes. Firstly, the models can be used for optimizing the fouling removal and prevention methods, such as backwash scheduling. Secondly, to enhance process understanding and interaction between different membrane filtration phenomena.

For typical linear time-invariant model-based control development, a sufficient model must fulfill a set of requirements; the model must be linear or be linearisable and ODE-based and have identifiable parameters. For advanced control design and optimization methods, such as MPC, the model is required to have relatively low computational load to facilitate online optimization calculation [120,121]. To ensure low computational load the chosen model must capture the essential dynamics and ignore insignificant details.

### 5.1. Blocking Laws

Early fouling models are developed in [122] and later extended in [123]. The models are based on constant TMP dead-end filtration and are divided into four types of blockage (illustrated in Figure 6) namely;

**Complete** assumes that every particle that reaches the membrane surface will cause sealing of a new pore.

**Intermediate** considers that every particle that reaches the membrane surface will be included in the fouling. This model includes the probability for the particles to settle on an already sealed pore.

**Standard** is derived based on the assumption that decreased pore volume is proportional to the permeate volume.

**Cake** assumes that not all fouling will occur inside the membrane, but rather on the surface of the membrane where a cake layer will accumulate.

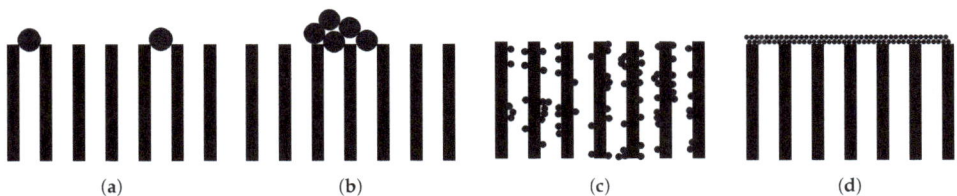

(a)          (b)          (c)          (d)

**Figure 6.** Four common types of blockage. (**a**) Complete blockage; (**b**) Intermediate blockage; (**c**) Standard blockage; (**d**) Cake blockage.

The blockage laws developed in [123] is summarized in Table 4, and can be simplified to into Equation (6).

**Table 4.** Hermia fouling models [123].

| Structure | Blockage Model |
|---|---|
| $\frac{d^2t}{dV^2} = \frac{a_k}{j^2}$ | Complete blockage |
| $\frac{d^2t}{dV^2} = \frac{a_i}{j}$ | Intermediate blockage |
| $\frac{d^2t}{dV^2} = \alpha_s j_0^{\frac{1}{2}} \left(\frac{1}{j}\right)^{\frac{2}{3}}$ | Standard blockage |
| $\frac{d^2t}{dV^2} = \alpha_c$ | Cake blockage |

$$\frac{d^2t}{dV^2} = k \left( \frac{dt}{dV} \right)^n \tag{6}$$

where $n$ is the type of fouling ($n = 0$ cake filtration, $n = 1$ intermediate blockage, $n = 3/2$ standard blockage, $n = 2$, for complete blockage). While it is unconventional to have the volume derivative of time, $\frac{dt}{dV}$ is an expression of the resistance. The equation can be rewritten into resistance and time-based derivative. In general the flow resistance through the membrane can be defined as Equation (7) [123].

$$\frac{dt}{dV} = \frac{1}{j} = \frac{R_t}{\Delta P}, \tag{7}$$

The second order derivative can be written as:

$$\frac{d^2t}{dV^2} = \frac{d}{dV} \left( \frac{R_t}{\Delta P} \right) = k \left( \frac{R_t}{\Delta P} \right)^n \tag{8}$$

Assuming constant TMP filtration, as assumed in the blocking law, and multiplying with $\frac{dv}{dt}$ on both sides results in Equation (9). Please note that given those assumptions the development of resistance over time, can be expressed exclusively with $R_t$ or $j$.

$$\frac{dR_t}{dt} = kR_t^n j \tag{9}$$

The models described in Table 4 do provide a common framework that is extensively deployed within membrane filtration [33,55,124–129]. The models were originally intended for dead-end filtration and did not consider flux recovery methods. Studies have been carried out to investigate the feasibility of applying the models to CF filtration, which is especially important for PW filtration systems as dead-end filtration is rarely deployed [87,130,131]. A limitation of Hermia's models is the assumption; that some process parameters remain constant throughout filtration operation, such as bulk concentration, TMP, and temperature. While it is possible to maintain certain parameters constant by deploying feedback control, the bulk concentration is for any offshore oil and water separation process uncontrollable. One approach for applying Hermia's models for CF filtration is to estimate the fouling coefficients for a specific operating condition. The resulting coefficients are only valid for a single CFV and model accuracy is significantly reduced if the system is operated away from the defined operating point [126].

Hermia's blocking laws have been widely deployed to model fouling behavior of an OiWM [37,38,132–134]. Reasonable model accuracy was achieved on by identifying model coefficients for each operating point and then using those coefficients for model prediction, indicating that the model structure can accurately describe the fouling behavior of an OiWM at a given operating point [132,133]. The cake blockage model provided the best fit to data, indicating that cake build-up is the main contributor to fouling when treating an OiWM [133]. In both [132,133] a relatively high TMP (above critical pressure) was applied without any significant reduction in model accuracy, hence droplets deformation require no model modification.

5.1.1. Critical Flux Extension

A critical flux-based model extension was proposed in [127]. The suggested extension introduces the shear rate created by CFV into Hermia's models. The modification, with respect to complete blocking, is written in Equation (10) [127].

$$A(t) = \underbrace{A_0 - \alpha V}_{\text{Hermia model}} + \underbrace{\int_0^t \lambda A_0 \, dt}_{\text{De Bruijn extension}} \tag{10}$$

Applying the shear rate to Hermia's blockage laws, results in Table 5.

**Table 5.** De Bruijn extended fouling models [127].

| Structure | Blockage Model |
|-----------|----------------|
| $\frac{dj_b}{dt} = -\alpha_b(j_b - j^*)$ | Complete blockage |
| $\frac{dj_i}{dt} = -j_i \cdot \alpha_i(j_i - j^*)$ | Intermediate blockage |
| $\frac{dj_c}{dt} = -j_c^2 \cdot \alpha_c(j_c - j^*)$ | Cake blockage |

The standard blockage law is unmodified as it occurs inside the membrane pores and is therefore assumed to be unaffected by CFV [127]. The proposed critical flux extension is validated in [127] using dextran T-70 as the solute and a reasonable prediction accuracy was achieved. A disagreement between the models proposed in [127,129] exists. The standard blockage model does not include critical flux since CF does not affect the fouling occurring in the pores [127], while it is included in [129].

Critical flux was introduced to adapt the blocking laws model from dead-end filtration to CF filtration, and as the standard blockage occurs inside the pores, this type of blockage is out of reach of CF, and CF should have no effect. However, the validity of this assumption is unconfirmed, and the CF introduced turbulence can affect the standard blockage to some degree.

The identification of critical flux is often based on flux stepping, a method that allows the critical flux to be approximated by visually inspecting the data [56,65,66], the method is good if the objective is to determine the critical flux for membrane performance evaluation, but for model parameter identification other methods should be investigated. Given the model structure and operating the system above the critical flux, system identification techniques can be deployed to identify both the fouling rate and the critical flux. If online identification is not used the critical flux must be identified across a range of operating conditions, and such experiments would be a time-consuming process.

5.1.2. Concentration Extension

The concentration of the feed does affect the fouling of the membrane [135], therefore the commonly used Hermia's models have been modified to incorporate the effect of the concentration directly [129]. This extension assumes that the probability of particles accumulating in or on the membrane is linearly proportional to the concentration. Experimental results obtained in [135] provided validation of the model structure, even across multiple concentration levels model accuracy remained good, but as polystyrene microspheres where used, it does not validate that the model assumptions are valid for a coalescing mixture, where size distribution and concentration is correlated.

The models proposed in [129] share a similar model structure to the models in [135], but is extended with the critical flux concept. The models for complete, standard, and cake blockage can be simplified as shown in Table 6.

**Table 6.** Kilduff fouling models [129].

| Structure | Blockage Model |
|---|---|
| $\frac{dj_b}{dt} = -\alpha_b(j_b - j^*) \cdot C$ | Complete blockage |
| $\frac{dj_s}{dt} = -\alpha_s j_s^{\frac{1}{2}}(j_s - j^*) \cdot C$ | Standard blockage |
| $\frac{dj_c}{dt} = -\alpha_c j_c^2(j_c - j^*) \cdot C$ | Cake blockage |

The assumption that fouling is linearly proportional to the concentration still needs to be validated for an OiWM before any conclusion can be made with respect to the model accuracy. Even though models have been explicitly developed to describe fouling for CF filtration, and to incorporate concentration into the models, recent studies for PWT continuously insist to use the original models developed for dead-end filtration [37,136]. If the extended models could be validated for PW, it would improve the accuracy of the models across different concentration levels, and thereby reduce the need to re-estimate the model for different concentration levels.

*5.2. Resistance-Based Models*

Another model approach for UF and MF is presented in [39], where the fouling is considered to be resistances in the series. The permeate flux is described with Darcy's law:

$$j = \frac{\Delta P}{\mu R_t},$$ (11)

The total membrane resistance is described as:

$$R_t = R_m + R_c,$$ (12)

where $R_m$ is the flow resistance through the membrane, defined as the clean membrane resistance combined with the resistances caused by pore blockage, as in Equation (13).

$$R_m = R_{m|c} + R_b$$ (13)

The cake layer resistance is described in terms of specific resistance and cake hight:

$$R_c = \hat{R}_c h_c$$ (14)

where the specific resistance and cake height is expressed as Equations (15) and (16):

$$\hat{R}_c = \frac{180(1 - \epsilon_c)^2}{d^2 \epsilon_c^3}$$ (15)

$$\frac{dh_c}{dt} = k_1 j - k_2 h_c$$ (16)

where the $k_1$ term is the transport of cake materials to the membrane wall, which is directly affected by the permeate flux. Theoretically, this effect should increase with concentration and decrease with larger particle size and CFV. The $k_2$ term is describing the back-transport (removal of cake materials) and should increase with CFV. The resistance of the membrane, partly blocked by pore blockage, can be expressed as in Equation (17).

$$R_{m|c} + R_b = \frac{8h_m}{f_o r_p^2}$$ (17)

The model is detailed and the relationship of the investigated parts are nicely described, but the models leave a few areas unexplained. Firstly, the development of the cake height is described to a degree where the exact correlation between CF, concentration, and cake development is unknown. Secondly, a complete description of how the pore blockage resistance develops over time is lacking, more precisely how $f_o$ and $r_p$ develops. Lastly, the model structure is not validated against experimental data, although an extended version is validated in [40].

From a control perspective, the model complexity is significantly increased compared to the blocking models described in Tables 4–6. The increased complexity is described in terms of 8 parameters that must be identified before the model can be used for control design purposes. The relatively high number of parameters that must be identified is a disadvantage, especially if the parameters are to be identified online.

### 5.2.1. Wiesner-Model Extension

An extension to the model, described in [39] (Equations (11)–(17)), is developed in [40]. The extension addresses some of the unexplained areas, such as cake transport and pore blockage resistance. The cake transport equations are modified to describe fouling in membrane bioreactors. The effective radius of the membrane pores is formulated as:

$$\frac{dr_p}{dt} = -\alpha_b C_w j \tag{18}$$

The fraction of open pore area $f_o$ is described similarly:

$$\frac{df_o}{dt} = -\alpha_{op} C_w j \tag{19}$$

The concentration at the membrane wall is less than the bulk concentration, as the passage through the cake layer causes some particles to settle in the cake layer itself:

$$C_w = C e^{-\frac{k_3 h_c}{j}} \tag{20}$$

The model for cake height is modified to a degree where concentration and air scouring is included with:

$$\rho_c \frac{dh_c}{dt} = j C - k_6 V_{air}^{k_5} \tag{21}$$

The model described in this section is validated and describes the fouling behavior of the membrane bioreactor with air scouring accurately. To apply the model for CF filtration, the back-transport term of Equation (21) must be modified to incorporate CFV instead of air scouring.

The suggested non-linear model structure requires nine parameters to be identified, which can be challenging. The advantage is that the identified parameters should remain constant even if the operating conditions, whereas the parameters of Hermia's models change with operating conditions. However, that is only if the model structure is able to completely capture the fouling behavior, in the likely case where the model structure cannot capture the fouling behavior across different conditions, the model parameters would require re-estimation.

### 5.2.2. Exponential Extension

The model proposed and developed in [39] and extended in [40] is further extended in [137]. An exponential term is added to the expressions for $R_c$ and $R_b$ to better explain behavior observed in the study. The extended equations are shown in Equations (22) and (23).

$$R_c = \hat{R}_c h_c \rho_c \cdot \underbrace{e^{N_c t}}_{\text{Extension}} \tag{22}$$

$$R_{m|c} + R_b = \frac{8h_m}{f_o r_p^2} \cdot \underbrace{e^{N_b t}}_{\text{Extension}} \tag{23}$$

Secondly, the cake layer expression shown in Equation (21) is modified as shown in Equation (24).

$$\rho_c \frac{dh_c}{dt} = jC - Cjk_4 \tag{24}$$

There is a slight difference between the model of the cake described in [39,40,137]. In [39] the cake layer density is mentioned as a possible extension to the model. In [40] the cake layer density is accounted for, as described in Equation (21). Lastly, in [137] the cake layer density is used in both the cake layer growth rate (Equation (24)) and to describe the cake resistance (Equation (22)). The cake layer density in Equation (22) is added to account for the compressibility of the cake, but how the density changes as a function of e.g., TMP and CF is not described, as such it is likely assumed constant in the validation experiments.

The dynamic exponential extensions (Equations (22) and (23)) are significantly slower than the valve, pump, and remaining fouling dynamics, and only improve the model accuracy when operating over several days [137]. Although the exponential extensions add dominating features, the short-term accuracy gain is insignificant, and the long-term accuracy gain can be addressed as process disturbances. Compared to the model suggested in [40], the exponential extended model also modifies the cake layer transport. The claimed improved accuracy is based on a comparison of the model with and without the exponential extension and naturally the additional degree of freedom improves model accuracy. The cake layer modification seems illogical, as both terms is dependent on both flux and concentration, compared to Equations (21) and (16) where only a single terms is dependent on the flux. As few details and no references are provided in the work, questions rise as to the logic behind it.

### 5.2.3. Length and Backwash Dependency

The model proposed in [138] is an extension of the work in [139]. The model deploys the methodology of resistances in series to described the fouling and is explicitly developed for process control. The resistances that are included are; membrane resistance ($R_m$), complete blocking resistance ($R_b$), cake layer resistance ($R_c$), and biofilm resistance ($R_{bo}$). The resistance caused by concentration polarization and scaling are concluded to be negligible after some initial study.

The resistance is considered over the entire length ($z$) of the membrane. The variation in resistance over the length of the membrane is mainly caused by uneven permeate flow throughout the length of the membrane. The total resistance for the proposed model is defined as:

$$R_t = R_{m|c} + R_c(z) + R_b(z) + R_{bo} \tag{25}$$

To model the fouling, the feed concentration is divided into two parts: Firstly, the part that is small enough to enter the membrane pores ($C_w$). Secondly, the remaining part that mainly consist of larger sizes, which often tend to settle as cake ($C_c$). The divided concentration is defined as:

$$C_w = G \cdot C \tag{26a}$$
$$C_c = (1 - G) \cdot C \tag{26b}$$

The change in the combined porosity of both the membrane and pore blockage is described as:

$$\rho_p \frac{d\epsilon_{m|b}(z)}{dt} = -\eta_{f,p} \cdot j(z) \cdot C_w \frac{A}{V} \tag{27}$$

The relationship between resistance and porosity is described using the Kozeny-Carman equation, similar to how cake resistance is calculated in [39].

$$R_{m|c} + R_b(z) = k_7 \frac{(1 - \epsilon_{m|b}(z))^2}{\epsilon_{m|b}(z)^3} \tag{28}$$

The cake resistance is found almost identical to Equations (14) and (21) using the Kozeny-Carman equation, except that the cake growth rate is modified to:

$$\rho_c \frac{dh_c(z)}{dt} = j(z)\alpha_c C_c \tag{29}$$

Removal of fouling inside the membrane pores during backwashing is described with Equation (30).

$$\rho_p \frac{d\epsilon_{m|b}(z)}{dt} = \begin{cases} -j(z)C_w \frac{A}{V}, & \text{if } \epsilon_{m|b} < \epsilon_{max}(z) \\ 0, & \text{if } \epsilon_{m|b} = \epsilon_{max}(z) \end{cases} \tag{30}$$

While $j(z)$ is negative the fouling is reversed, and the fouling removal persists while the porosity is below $\epsilon_{max}(z)$, where $\epsilon_{max}(z)$ is defined as Equation (31).

$$\frac{d\epsilon_{max}(z)}{dt} = \alpha_{ir} \frac{\epsilon_{m|b}(z)}{dt}, \qquad \epsilon_{max}(z)|_{t=0} = \epsilon_0, \tag{31}$$

The increased porosity caused by backwashing will only effect the complete blocking resistance. A second expression is introduced in [139], to model the removal of cake thickness caused by backwashing:

$$\frac{dh_c(z)}{dt} = -\frac{h_c(z)}{\tau_r}, \qquad \tau_r = \tau_n \left(\frac{j_r}{j_n}\right)^{k_8} \tag{32}$$

where $\tau_n$ is the observed backwashing time constant at $j_n$ flux, $\tau_b$ is the backwashing time constant at the current applied flux. This assumed that the fouling removal time constant scales linear with the applied backwashing flux. The model shares similarities with the previously described resistance-based models (Equations (11)–(24)) and considerably extends the models in multiple areas such as concentration polarization, non-linear TMP gradient, and backwashing.

The model emphasizes that several resistances are described as a function of the axial length coordinate. No experimental results are presented to validate the axial length and backwashing extensions developed by [138], as such, the significance of the extension is indecisive. The axial length extensions may provide insight into the fouling behavior, but for control purpose the extensions is of little use.

The general model structure without the axial length extension, and where backwashing only affects the cake layer was originally developed in [139], where it was validated. This validation shows good accuracy at the flux where the coefficient was identified. Shifting the operating flux affects model accuracy to a relatively low degree, indicating that the model structure is unable to fully capture the fouling behavior within a large operating range. The models described with the resistance in series methodology do in some cases consider submerged membrane filtration, which deploys alternatives to CF such as air scouring. None of the resistance-based covered in this work have been validated for an OiWM, but the resistance-based model methodology has been used to model oil fouling with good success [140].

### 5.3. Model Discussion

All the models investigated in this work are summarized in Table 7. The complexities of the different models are quite diverse, from simple models [123,127,129] to more advanced models [137,138]. All the models described throughout this study have the potential for supporting control design. Model complexity is often considered an advantage in terms of accuracy, but high complexity can cause problems when trying to identify the model parameters. Some models, such as the model proposed in [138], have a high complexity and a large set of parameters that must be identified before the model can be deployed for control design. The complexity combined with the number of parameters will complicate the parameter identification process and may not benefit the control design.

A challenge with applying fouling models for filtration of PW is that the system is rarly in steady state, as such perssure, flow rates and feed properties changes the fouling behavior and model parameters. In [30] it was suggested to enhance understanding of the fouling and fouling models, and thereby avoid the need for a pilot plant for experimental pre-investigation, but achieving such a deep understanding of the filtration processes across different application is not an easy task. Furthermore, for PWT it would require not only a fouling model but a reservoir model to predict how the reservoir changes affect the fouling.

An alternative to increased model complexity, and a probably more feasible approach, is to apply online automatic estimation to adapt to the varying conditions. Estimating the model online would provide a model that is continuously updated as conditions, such as feed properties, change. Additional benefits of online identification are that model parameters, such as fouling rate parameter can be observed, while changing system conditions, and that the functional range of the model is extended. However, online model identification can be difficult, especially if the model has a high degree of complexity.

**Table 7.** Model comparison overview.

| Models | Model Type | Advantages | Limitations | System Variables | Extends | Year |
|---|---|---|---|---|---|---|
| Hermia [123] | Blocking laws | Simple structure, widely used and validated | Dead-end filtration, constant TMP assumption | TMP, flux | - | 1982 |
| Wiesner [39] | Resistance | Good insight into fouling process, Simple resistance-based structure | No experimental validation, without consideration of CF, concentration and backwash | TMP, flux | - | 1996 |
| Kilduff [129] | Blocking laws | Includes critical flux concept and concentration | No direct link between CF and critical flux | TMP, flux, concentration, critical flux (CF), | Hermia | 2002 |
| De Bruijn [127] | Blocking laws | Includes critical flux concept | No concentration and direct link between CF and critical flux | TMP, flux, critical flux | Hermia | 2005 |
| Busch [138] | Resistance | Includes backwash, resistance as a function of membrane length | High computational complexity, no CF | TMP, flux, concentration, air scouring | - | 2007 |
| Giraldo [40] | Resistance | Includes pore blockage and flux recovery from CFV | No backwash and CF | TMP, flux, concentration, air scouring | Wiesner | 2014 |
| Fazana [137] | Resistance | Good long term model accuracy | No air scouring nor CF, No backwash | TMP, flux, concentration | Wiesner | 2017 |

## 6. Conclusions

This study has investigated membrane filtration of an OiWM from an oil and gas process control point of view. The effectiveness of membrane filtration is greatly reduced by fouling, and thus fouling remains a major complication for deploying membrane filtration in PWT [49,108,141]. This is especially true of offshore installations, where the problem is intensified by the immense cost of expanding current platforms to incorporate membrane filtration.

Studies on membrane filtration performance for OIW removal, shows that fouling is sensitive to steady state operating conditions [49,51], but how the transient performance in operating conditions affect the fouling rate of the membrane is unclear at this moment. Most industrial filtration systems are operated in steady state, but for PWT the feed flow, pressure, and feed properties, to some degree are governed by the specific well and vary with respect to time. Since steady state operation is rare with respect to PWT it is necessary to consider how dynamic variations affect the fouling rate, a topic which has not received much attention in the literature. If the dynamic variations significantly affect the fouling rate, advanced control could improve the robustness to disturbance, ensuring the operational conditions remains constant and thereby maintains a relatively low fouling degree. Such advanced control solutions, which address the transient performance, would require pump, valve, and possibly hydrodynamics to be considered in the model. Improving the currently deployed process control and thereby increase the capacity and reduce cost, would increase the attractiveness of membrane filtration for PWT.

The described features, advantages, and limitations of the fouling models are summarized in Table 7. In general, several models that describe fouling behavior are inspired by Hermia, and the models have extensively and successfully been used to model membrane fouling of an OiWM [37,132–134]. The resistance-based membrane fouling models described in this paper are not explicitly developed or validated for an OiWM and further studies are required to confirm that the model structure is suited for PW. It is observed that the resistance-based membrane fouling model methodology can capture the fouling behavior of an OiWM [140]. The variety of models and deviations between cases of filtration demands a unified model structure, that can automatically adapt to each individual filtration case.

Online identification of model coefficients, can be a natural progression to adapt control and scheduling of backwash to system changes. Hermia's models or modifications thereof (Tables 4–6) are suitable candidates, as the models have been proven to be sufficiently accurate for an OiWM. Furthermore, the limited consideration for changes in the operating conditions could be compensated for with online identification.

Financial incentives for deploying membrane filtration for PWT are negligible as improved separation has no financial impact if regulations are complied with. It is predicted that the successful implementation of membrane technology into the offshore PWT processes are dependent on future regulations, as the existing technologies in the vast majority of cases sufficiently comply with current regulations. As future regulations and industrial tendencies for the North Sea move towards a zero discharge policy, that could lead to the possibility of the membrane filtration technology to become an integral part of PWT [142].

**Author Contributions:** K.L.J. carried out the literature study, model comparison, and wrote the paper. M.V.B. assisted with the investigation of the fluid dynamics. M.V.B., S.P., and Z.Y. provided a forum for technical and valuable discussions. M.V.B., S.P., and Z.Y. reviewed and correct some parts of the paper.

**Funding:** This research was funded by Danish Hydrocarbon Research and Technology Centre (DHRTC) and AAU joint project—Smart Water Management Systems (AAU Pr-no:870051). **Acknowledgments:** The authors

thanks the support from the DTU-DHRTC and AAU joint project—Smart Water Management Systems (AAU Pr-no: 870051). Thanks go to DTU colleagues: E. Bek-Pedersen, T. M. Jørgensen and M. Lind. Thanks go to AAU colleagues: P. Durdevic, L. Hansen, D. Hansen, S. Jepsersen for many valuable discussions and technical support.

**Conflicts of Interest:** The authors declare no conflict of interest.

## Nomenclature

| | |
|---|---|
| $\alpha$ | Fouling rate |
| $\alpha_{op}$ | Open pore reduction rate |
| $\beta$ | Interface contact angle |
| $\Delta P$ | Transmembrane pressure |
| $\epsilon$ | Porosity |
| $\epsilon_{max}$ | Highest recoverable porosity thought backwashing |
| $\eta_{f,p}$ | Fraction of stuck and leaving particles |
| $\gamma$ | Interfacial tension |
| $\hat{R}$ | Specific resistance |
| $\lambda$ | Shear rate |
| $\mu$ | Dynamic viscosity of the permeate flow |
| $\rho$ | Density |
| $\rho_p$ | Density of the removed fouling |
| $\tau_n$ | Backwashing time constant, given $j_n$ flux |
| $\tau_r$ | Backwashing time constant, given $j_r$ flux |
| $A$ | Total membrane area |
| $C$ | Bluk concentration |
| $c$ | Oil-water interface mean curvature |
| $C_c$ | Bulk concentration retained by the cake layer |
| $C_w$ | Concentration below the cake layer |
| $d$ | Particle and droplet diameter |
| $F$ | Pressure gradient threshold |
| $f_o$ | Fraction of open pore area |
| $G$ | Ratio of particles that permeate the cake |
| $h_c$ | Cake layer height |
| $h_m$ | Effective membrane pore length |
| $j_n$ | Flux at which $\tau_n$ is found |
| $k$ | Generalized fouling coefficient |
| $k_1$ | Cake transport coefficient |
| $k_2$ | Back transport coefficient |
| $k_3$ | Cake layer filtration coefficient |
| $k_4$ | Cake removal rate |
| $k_5$ | Air scouring exponent coefficient |
| $k_6$ | Air scouring coefficient |
| $k_7$ | Membrane specific constant |
| $k_8$ | Backwashing constant |
| $L$ | Membrane length |
| $N$ | Exponential coefficient |
| $n$ | Type of fouling, $n \in 0, 1, 3/2, 2$ |
| $P$ | Pressure |
| $Q$ | Flow rate |
| $R$ | Resistance |
| $r_p$ | Effective pore radius |
| $R_{bo}$ | Resistance from biofilm |
| $t$ | Time |
| $V$ | Total processed volume of water |
| $V_{air}$ | Air scouring velocity |
| $z$ | Longitude position on the membrane |
| Subscripts | |
| 0 | Initial |
| $avg$ | Average |
| $b$ | Complete blockage |
| $c$ | Cake blockage |
| $d$ | Downstream |

| $f$ | Forward filtration operation |
|---|---|
| $i$ | Intermediate blockage |
| $ir$ | Irreversible |
| $m$ | Membrane |
| $m \mid c$ | Clean membrane |
| $r$ | Backward filtration operation |
| $s$ | Standard blockage |
| $t$ | Total |
| $u$ | Upstream |
| $m \mid b$ | Membrane given complete blockage |
| $m \mid p$ | Membrane given pore blockage |
| Superscripts | |
| $*$ | Critical |

## References

1. Danish Energy Agency. Available online: http://www.webcitation.org/70Qv4kLDk (accessed on 22 June 2018).
2. Tellez, G.T.; Nirmalakhandan, N.; Gardea-Torresdey, J.L. Evaluation of biokinetic coefficients in degradation of oilfield produced water under varying salt concentrations. *Water Res.* **1995**, *29*, 1711–1718. [CrossRef]
3. Danish Environmental Protection Agency. Available online: http://www.webcitation.org/70QvEretB (accessed on 22 June 2018).
4. Danish Environmental Protection Agency. Available online: http://www.webcitation.org/70QvOYGju (accessed on 22 June 2018).
5. Coca-Prados, J.; Gutiérrez-Cervelló, G. *Water Purification and Management*; Springer: Dordrecht, The Netherlands, 2010; ISBN 978-90-481-9775-0
6. Judd, S.; Qiblawey, H.; Al-marri, M.; Clarkin, C.; Watson, S.; Ahmed, A.; Bach, S. The size and performance of offshore produced water oil-removal technologies for reinjection. *Sep. Purif. Technol.* **2014**, *134*, 241–246. [CrossRef]
7. Amini, S.; Mowla, D.; Golkar, M.; Esmaeilzadeh, F. Mathematical modelling of a hydrocyclone for the down-hole oil-water separation (DOWS). *Chem. Eng. Res. Des.* **2012**, *90*, 2186–2195. [CrossRef]
8. Wolbert, D.; Ma, B.F.; Aurelle, Y.; Seureau, J. Efficiency estimation of liquid-liquid Hydrocyclones using trajectory analysis. *AIChE J.* **1995**, *41*, 1395–1402. [CrossRef]
9. Cengel, Y.A.; Cimbala, J.M.; Turner, R.H. *Fundamentals of Thermal-Fluid Sciences*, 3rd ed.; Mcgraw-Hill Education: New York, NY, USA, 2008; ISBN 978-007-126631-4.
10. Fakhru'l-Razi, A.; Pendashteh, A.; Abdullah, L.C.; Biak, D.R.A.; Madaeni, S.S.; Abidin, Z.Z. Review of technologies for oil and gas produced water treatment. *J. Hazard. Mater.* **2009**, *170*, 530–551. [CrossRef] [PubMed]
11. Sadrzadeh, M.; Hajinasiri, J.; Bhattacharjee, S.; Pernitsky, D. Nanofiltration of oil sands boiler feed water: Effect of pH on water flux and organic and dissolved solid rejection. *Sep. Purif. Technol.* **2015**, *141*, 339–353. [CrossRef]
12. Yi, H.; Zhu-Wu, J. Technology review: Treating oilfield wastewater. *Filtr. Sep.* **2008**, *45*, 14–16.
13. Cheryan, M.; Rajagopalan, N. Membrane processing of oil streams. Wastewater treatment and waste reduction. *J. Membr. Sci.* **1998**, *151*, 13–28. [CrossRef]
14. Santos, S.M.; Weisner, M.R.; Wiesner, M.R. Ultrafiltration of water generated in oil and gas production. *Water Environ. Res.* **1997**, *69*, 1120–1127. [CrossRef]
15. Ashaghi, K.S.; Ebrahimi, M.; Czermak, P. Ceramic ultra-and nanofiltration membranes for oilfield produced water treatment: A mini review. *Open Environ. Sci.* **2007**, *1*, 1–8. [CrossRef]
16. Igunnu, E.T.; Chen, G.Z. Produced water treatment technologies. *Int. J. Low-Carbon Technol.* **2014**, *9*, 157–177. [CrossRef]
17. Chen, A.; Flynn, J.; Cook, R.; Casaday, A. Removal of oil, grease, and suspended solids from produced water with ceramic crossflow microfiltration. *Soc. Pet. Eng.* **1991**, *6*, 131–136. [CrossRef]
18. Srijaroonrat, P.; Julien, E.; Aurelle, Y. Unstable secondary oil/water emulsion treatment using ultrafiltration: Fouling control by backflushing. *J. Membr. Sci.* **1999**, *159*, 11–20. [CrossRef]

19. Shi, X.; Tal, G.; Hankins, N.P.; Gitis, V. Fouling and cleaning of ultrafiltration membranes: A review. *J. Water Process Eng.* **2014**, *1*, 121–138. [CrossRef]
20. Guo, W.; Ngo, H.H.; Li, J. A mini-review on membrane fouling. *Bioresour. Technol.* **2012**, *122*, 27–34. [CrossRef] [PubMed]
21. Ng, A.N.L.; Kim, A.S. A mini-review of modeling studies on membrane bioreactor (MBR) treatment for municipal wastewaters. *Desalination* **2007**, *212*, 261–281. [CrossRef]
22. Kim, J.; DiGiano, F.A. Fouling models for low-pressure membrane systems. *Sep. Purif. Technol.* **2009**, *68*, 293–304. [CrossRef]
23. Jaffrin, M.Y. Dynamic shear-enhanced membrane filtration: A review of rotating disks, rotating membranes and vibrating systems. *J. Membr. Sci.* **2008**, *324*, 7–25. [CrossRef]
24. Baker, R.W. *Membrane Technology and Applications*, 3rd ed.; John Wiley & Sons, Ltd.: Chichester, UK, 2012; ISBN 978-0-470-74372-0.
25. Bilstad, T.; Espedal, E. Membrane separation of produced water. *Water Sci. Technol.* **1996**, *34*, 239–246. [CrossRef]
26. Alzahrani, S.; Mohammad, A.W. Challenges and trends in membrane technology implementation for produced water treatment: A review. *J. Water Process Eng.* **2014**, *4*, 107–133. [CrossRef]
27. Padaki, M.; Murali, R.S.; Abdullah, M.S.; Misdan, N.; Moslehyani, A.; Kassim, M.A.; Hilal, N.; Ismail, A.F. Membrane technology enhancement in oil–water separation. A review. *Desalination* **2015**, *357*, 197–207. [CrossRef]
28. Munirasu, S.; Haija, M.A.; Banat, F. Use of membrane technology for oil field and refinery produced water treatment—A review. *Process Saf. Environ. Prot.* **2016**, *100*, 183–202. [CrossRef]
29. Yu, L.; Han, M.; He, F. A review of treating oily wastewater. *Arab. J. Chem.* **2017**, *10*, S1913–S1922. [CrossRef]
30. Dickhout, J.M.; Moreno, J.; Biesheuvel, P.M.; Boels, L.; Lammertink, R.G.; de Vos, W.M. Produced water treatment by membranes: A review from a colloidal perspective. *J. Colloid Interface Sci.* **2017**, *487*, 523–534. [CrossRef] [PubMed]
31. Nazirah Wan Ikhsan, S.; Yusof, N.; Aziz, F.; Misdan, N. Malaysian Journal of Analytical Sciences a Review of Oilfield Wastewater Treatment Using Membrane Filtration over Conventional Technology. *Malays. J. Anal. Sci.* **2017**, *21*, 643–658. [CrossRef]
32. Guo, X.; Zhang, Z.; Fang, L.; Su, L. Study on ultrafiltration for surface water by a polyvinylchloride hollow fiber membrane. *Desalination* **2009**, *238*, 183–191. [CrossRef]
33. Maiti, A.; Sadrezadeh, M.; Guha Thakurta, S.; Pernitsky, D.J.; Bhattacharjee, S. Characterization of boiler blowdown water from steam-assisted gravity drainage and silica-organic coprecipitation during acidification and ultrafiltration. *Energy Fuels* **2012**, *26*, 5604–5612. [CrossRef]
34. Hayatbakhsh, M.; Sadrzadeh, M.; Pernitsky, D.; Bhattacharjee, S.; Hajinasiri, J. Treatment of an in situ oil sands produced water by polymeric membranes. *Desalin. Water Treat.* **2016**, *57*, 14869–14887. [CrossRef]
35. Le, N.L.; Nunes, S.P. Materials and membrane technologies for water and energy sustainability. *Sustain. Mater. Technol.* **2016**, *7*, 1–28. [CrossRef]
36. Coday, B.D.; Xu, P.; Beaudry, E.G.; Herron, J.; Lampi, K.; Hancock, N.T.; Cath, T.Y. The sweet spot of forward osmosis: Treatment of produced water, drilling wastewater, and other complex and difficult liquid streams. *Desalination* **2014**, *333*, 23–35. [CrossRef]
37. Abbasi, M.; Mohammad, R.S.; Salahi, A.; Mirza, B. Modeling of membrane fouling and flux decline in microfiltration of oily wastewater using ceramic membranes. *Chem. Eng. Commun.* **2012**, *199*, 78–93. [CrossRef]
38. Vasanth, D.; Pugazhenthi, G.; Uppaluri, R. Cross-flow microfiltration of oil-in-water emulsions using low cost ceramic membranes. *Desalination* **2013**, *320*, 86–95. [CrossRef]
39. Mallevialle, J.; Odendaal, P.E.; Wiesner, M.R. *Water Treatment Membrane Processes*, 1st ed.; McGraw-Hil: New York, NY, USA, 1996; ISBN 0-07-0011559-7
40. Giraldo, E.; Systems, N.; Water, L.A.; Giraldo, E. Dynamic Mathematical Modeling of Membrane Fouling in Submerged Membrane Bioreactors. *Proc. Water Environ. Fed.* **2006**, *2006*, 4895–4913. [CrossRef]
41. Darvishzadeh, T.; Priezjev, N.V. Effects of crossflow velocity and transmembrane pressure on microfiltration of oil-in-water emulsions. *J. Membr. Sci.* **2012**, *423–424*, 468–476. [CrossRef]
42. Bavière, M. *Basic Concepts in Enhanced Oil Recovery Processes*; Elsevier Applied Science: Essex, UK, 1991; Volume 33, ISBN 1-85166-617-6.

43. Nazzal, F.F.; Wiesner, M.R. Microfiltration of oil-in-water emulsions. *Water Environ. Res.* **1996**, *68*, 1187–1191. [CrossRef]
44. Cumming, I.W.; Holdich, R.G.; Smith, I.D. The rejection of oil by microfiltration of a stabilised kerosene/water emulsion. *J. Membr. Sci.* **2000**, *169*, 147–155. [CrossRef]
45. Guo, J.; Cao, J.; Li, M.; Xia, H. Influences of water treatment agents on oil-water interfacial properties of oilfield produced water. *Pet. Sci.* **2013**, *10*, 415–420. [CrossRef]
46. Darvishzadeh, T.; Tarabara, V.V.; Priezjev, N.V. Oil droplet behavior at a pore entrance in the presence of crossflow: Implications for microfiltration of oil-water dispersions. *J. Membr. Sci.* **2013**, *447*, 442–451. [CrossRef]
47. Monfared, M.A.; Kasiri, N.; Mohammadi, T. Microscopic modeling of critical pressure of permeation in oily waste water treatment via membrane filtration. *RSC Adv.* **2016**, *6*, 71744–71756. [CrossRef]
48. Tummons, E.N.; Tarabara, V.V.; Chew, J.W.; Fane, A.G. Behavior of oil droplets at the membrane surface during crossflow microfiltration of oil-water emulsions. *J. Membr. Sci.* **2016**, *500*, 211–224. [CrossRef]
49. Abadi, S.R.H.; Sebzari, M.R.; Hemati, M.; Rekabdar, F.; Mohammadi, T. Ceramic membrane performance in microfiltration of oily wastewater. *Desalination* **2011**, *265*, 222–228. [CrossRef]
50. Zhong, J.; Sun, X.; Wang, C. Treatment of oily wastewater produced from refinery processes using flocculation and ceramic membrane filtration. *Sep. Purif. Technol.* **2003**, *32*, 93–98. [CrossRef]
51. Hua, F.L.; Tsang, Y.F.; Wang, Y.J.; Chan, S.Y.; Chua, H.; Sin, S.N. Performance study of ceramic microfiltration membrane for oily wastewater treatment. *Chem. Eng. J.* **2007**, *128*, 169–175. [CrossRef]
52. Kwon, D.Y.; Vigneswaran, S.; Fane, A.G.; Aim, R.B. Experimental determination of critical flux in cross-flow microfiltration. *Sep. Purif. Technol.* **2000**, *19*, 169–181. [CrossRef]
53. Van der Marel, P.; Zwijnenburg, A.; Kemperman, A.; Wessling, M.; Temmink, H.; van der Meer, W. An improved flux-step method to determine the critical flux and the critical flux for irreversibility in a membrane bioreactor. *J. Membr. Sci.* **2009**, *332*, 24–29. [CrossRef]
54. Bacchin, P.; Aimar, P.; Field, R.W. Critical and sustainable fluxes: Theory, experiments and applications. *J. Membr. Sci.* **2006**, *281*, 42–69. [CrossRef]
55. Field, R.W.; Wu, D.; Howell, J.A.; Gupta, B.B. Critical flux concept for microfiltration fouling. *J. Membr. Sci.* **1995**, *100*, 259–272. [CrossRef]
56. Wicaksana, F.; Fane, A.G.; Pongpairoj, P.; Field, R. Microfiltration of algae (Chlorella sorokiniana): Critical flux, fouling and transmission. *J. Membr. Sci.* **2012**, *387–388*, 83–92. [CrossRef]
57. Gander, M.; Jefferson, B.; Judd, S. Aerobic MBRs for domestic wastewater treatment: A review with cost considerations. *Sep. Purif. Technol.* **2000**, *18*, 119–130. [CrossRef]
58. Tanudjaja, H.J.; Tarabara, V.V.; Fane, A.G.; Wei, J.; Chew, J.W. Effect of cross-flow velocity , oil concentration and salinity on the critical flux of an oil-in-water emulsion in micro filtration. *J. Membr. Sci.* **2017**, *530*, 11–19. [CrossRef]
59. Yi, X.; Wang, Y.; Jin, L.; Shi, W. Critical flux investigation in treating o/w emulsion by $TiO_2/Al_2O_3$-PVDF UF membrane. *Water Sci. Technol.* **2017**, *76*, 2785–2792. [CrossRef] [PubMed]
60. Howell, J.A. Sub-critical flux operation of microfiltration. *J. Membr. Sci.* **1995**, *107*, 165–171. [CrossRef]
61. Guglielmi, G.; Chiarani, D.; Judd, S.J.; Andreottola, G. Flux criticality and sustainability in a hollow fibre submerged membrane bioreactor for municipal wastewater treatment. *J. Membr. Sci.* **2007**, *289*, 241–248. [CrossRef]
62. Ognier, S.; Wisniewski, C.; Grasmick, A. Membrane bioreactor fouling in sub-critical filtration conditions: A local critical flux concept. *J. Membr. Sci.* **2004**, *229*, 171–177. [CrossRef]
63. Bacchin, P.; Espinasse, B.; Aimar, P. Distributions of critical flux: Modelling, experimental analysis and consequences for cross-flow membrane filtration. *J. Membr. Sci.* **2005**, *250*, 223–234. [CrossRef]
64. Metsämuuronen, S.; Howell, J.; Nyström, M. Critical flux in ultrafiltration of myoglobin and baker's yeast. *J. Membr. Sci.* **2002**, *196*, 13–25. [CrossRef]
65. Ye, Y.; Clech, P.L.; Chen, V.; Fane, A.G. Evolution of fouling during crossflow filtration of model EPS solutions. *J. Membr. Sci.* **2005**, *264*, 190–199. [CrossRef]
66. Chan, R.; Chen, V. The effects of electrolyte concentration and pH on protein aggregation and deposition: Critical flux and constant flux membrane filtration. *J. Membr. Sci.* **2001**, *185*, 177–192. [CrossRef]
67. Espinasse, B.; Bacchin, P.; Aimar, P. On an experimental method to measure critical flux in ultrafiltration. *Desalination* **2002**, *146*, 91–96. [CrossRef]

68. Espinasse, B.; Bacchin, P.; Aimar, P. Filtration method characterizing the reversibility of colloidal fouling layers at a membrane surface: Analysis through critical flux and osmotic pressure. *J. Colloid Interface Sci.* **2008**, *320*, 483–490. [CrossRef] [PubMed]

69. Li, H.; Fane, A.G.; Coster, H.G.L.; Vigneswaran, S. An assessment of depolarisation models of crossflow microfiltration by direct observation through the membrane. *J. Membr. Sci.* **2007**, *172*, 135–147. [CrossRef]

70. Sui, P.; Wen, X.; Huang, X. Membrane fouling control by ultrasound in an anaerobic membrane bioreactor. *Front. Environ. Sci. Eng. China* **2007**, *1*, 362–367. [CrossRef]

71. Aliasghari Aghdam, M.; Mirsaeedghazi, H.; Aboonajmi, M.; Kianmehr, M.H. Effect of ultrasound on different mechanisms of fouling during membrane clarification of pomegranate juice. *Innov. Food Sci. Emerg. Technol.* **2015**, *30*, 127–131. [CrossRef]

72. Li, Q.; Elimelech, M. Organic fouling and chemical cleaning of nanofiltration membranes: Measurements and mechanisms. *Environ. Sci. Technol.* **2004**, *38*, 4683–4693. [CrossRef] [PubMed]

73. James, P.; Vigneswaran, S.; Hao, H.; Ben-aim, R.; Nguyen, H. A new approach to backwash initiation in membrane systems. *J. Membr. Sci.* **2007**, *278*, 381–389. [CrossRef]

74. Gao, W.; Liang, H.; Ma, J.; Han, M.; Chen, Z.L.; Han, Z.S.; Li, G.B. Membrane fouling control in ultrafiltration technology for drinking water production: A review. *Desalination* **2011**, *272*, 1–8. [CrossRef]

75. Zhao, Y.J.; Wu, K.F.; Wang, Z.J.; Zhao, L.; Li, S.S. Fouling and Cleaning of Membrane—A Literature Review. *J. Environ. Sci.* **2000**, *12*, 241–251.

76. Gopal, R.; Kaur, S.; Feng, C.Y.; Chan, C.; Ramakrishna, S.; Tabe, S.; Matsuura, T. Electrospun nanofibrous polysulfone membranes as pre-filters: Particulate removal. *J. Membr. Sci.* **2007**, *289*, 210–219. [CrossRef]

77. Kharoua, N.; Khezzar, L.; Nemouchi, Z. Hydrocyclones for de-oiling applications-a review. *Pet. Sci. Technol.* **2010**, *28*, 738–755. [CrossRef]

78. Durdevic, P.; Pedersen, S.; Yang, Z. Evaluation of OiW Measurement Technologies for Deoiling Hydrocyclone Efficiency Estimation and Control. In Proceedings of the OCEANS'16 MTS/IEEE, Shanghai, China, 10–13 April 2016. [CrossRef].

79. Durdevic, P. Real-Time Monitoring and Robust Control of Offshore De-Oiling Processes. Ph.D. Thesis, Aalborg University, Aalborg, Denmark, 2017.

80. Arnot, T.C.; Field, R.W.; Koltuniewicz, A.B. Cross-flow and dead-end microfiltration of oily-water emulsions. Part II. Mechanisms and modelling of flux decline. *J. Membr. Sci.* **2000**, *169*, 1–15. [CrossRef]

81. Foley, G.; Malone, D.M.; Macloughlin, F. Modelling the effects of particle polydispersity in crossflow filtration. *Membr. Sci.* **1995**, *99*, 77–88. [CrossRef]

82. Beier, S.P.; Guerra, M.; Garde, A.; Jonsson, G. Dynamic microfiltration with a vibrating hollow fiber membrane module: Filtration of yeast suspensions. *J. Membr. Sci.* **2006**, *281*, 281–287. [CrossRef]

83. Belfort, G.; Pimbley, J.M.; Greiner, A.; Chun, K.Y. Diagnosis of membrane Cell culture media fouling using a rotating annular 1. Cell culture media. *J. Membr. Sci.* **1993**, *77*, 1–22. [CrossRef]

84. Zsirai, T.; Qiblawey, H.; A-Marri, M.J.; Judd, S. The impact of mechanical shear on membrane flux and energy demand. *J. Membr. Sci.* **2016**, *516*, 56–63. [CrossRef]

85. Finnigan, S.M.; Howell, J.A. The effect of pulsed flow on ultrafiltration fluxes in a baffled tubular membrane system. *Desalination* **1990**, *79*, 181–202. [CrossRef]

86. Gupta, B.B.; Howell, J.A.; Wu, D.; Field, R.W. A helical baffle for cross-flow microfiltration. *J. Membr. Sci.* **1995**, *102*, 31–42. [CrossRef]

87. Chakrabarty, B.; Ghoshal, A.K.; Purkait, M.K. Cross-flow ultrafiltration of stable oil-in-water emulsion using polysulfone membranes. *Chem. Eng. J.* **2010**, *165*, 447–456. [CrossRef]

88. Mackley, M.R.; Sherman, N.E. Cross-flow cake filtration mechanisms and kinetics. *Chem. Eng. Sci.* **1992**, *47*, 3067–3084. [CrossRef]

89. Wakeman, R.; Tarleton, E. Colloidal Fouling of Microfiltration Membranes during the Treatment of Aqueous Feed Streams. *Desalination* **1991**, *83*, 35–52. [CrossRef]

90. Abbasi, M.; Salahi, A.; Mirfendereski, M.; Mohammadi, T.; Pak, A. Dimensional analysis of permeation flux for microfiltration of oily wastewaters using mullite ceramic membranes. *Desalination* **2010**, *252*, 113–119. [CrossRef]

91. Koltuniewicz, A.B.; Field, R.W. Process factors during removal of oil-in-water emulsions with cross-flow microfiltration. *Desalination* **1996**, *105*, 79–89. [CrossRef]

92. Seyed Shahabadi, S.M.; Reyhani, A. Optimization of operating conditions in ultrafiltration process for produced water treatment via the full factorial design methodology. *Sep. Purif. Technol.* **2014**, *132*, 50–61. [CrossRef]

93. Reyhani, A.; Mashhadi Meighani, H. Optimal operating conditions of micro- and ultra-filtration systems for produced-water purification: Taguchi method and economic investigation. *Desalin. Water Treat.* **2016**, *57*, 19642–19654. [CrossRef]

94. James, P.; Vigneswaran, S.; Hao, H.; Ben-aim, R.; Nguyen, H. Design of a generic control system for optimising back flush durations in a submerged membrane hybrid reactor. *J. Membr. Sci.* **2005**, *255*, 99–106. [CrossRef]

95. Busch, J.; Marquardt, W. Run-to-run control of membrane filtration in wastewater treatment—An experimental study. *IFAC Proc. Vol.* **2007**, *40*, 195–200. [CrossRef]

96. Bartman, A.R.; Zhu, A.; Christofides, P.D.; Cohen, Y. Minimizing energy consumption in reverse osmosis membrane desalination using optimization-based control. *J. Process Control* **2010**, *20*, 1261–1269. [CrossRef]

97. Ferrero, G.; Monclús, H.; Buttiglieri, G.; Comas, J.; Rodriguez-Roda, I. Automatic control system for energy optimization in membrane bioreactors. *Desalination* **2011**, *268*, 276–280. [CrossRef]

98. Lee, E.K.; Chen, V.; Fane, A.G. Natural organic matter (NOM) fouling in low pressure membrane filtration—Effect of membranes and operation modes. *Desalination* **2008**, *218*, 257–270. [CrossRef]

99. Miller, D.J.; Kasemset, S.; Paul, D.R.; Freeman, B.D. Comparison of membrane fouling at constant flux and constant transmembrane pressure conditions. *J. Membr. Sci.* **2014**, *454*, 505–515. [CrossRef]

100. Defrance, L.; Jaffrin, M.Y. Comparison between filtrations at fixed transmembrane pressure and fixed permeate flux: Application to a membrane bioreactor used for wastewater treatment. *J. Membr. Sci.* **1999**, *152*, 203–210. [CrossRef]

101. Pedersen, S. Plant-Wide Anti-Slug Control for Offshore Oil and Gas Processes. Doctoral dissertation, Ph. D. Thesis, Aalborg University, Aalborg, Denmark, 2016. [CrossRef]

102. Abbas, A. Model predictive control of a reverse osmosis desalination unit. *Desalination* **2006**, *194*, 268–280. [CrossRef]

103. Weschenfelder, S.E.; Louvisse, A.M.T.; Borges, C.P.; Meabe, E.; Izquierdo, J.; Campos, J.C. Evaluation of ceramic membranes for oilfield produced water treatment aiming reinjection in offshore units. *J. Pet. Sci. Eng.* **2015**, *131*, 51–57. [CrossRef]

104. Katsoufidou, K.; Yiantsios, S.; Karabelas, A. A study of ultrafiltration membrane fouling by humic acids and flux recovery by backwashing: Experiments and modeling. *J. Membr. Sci.* **2005**, *266*, 40–50. [CrossRef]

105. Hofs, B.; Ogier, J.; Vries, D.; Beerendonk, E.F.; Cornelissen, E.R. Comparison of ceramic and polymeric membrane permeability and fouling using surface water. *Sep. Purif. Technol.* **2011**, *79*, 365–374. [CrossRef]

106. Liang, H.; Gong, W.; Chen, J.; Li, G. Cleaning of fouled ultrafiltration (UF) membrane by algae during reservoir water treatment. *Desalination* **2008**, *220*, 267–272. [CrossRef]

107. Weschenfelder, S.E.; Borges, C.P.; Campos, J.C. Oilfield produced water treatment by ceramic membranes: Bench and pilot scale evaluation. *J. Membr. Sci.* **2015**, *495*, 242–251. [CrossRef]

108. Weschenfelder, S.E.; Louvisse, A.M.T.; Borges, C.P.; Campos, J.C. Preliminary Studies on the Application of Ceramic Membranes for Oilfield Produced Water Management. In Proceedings of the Offshore Technology Conference, OTC Brasil, Rio de Janeiro, Brazil, 29–31 October 2013; pp. 1–10.

109. Slimane, F.Z.; Ellouze, F.; Ben miled, G.; Ben Amar, N. Physical backwash optimization in membrane filtration processes: Seawater Ultrafiltration case. *J. Membr. Process. Res.* **2018**, *4*, 63–68. [CrossRef]

110. Cakl, J.; Bauer, I.; Dole, P.; Mikul, P. Effects of backflushing conditions on permeate flux in membrane crossflow microfiltration of oil emulsion. *Desalination* **2000**, *127*, 189–198. [CrossRef]

111. Ebrahimi, M.; Willershausen, D.; Ashaghi, K.S.; Engel, L.; Placido, L.; Mund, P.; Bolduan, P.; Czermak, P. Investigations on the use of different ceramic membranes for efficient oil-field produced water treatment. *Desalination* **2010**, *250*, 991–996. [CrossRef]

112. Mi, B.; Elimelech, M. Organic fouling of forward osmosis membranes: Fouling reversibility and cleaning without chemical reagents. *J. Membr. Sci.* **2010**, *348*, 337–345. [CrossRef]

113. Lamminen, M.O.; Walker, H.W.; Weavers, L.K. Mechanisms and factors influencing the ultrasonic cleaning of particle-fouled ceramic membranes. *J. Membr. Sci.* **2004**, *237*, 213–223. [CrossRef]

114. Yutaka, M.; Tan, M.; Shin-ichi, N.; Shoji, K. Improvement of membrane permeation performance by ultrasonic microfiltration. *J. Chem. Eng. Jpn.* **1996**, *29*, 561–567.

115. Thiam Teik Wan, W.L.L. Experimental Study of the Separation of Oil in Water Emulsions by Tangential Flow Microfiltration Process. Part 2: The Use of Ultrasound for In-Situ Controlling of the Membrane Fouling. *J. Membr. Sci. Technol.* **2014**, *5*, 1–6. [CrossRef]

116. Kobayashi, T.T.; Kobayashi, T.T.; Hosaka, Y.; Fujii, N. Ultrasound-enhanced membrane-cleaning processes applied water treatments: Influence of sonic frequency on filtration treatments. *Ultrasonics* **2003**, *41*, 185–190. [CrossRef]

117. Zhu, C.; Liu, G. Modeling of ultrasonic enhancement on membrane distillation. *J. Membr. Sci.* **2000**, *176*, 31–41. [CrossRef]

118. Grundfos. Available online: https://product-selection.grundfos.com/product-detail.product-detail.html?custid=GMA&productnumber=98390030&qcid=393096734 (accessed on 22 June 2018).

119. Borea, L.; Naddeo, V.; Shalaby, M.S.; Zarra, T.; Belgiorno, V.; Abdalla, H.; Shaban, A.M. Wastewater treatment by membrane ultrafiltration enhanced with ultrasound: Effect of membrane flux and ultrasonic frequency. *Ultrasonics* **2018**, *83*, 42–47. [CrossRef] [PubMed]

120. Amir, M.; Givargis, T. Hybrid State Machine Model for Fast Model Predictive Control: Application to Path Tracking. In Proceedings of the 2017 IEEE/ACM International Conference on Computer-Aided Design (ICCAD), Irvine, CA, USA, 12–16 November 2017; pp. 185–192.

121. Seborg, D.E.; Mellichamp, D.A.; Edgar, T.F.; Doyle, F.J., III. *Process Dynamics and Control*, 3rd ed.; John Wiley & Sons: Hoboken, NJ, USA, 2010; ISBN 978-0-470-12867-1.

122. Hermans, P.H.; Bredée, H.L. Zur Kenntnis der Filtrationsgesetze. *Recueil des Travaux Chimiques des Pays-Bas* **1935**, *54*, 680–700. [CrossRef]

123. Hermia, J. Constant pressure blocking filtration law: Application to power law non-Newtonian fluids. *Trans. Inst. Chem. Eng.* **1982**, *60*, 183–187.

124. Sampath, M.; Shukla, A.; Rathore, A. Modeling of Filtration Processes—Microfiltration and Depth Filtration for Harvest of a Therapeutic Protein Expressed in Pichia pastoris at Constant Pressure. *Bioengineering* **2014**, *1*, 260–277. [CrossRef] [PubMed]

125. Bowen, W.R.; Calvo, J.I.; Hernández, A. Steps of membrane blocking in flux decline during protein microfiltration. *J. Membr. Sci.* **1995**, *101*, 153–165. [CrossRef]

126. Vincent Vela, M.C.; Álvarez Blanco, S.; Lora García, J.; Bergantiños Rodríguez, E. Analysis of membrane pore blocking models adapted to crossflow ultrafiltration in the ultrafiltration of PEG. *Chem. Eng. J.* **2009**, *149*, 232–241. [CrossRef]

127. De Bruijn, J.; Salazar, F.; Bórquez, R. Membrane Blocking In Ultrafiltration. *Food Bioprod. Process.* **2005**, *83*, 211–219. [CrossRef]

128. Bolton, G.; LaCasse, D.; Kuriyel, R. Combined models of membrane fouling: Development and application to microfiltration and ultrafiltration of biological fluids. *J. Membr. Sci.* **2006**, *277*, 75–84. [CrossRef]

129. Kilduff, J.E.; Mattaraj, S.; Sensibaugh, J.; Pieracci, J.P.; Yuan, Y.; Belfort, G. Modeling flux decline during nanofiltration of NOM with poly(arylsulfone) membranes modified using UV-assisted graft polymerization. *Environ. Eng. Sci.* **2002**, *19*, 477–495. [CrossRef]

130. Ebrahimi, M.; Ashaghi, K.S.; Engel, L.; Willershausen, D.; Mund, P.; Bolduan, P.; Czermak, P. Characterization and application of different ceramic membranes for the oil-field produced water treatment. *Desalination* **2009**, *245*, 533–540. [CrossRef]

131. Silalahi, S.H.D.; Leiknes, T. Cleaning strategies in ceramic microfiltration membranes fouled by oil and particulate matter in produced water. *Desalination* **2009**, *236*, 160–169. [CrossRef]

132. Peng, H.; Tremblay, A.Y. Membrane regeneration and filtration modeling in treating oily wastewaters. *J. Membr. Sci.* **2008**, *324*, 59–66. [CrossRef]

133. Salahi, A.; Abbasi, M.; Mohammadi, T. Permeate flux decline during UF of oily wastewater: Experimental and modeling. *Desalination* **2010**, *251*, 153–160. [CrossRef]

134. Mohammadi, T.; Kazemimoghadam, M.; Saadabadi, M. Modeling of membrane fouling and flux decline in reverse osmosis during separation of oil in water emulsions. *Desalination* **2003**, *157*, 369–375. [CrossRef]

135. Duclos-Orsello, C.; Li, W.; Ho, C.C.C. A three mechanism model to describe fouling of microfiltration membranes. *J. Membr. Sci.* **2006**, *280*, 856–866. [CrossRef]

136. Badrnezhad, R.; Beni, A.H. Ultrafiltration membrane process for produced water treatment: Experimental and modeling. *J. Water Reuse Desalin.* **2013**, *3*, 249. [CrossRef]

137. Fazana, M.; Zuthi, R.; Guo, W.; Hao, H.; Long, D.; Hai, F.I.; Xia, S.; Li, J.; Li, J.; Liu, Y. Bioresource Technology New and practical mathematical model of membrane fouling in an aerobic submerged membrane bioreactor. *Bioresour. Technol.* **2017**, *238*, 86–94. [CrossRef]

138. Busch, J.; Cruse, A.; Marquardt, W. Modeling submerged hollow-fiber membrane filtration for wastewater treatment. *J. Membr. Sci.* **2007**, *288*, 94–111. [CrossRef]

139. Broeckmann, A.; Busch, J.; Wintgens, T.; Marquardt, W. Modeling of pore blocking and cake layer formation in membrane filtration for wastewater treatment. *Desalination* **2006**, *189*, 97–109. [CrossRef]

140. Lobo, A.; Benito, M.; Pazos, C. Ultrafiltration of oil-in-water emulsions with ceramic membranes: Influence of pH and crossflow velocity. *J. Membr. Sci.* **2006**, *278*, 328–334. [CrossRef]

141. Pedenaud, P.; Heng, S.; Evans, W.; Bigeonneau, D. Ceramic membrane and core pilot results for produced water management. In Proceedings of the Annual Offshore Technology Conference, Rio de Janeiro, Brazil, 4–6 October 2011; pp. 385–400.

142. OSPAR. Available online: https://www.ospar.org/documents?d=33828 (accessed on 22 June 2018).

*water*

MDPI

*Article*

# Multivariate Chemometric Analysis of Membrane Fouling Patterns in Biofilm Ceramic Membrane Bioreactor

**Olga Kulesha [1,2,\*], Zakhar Maletskyi [1] and Harsha Ratnaweera [1]**

[1]  Faculty of Science and Technology (REALTEK), Norwegian University of Life Sciences, P.O. Box 5003, 1432 Aas, Norway; zakhar.maletskyi@nmbu.no (Z.M.); harsha.ratnaweera@nmbu.no (H.R.)

[2]  Department of General and Inorganic Chemistry, Faculty of Chemical Technology, National Technical University of Ukraine "Igor Sikorsky Kyiv Polytechnic Institute", Peremohy 37, 03056 Kyiv, Ukraine

\*  Correspondence: olga.kulesha@nmbu.no; Tel.: +47-406-755-92

Received: 27 June 2018; Accepted: 22 July 2018; Published: 26 July 2018

**Abstract:** Membrane fouling highly limits the development of Membrane bioreactor technology (MBR), which is among the key solutions to water scarcity. The current study deals with the determination of the fouling propensity of filtered biomass in a pilot-scale biofilm membrane bioreactor to enable the prediction of fouling intensity. The system was designed to treat domestic wastewater with the application of ceramic microfiltration membranes. Partial least squares regression analysis of the data obtained during the long-term operation of the biofilm-MBR (BF-MBR) system demonstrated that Mixed liquor suspended solids (MLSS), diluted sludge volume index (DSVI), chemical oxygen demand (COD), and their slopes are the most significant for the estimation and prediction of fouling intensity, while normalized permeability and its slope were found to be the most reliable fouling indicators. Three models were derived depending on the applied operating conditions, which enabled an accurate prediction of the fouling intensities in the system. The results will help to prevent severe membrane fouling via the change of operating conditions to prolong the effective lifetime of the membrane modules and to save energy and resources for the maintenance of the system.

**Keywords:** water crisis; biofilm membrane bioreactor; membrane fouling; operation; ceramic membranes; multivariate statistics

---

## 1. Introduction

The World Economic Forum (WEF) includes water crises in the group of risks with the highest likelihood and impact, which are strongly interconnected with the trends in climate change that can degrade the environment and cause food crises [1]. According to the WEF, the main reason for a water crisis is a significant decline in the available quality and quantity of fresh water, thus resulting in harmful effects to human health and economic activity. Competition for water between agriculture, industry and municipal supply is being complicated by political tension around water in stressed regions, thus leading to the future shock of so-called "grim reaping" [2].

Water reuse is gaining momentum as a reliable alternative source of freshwater in the face of growing water demand, which is shifting the paradigm of wastewater management from "disposal" to "reuse and resource recovery" [3]. Growing globally [4], water reuse plays a key role in bringing significant environmental, social and economic benefits [5]. Advanced tertiary treatment is a rule of thumb in water reuse and is an important factor of system resilience in the case of wastewater reuse as a part of a decentralized water supply [6]. However, of all the wastewater produced worldwide, only a very small fraction actually undergoes tertiary treatment [3]. Efficient, reliable, sustainable and economically feasible technologies are highly demanded when it comes to potential cost recovery by treating wastewater to a water quality standard acceptable to users.

Membrane bioreactor technology (MBR) is a highly competitive technology when applied in water reuse schemes. It provides excellent nutrient removal efficiency, compactness, complete biomass retention with no use of a secondary clarifier, and produces a low carbon footprint [7–9]. Additionally, strengthening requirements for reclaimed water quality is expected to drive the MBR market to USD 8.27 billion by 2025 [10].

However, membrane fouling is the main restraint to further penetration of MBR into cost-sensitive markets, including the water reuse market in small communities and developing countries, which is primarily due to the occurrence of unplanned high operating costs [11–14]. Several approaches to detect, control and prevent membrane fouling in MBR have been developed during the last decades, focusing on pre-treatment or modification of mixed liquor, membrane properties, operating conditions, etc. [15–19]. Considering the pros and cons of the aforementioned, there is no unified approach to dealing with membrane fouling.

Several types of research demonstrated that a combination of two or more fouling prevention factors gives the best practical results through the synergy of anti-fouling mechanisms [20–22]. Therefore, the current research considers the use of a combination of biofilm-MBR (BF-MBR) process configuration with the application of ceramic flat-sheet membranes.

BF-MBR combines membrane separation, biological contact oxidation and fluidized bed wastewater treatment (as in the moving-bed-biofilm reactor (MBBR) process). This results in better effluent quality due to reliable degradation of organics and nutrients, a lower sludge production rate and a smaller footprint, together with stable and reliable operation, strong resistance to shock loading, and adaptability due to high biomass concentration and diversity in bacterial population [23]. The BF-MBR process has demonstrated lower membrane fouling rates and better settling ability of suspended biomass than in conventional MBR and MBBR processes separately [12,24].

In another study [25], porous suspended biofilm carriers were introduced to a submerged ceramic membrane bioreactor to explore their effectiveness in membrane flux enhancement. Alleviation of membrane fouling, in this case, is anticipated via mechanical scouring of the cake layer on the membrane surface and modification of mixed liquor characteristics. It has been shown that a combination of biofilm carriers with the ceramic membrane in MBR leads to 2.7 times lower cake resistance and 1.5 times lower total resistance.

Mixed liquor suspended solids (MLSS), chemical oxygen demand (COD) and sludge relative hydrophobicity (RH) are among the main characteristic parameters of activated sludge suspension that are traditionally monitored in an MBR system [22,26–31].

MLSS provides information about mixed liquor fouling propensity, apart from indicating a biomass potential to decompose wastewater impurities, determining an aeration tank volume, and affecting the aeration demand and sludge production [28,32]. Several researchers acknowledged there was a complex relationship between MLSS and membrane fouling [9,29,33].

The COD parameter accounts for the organic load and the biological treatment efficiency in terms of the degradation of organic contaminants [34]. In addition, as specified by Le-Clech et al. [29], Ji and Zhou [35], Meng et al. [36], in MBR systems, soluble COD is an indicator of the soluble microbial product (SMP) level. SMP is generally considered to be one of the major foulants in MBR [37–39].

Biomass RH is one of the key parameters used to estimate the resistance caused by microbial aggregates. RH determines flocculation ability of the sludge flocs based on their hydrophobic interactions with each other, which in turn controls their dewaterability [32,40,41]. RH of the activated sludge influences initial biomass attachment to the membrane and, therefore, membrane permeability (i.e., determines whether a membrane can be more or less sensitive to different foulants).

The sludge volume index (SVI)/diluted sludge volume index (DSVI) is another characteristic that is monitored in MBR systems. Although this parameter primarily characterizes the activated sludge settling properties, it is also widely applied in MBRs, since it indicates the flocculation characteristics of the activated sludge and is associated with filamentous bacteria. The latter induces membrane fouling through the release of SMPs from the sludge flocs, thus increasing their concentration via viscosity increase and by fixing the foulants on the membrane surface, thus forming practically a non-porous cake layer [9,33,42–44].

In general, a number of studies indicated that the above-mentioned biomass characteristics exhibit specific tendencies in influencing fouling in MBR (Table 1).

**Table 1.** The influence of activated sludge parameters on the biomass fouling propensity.

| Parameter | Correlation with the Fouling | Possible Fouling Mechanism | References |
|---|---|---|---|
| MLSS [1] | Positive | Intense cake layer formation on the membrane surface. Increase in the suspension viscosity. Excessive growth of filamentous bacteria. Increase in microbial metabolic products such as SMP [2] and EPS [3], which are the major foulants. | [34,45–51] |
| MLSS [1] | Negative (irreversible fouling) | MLSS [1] 12–18 g/L: The formed cake layer causes the prevention of the pore blocking development and induces an increased porosity of the cake layer. | [15,45] |
| COD [4] | Positive | COD [4] in the form of colloids proteins (adsorption mechanism) and other soluble organic fractions, causing irreversible fouling; higher organic load causes an increase in the production of specific EPS [3] and macromolecules in the SMP [2]/EPS [3] fractions, defloculation of the mixed liquor, and a fast formation of cake layers. | [9,29,35,52–56] |
| RH [6] (mostly hydrophilic membranes) | Negative | RH [6] increase: Enhanced AS [5] flocculation due to more intense hydrophobic interactions between sludge flocs, resulting in the formation of larger aggregates with less water content, and decreased interaction between the flocs and membrane surface. RH [6] decrease: Floc deterioration. | [57–62] |
| RH [6] (mostly hydrophilic membranes) | Positive | RH [6] increase: A formation of a thin cake layer, promoting the adhesion of proteins and carbohydrates in the form of SMP [2] on the membrane surface and its pores, resulting in irreversible and irrecoverable fouling. | [26,63] |
| SVI (DSVI) [7] | Positive | High DSVI [7]: Evolution of the flocs to the more irregular rougher shapes which more likely adhere to the surface of the membrane, intertwisting with the fibers. This forms a dense, non-porous cake with large thickness. The possible decrease of the bound protein and release of SMP [2] triggers deflocculation and the increase in fouling intensity. | [64–69] |

Notes: [1] Mixed liquor suspended solids; [2] Soluble microbial products; [3] Extracellular polymeric substances; [4] Chemical oxygen demand; [5] Activated sludge; [6] Relative hydrophobicity; [7] Sludge volume index (diluted sludge volume index).

It is worth noting that application of ceramic membranes in MBR started from a niche where polymer membranes either failed or provided insufficient results: The cases when high effluent quality is required or the process depends on ceramic membrane robustness [70]. Compared to their polymeric counterparts, ceramic membranes have the following advantages:

1. Higher mechanical strength and chemical resistance to oxidants and solvents. The modules are backwashable with the possible application of high backwash pressure/flux [71,72] and can withstand much more aggressive operation and chemical cleaning conditions (i.e., can be used in combination with ultrasonic irradiation and undergo a soaking in more concentrated NaClO, NaOH, and acidic solutions). In addition, they can undergo the influence of higher temperatures and pH without damaging the active layer [73–77].

2. Higher hydrophilicity, thus no affinity to organic foulants which are mostly of a hydrophobic nature [70,78,79].

The outcomes are: High permeability recovery [80]; a longer period of operation between the chemical cleanings due to more efficient removal of reversible and irreversible fouling [29,79]; enhanced concentration polarization control; and, higher applicable net permeate fluxes and permeabilities are sustained [81–83], consequently leading to a long lifespan.

Ceramic membranes proved to be an effective and reliable MBR component, leading to higher treatment efficiencies of COD, ammonium, and phosphorus elimination [84,85]. In addition, higher treatment performance in terms of COD and MLSS removal, more stable operation and less transmembrane pressure (TMP) increase was exhibited by the MBR with ceramic modules, compared to the system with the polymeric units [86]. Lower TMP increase, higher removal of non-purgeable organic compounds and lower UV absorbance of the permeate was demonstrated by Hofs et al. [87] in relation to the surface water samples being treated by ceramic modules.

From an economic point of view, the tremendously higher cost of the application of the MBR systems with the ceramic membranes in comparison to the use of the systems with the polymeric modules is rather a stereotype than a reality at present. According to a study by Park et al. [83], the incorporation of membrane modules into the water treatment plant (WTP) makes up 13% and 24% of the total capital cost for polymeric and ceramic WTP, respectively. The comparative analysis demonstrated that the polymeric WTP (with capacity 30,000 m$^3$/day) are indeed cheaper in terms of the capital costs than their ceramic counterparts, but the difference is not significant: USD 28,019 vs. USD 32,634, respectively. Moreover, the annual operating expenses of the filtration process are more than twice as high for the polymeric modules (USD 562,717) as for the ceramic modules (USD 217,725). This is mainly due to the membrane replacement costs for polymeric WTP, which constitute 61% of the operational expenses. Low operation costs of the systems with ceramic membranes were also acknowledged by Jin et al. [74]. As specified by Park et al. [83], the assessed life cycle costs (LCC) of water from the ceramic and polymeric membrane WTPs are, USD 0.28/m$^3$ and USD 0.274/m$^3$, respectively (at the flux of 41.7 LMH). If fluxes of 63 LMH and higher are applied, the LCC of the produced water decreases for the ceramic membranes, thus increasing their feasibility.

In addition, since the manufacturing of the ceramic membranes is an energy-consuming process, a number of recent studies have successfully developed and evaluated the performance of low-cost ceramic membranes [88–93].

Despite many studies on membrane fouling in general, and on BF-MBR or the application of ceramic membranes in particular, only a few findings that are relevant to detection and control of membrane fouling in submerged ceramic BF-MBR come from a pilot or full-scale product. Nevertheless, understanding, detection, and control of membrane fouling via applying advanced statistics and mathematical modelling represents a significant potential for improvement of the cost-efficiency of the process and provides the instruments for dynamic and real-time process control.

Chemometrics serves as a bridge between the state of a chemical system and its measured characteristics, which enhances the efficiency of automatic control systems. Chemometric analysis is based on the application of mathematical and statistical techniques to improve comprehension of the system properties and to link them to analytical measurements. The modelling of the patterns in the dataset results in model derivation. This model can be further used to predict identical parameters as in the initial model but in application to new data [94]. The following multivariate statistical data analysis methods are commonly used as chemometric tools for the interpretation of the acquired data: Cluster analysis (CA), discriminant analysis (DA), principal component analysis (PCA), partial least squares analysis (PLS), multiple linear regression (MLR), principal component regression (PCR), and partial least squares discriminant analysis (PLS-DA) [94–96].

It is worth mentioning that PLS is an advanced statistical technique due to the applied validation tools, noise elimination, and the ability to determine the independent influence of each input variable, even if there is a collinearity between them [59].

A number of recent studies were devoted to the application of modelling using multivariate data analysis for fouling control in MBR. In the study by Philippe et al. [97], the authors performed a PCA to distinguish a correlation between the operational parameters and the characteristics of filtered biomass in a full-scale municipal MBR. Among all the variables, solids retention time (SRT), MLSS, the food to microorganism ratio (F:M), pH and temperature (T) were found to be representative for describing the fouling behaviour. According to the plot of weighted variables, SRT, MLSS and pH positively contributed to the principal components (PCs) one and two, while the F:M ratio exhibited a negative influence. Temperature has a controversial contribution to the PCs in the model. However, the attained models managed to predict the development of permeability merely in one membrane tank and failed while applying them at different operation stages for all four membrane tanks in the system.

In the work by Kaneko and Funatsu [98], wastewater temperature, the duration of filtration, water temperature, and the inverse of flux and TMP were inputted into the model. PCA was applied as a visualization tool for the discriminant model. As concluded, the accuracy and the predictive ability of the derived model can be increased if the additional parameters related to the water quality and operating conditions are used.

A similar choice of variables was made in the study by De Temmerman et al. [99], where PCA was based on temperature, flux, TMP slope, and pressure peaks during the filtration and chemically enhanced backwash (CEB) for the full-scale MBR. The detection of the fouling types (reversible/irreversible and irrecoverable) was among the prime research goals. The TMP slope and pressure peak during the filtration were found to have a positive relationship. Meanwhile, they were negatively linked to the temperature and the CEB pressure peak along the PC-1 axis. Along the PC-3 axis, flux exhibited a negative correlation with water temperature and the backwash pressure peak. The variance of the CEB pressure peak was attributed to irrecoverable fouling, while pressure peaks during the filtration were attributed to reversible and irreversible fouling types. However, the scores plot indicated no clear trends.

Partial least squares regression analysis applying leave-one-out cross-validation was performed in the work by Van den Broeck [59] to find the influence of the activated sludge parameters on filterability in industrial and municipal MBRs. A relatively deep analysis of the biomass characteristics was conducted. The content of proteins and polysaccharides, sludge relative hydrophobicity, sludge dissociation constant, mean particle size, and the surface fraction of activated sludge particles equal to 1 pixel were used to predict any change of filtration resistance. An accurate estimation of the filtration resistance was observed, which was characterized by the sum of square errors equal to 0.076 (R-squared = 0.99). However, a number of factors (latent variables) exceeded 9, indicating a complexity of the derived model. As concluded, a combination of chosen activated sludge parameters succeeded in predicting sludge filterability, while, when taken individually, they were poor indicators of the biomass fouling propensity.

Consequently, the following knowledge gaps can be identified: The studies which are focused on the modelling of the relationship between operating parameters and filterability do not typically take into consideration biomass characteristics as potential fouling indicators, despite the fact that these are among the main factors affecting the fouling process [9,100,101]. Meanwhile, those studying the statistical evaluation of the relationship between mixed liquor parameters and biomass fouling propensity do not provide the information on the influence of the operating parameters on the fouling intensity. Most importantly, there is also still a need to study the application of the PLS regression to the processes in the biofilm membrane bioreactor due to the lack of research data. In addition, there is a controversy over the influence of the selected biomass parameters on the fouling intensity (Table 1), whereas the development of a reliable BF-MBR system requires concrete patterns.

Applying PLS analysis, the current work encapsulates the relationship between the mixed liquor characteristics, fouling indicators and the operation conditions in BF-MBR with ceramic modules, and thus provides a comprehensive analysis of the system performance and the mechanisms for influencing it.

The purpose of this research was to develop and validate a PLS regression model based on the mixed liquor characteristics and the indicators of fouling intensity, considering the influence of the operation parameters on the filtration performance in the BF-MBR with ceramic membranes, in order to detect membrane fouling patterns and to develop process control and a fouling mitigation approach.

## 2. Materials and Methods

In general, this study consists of the acquisition of operational data from a BF-MBR pilot plant at various sets of operating conditions followed by statistical analysis.

The BF-MBR pilot plant had a four-stage design (Figure 1) comprising equalization (I) and treated water (IV) compartments, and a MBBR chamber (II) and a separation chamber (III) with the submerged membranes being in contact with suspended biofilm carriers. Compartments I, II and III were interconnected through overflow, while the separation process from chamber III to chamber IV was driven by a reversible peristaltic pump (Verderflex, Castleford, UK), controlled from the programmable logic controller (PLC) (MoreControl, Aas, Norway). A return activated sludge (RAS) line was incorporated into the system between chambers III and II, and was controlled by RAS pumping intervals: With lower RAS intervals, more sludge is returned.

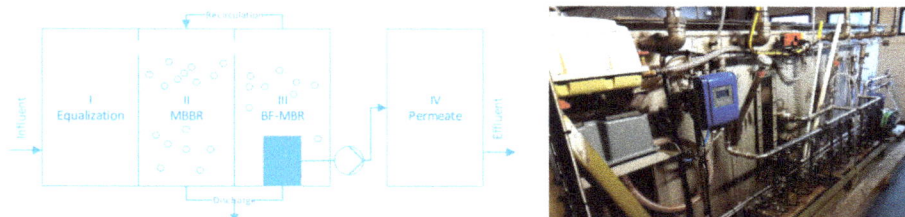

**Figure 1.** The BF-MBR pilot plant: Schematic diagram (**left**) and the photo of installation (**right**).

Wastewater was supplied at 0.3 m$^3$/day through the screens to the equalization tank (I) from the source-separated sewer network, keeping the ratio of black to grey water at 1:9. Black water was collected from the toilets and grey water from all other discharge points of the households around the pilot site [102]. This allowed maintenance of the influent quality at 1–1.3 g/L by suspended solids and 100–350 mg-$O_2$/L by COD.

Flat sheet SiC microfiltration membranes with 0.1 µm pore size (Cembrane, Lynge, Denmark) were used in the separation chamber (III), providing total filtering area of 0.828 m$^2$. Aeration was organized in chambers II and III by a MEDO LA-60E air compressor at 60 L/min.

Initial biological activity in the system was provided by inoculation with sludge from the municipal MBBR wastewater treatment plant (BEVAS, Oslo, Norway).

The BF-MBR pilot plant was operated in automatic mode under constant flux conditions, controlled through the PLC. The initial filtration settings were: 300 s of filtration at net-flux 8.2 LMH, 60 s relaxation, 15 s backwash with permeate at net-flux 180 LMH, 120 s relaxation. Further changes were introduced into the plant operation settings in order to reach different operation states (Table 2), which divided full operation time of 114 days into 8 relevant periods.

Plant operation data was continuously recorded every 3 s to the data-logger, in-built in the PLC. Values of system inflow, level in the separation chamber, TMP and permeate flow were stored and recalculated further to analytical values.

Filtration settings were programmed as $t_{filtr/relax/BW}$, filtration/relaxation/backwash time, and $RAS_{pulse\ interval} = RAS_{PI}$, the pulse interval of the return activated sludge. For every period of operation, normalized net membrane flux was calculated ($J_{n(net)}$). The normalized permeability, $P_n$, and permeability slope, $dP_n/dt$, were determined.

Permeate flow was used to calculate membrane flux J (LMH; Equation (1)), normalized to 20 °C as $J_n$ (Equation (2)), and used to calculate normalized permeability, $P_n$ (Equation (3)), and the fouling rate in terms of membrane permeability decrease, $dP_n/dt$ (Equation (4)):

$$J = \frac{F}{S_f}$$

(1)

$$J_n = J \cdot e^{(-0.032 \cdot (t-20))} \tag{2}$$

$$P_n = \frac{J_n}{TMP} \tag{3}$$

$$\frac{dP_n}{dt} = \frac{P_{ni} - P_{ni-1}}{t_i - t_{i-1}} \tag{4}$$

where F is permeate flow, L/h, and $S_f$ is the active filtration surface ($m^2$).

**Table 2.** BF-MBR pilot plant operation settings.

| Period | Days | Adjustments in Settings | Processes and Changes in the System |
|--------|------|-------------------------|-------------------------------------|
| I | 1–20 | $J_{n(net)}$ [1] = 8.2 LMH, $J_{n(gross)}$ [2] = 37.6 LMH Filtration cycle settings: $t_{filtr}$ = 300 s, $t_{relaxI}$ = 60 s, $t_{relaxII}$ = 120 s, $t_{BW}$ = 15 s $RAS_{pulse\ interval}$ [3] = 1620 s, $SRT_{av}$ [4] =20 days | Conditions for sludge adaptation and conditional fouling of fresh membranes. |
| II | 21–34 | $J_{n(net)}$ [1] = 5.3 LMH, $J_{n(gross)}$ [2] = 32.6 LMH, $RAS_{pulse\ interval}$ [3] = 740 s, $SRT_{av}$ [4] =20 days | System stabilization and an increase of sludge recirculation between separation and MBBR [5] chambers through the decrease of RAS [6] interval. |
| III | 35–36 | $J_{n(net)}$ [1] = 12.2 LMH, $J_{n(gross)}$ [2] = 44.0 LMH | Increase of net-flux in order to get close to TMP [7] jump. |
| IV | 37–44 | $J_{n(net)}$ [1] = 10.0 LMH, $J_{n(gross)}$ [2] = 43.7 LMH, $t_{BW}$ = 19.5, $t_{relaxI}$ = 30 | Prolongation of backwash in order to stabilize the system and TMP [7] jump. |
| V | 45–47 | CIP [8] I, 1% NaOCl, 2% Citric acid | TMP [7] ↓; $P_n$ ↑ (58%), $dP_n/dt$ ↑ (88%)—removal of reversible and irreversible fouling. |
| VI | 48–77 | Same as in period IV, SRT = 31 days | Reproduction of last stable operation. |
| VII | 78–85 | CIP [8] II | TMP [7] ↓ (82%), $P_n$ ↑ (82%), $dP_n/dt$ ↑. |
| VIII | 86–114 | $J_{n(net)}$ [1] = 4.5 LMH, $J_{n(gross)}$ [2] = 30.4 LMH, Infinite SRT (no wastage/sludge discharge) | Lower hydraulic loading. |

Notes: [1] Normalized net flux; [2] Normalized gross flux; [3] The pulse interval of the return activated sludge; [4] Average solids retention time; [5] Moving-bed-biofilm reactor; [6] Return activated sludge; [7] Transmembrane pressure; [8] Cleaning-in-place.

The data array of hydraulic parameters was statistically treated and expressed in the form of 8 representative filtration cycles for every day. For a single cycle, a set of average initial ($TMP_i$, $J_{Ni}$, $P_{ni}$) and final parameters ($TMP_{i-1}$, $J_{Ni-1}$, $P_{ni-1}$) was calculated, excluding ramp and relaxation periods of the peristaltic pump.

Recovery of membrane permeability was expressed as the ratio of permeability after chemical cleaning and before chemical cleaning [103]:

$$Recovery_{P_n} = \frac{P_{CIP/BW_{fin}} - P_{CIP/BW_{in}}}{P_{in} - P_{fin}} \tag{5}$$

where: $P_{CIP/BW_{fin}}$ is a permeability of the new filtration cycle after the backwash/Chemical cleaning (CIP); $P_{CIP/BW_{in}}$ is the initial permeability before the cleaning, which is equal to $P_{fin}$, the permeability at the end of previous filtration cycle; and, $P_{in}$ is the initial permeability at the beginning of the previous filtration cycle.

In other words, recovery of permeability expresses the extent to which membrane permeability is restored after the application of different types of cleaning to remove the foulants [104].

A sampling of mixed liquor, and raw and treated wastewater was organized on a daily basis. Samples of raw wastewater (chamber I), MBBR mixed liquor (chamber II), BF-MBR mixed liquor (chamber III) and permeate (chamber IV) were analyzed accordingly for suspended solids (SS, MLSS), COD of the filtrates, DSVI, and RH. COD was measured by COD-cuvette test (HACH, Manchester, UK) applying the dichromate method, DSVI was measured by a settleability test. RH was determined by the MATH (microbial adherence to hydrocarbons) method. The analyses were conducted in accordance with SMWW (Standard Methods for the Examination of Water and Wastewater) (22nd edition) and the MATH test [59,105]. Flow in permeate line and TMP were measured through flow and pressure sensors (Krohne, Dilling, Norway) and logged every second to the PLC together with filtration cycle settings.

PLS regression was used to distinguish the relationship between the parameters of the mixed liquor and the fouling indicators and to predict the fouling intensity. The statistical software, The Unscrambler® X10.3 (CAMO Software AS, Oslo, Norway), was used to perform the analysis of the monitored data.

## 3. Results and Discussion

### 3.1. Pilot Plant Operation Results

During 114 days of operation of the BF-MBR pilot plant, notable trends in TMP, permeability, permeability slope, MLSS in the membrane separation chamber (MLSS-III) and COD removal were observed (Figure 2), allowing the development of the qualitative description of the biological activity and its influence on membrane separation process.

The first period (1–20 days) can be described as the period of biological adaptation and biomass development. It is characterized by moderate growth of biomass up to MLSS-III 5–6 g/L and increasing biodegradation of organics in the range of 67–81%, together with a steep TMP growth and a respective decrease of permeability at a relatively high rate of 0.35–0.47 LMH/bar/s. This state can be identified as conditioning fouling.

After reaching the conditionally critical value of 1.7 times permeability decrease, the return of suspended solids from separation chamber (III) to MBBR chamber (II) was doubled, leading to stabilization of permeability and MLSS-III in the next period (21–34 days) and decreasing the membrane fouling rate to 0.25–0.27 LMH/bar/s by permeability, which is considered steady fouling.

In order to increase the system productivity in terms of permeate, membrane flux was increased, entailing the TMP jump during the third period (35–36 days), which indicates a severe fouling. Following that, backwash and relaxation times were adjusted in order to stabilize rapid fouling development during 37–44 days.

Chemical cleaning (CIP), applied in the fifth period, exhibited relatively high values of the recovered membrane permeability. While recovery of the permeability between the backwashes performed at the end of every filtration cycle was in the range 88–126%, recovery of the permeability after CIP was in the range of 158–182%.

The sixth period (48–77 days) was another steady fouling state. It reproduced the same trends from the second period (21–34 days), except for a more stable COD degradation due to well-developed biofilms in MBBR part and on carriers in the separation chamber (III). After reaching 400 mbar of TMP, a second chemical cleaning was provided, applying higher backwash pressure with the subsequent soaking of the membrane elements in the cleaning solutions, which caused the permeability to recover to the initial value.

The last, eighth period of system operation is a control period which is characterized by both conditional and steady fouling in the permeability pattern.

In general, in the way described above, the operation of the BF-MBR pilot plant was observed during all the states, which is important for the determination of membrane fouling patterns: Conditional fouling, steady fouling, and TMP jump at different fluxes. Two chemical cleaning procedures were conducted to estimate the recovery of permeability. Data, which were recorded during these states, were taken as the basis for further statistical analysis.

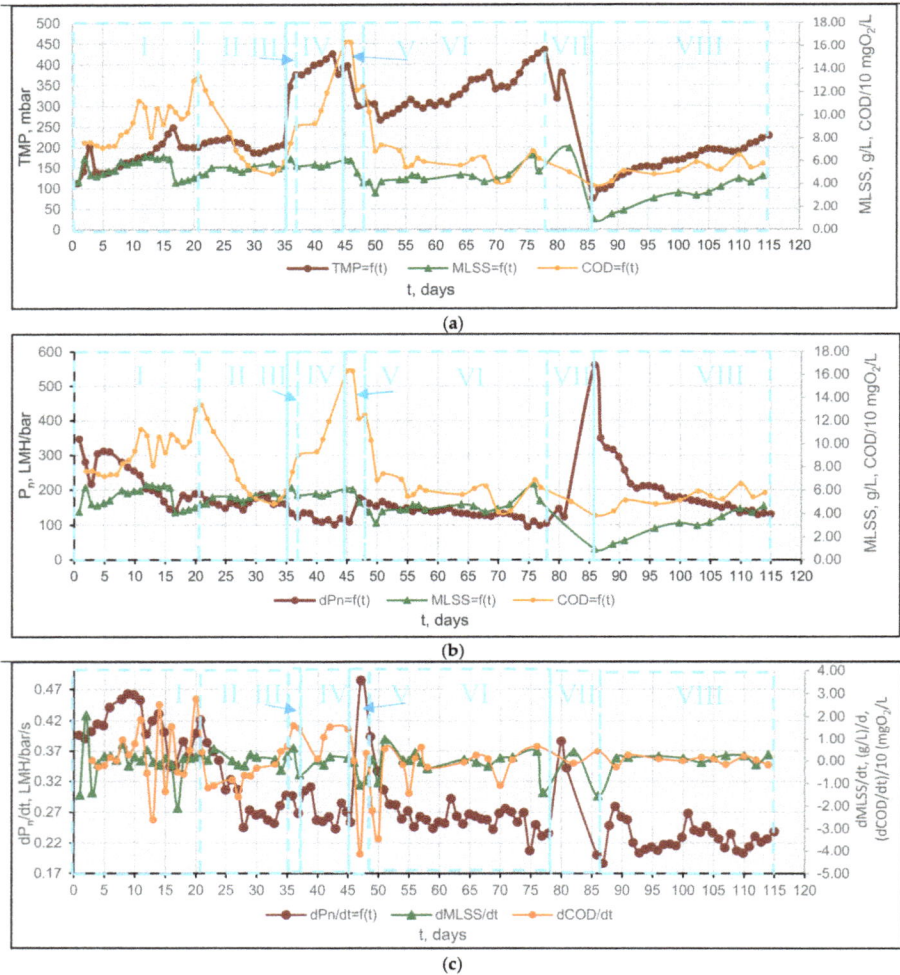

**Figure 2.** BF-MBR pilot plant operation profile: (**a**) TMP, MLSS-III, COD-III change within operation time; (**b**) normalized permeability (P$_n$), MLSS-III, COD-III change within operation time; (**c**) first derivative of normalized permeability (dP$_n$/dt), MLSS-III (dMLSS/dt), COD-III (dCOD/dt) within operation time.

### 3.2. Statistical Determination of Membrane Fouling Patterns

According to the literature, the influence of the mixed liquor parameters (i.e., MLSS, SVI (DSVI), COD, and RH) on the filtration performance and fouling intensity is controversial. Indeed, a positive impact of higher MLSS concentration on MBR hydraulic performance has been indicated [15,106]. On the contrary, Chang et al. [46] observed a positive link between the MLSS increase and the flux decline, which is the opposite of its effect on the specific cake resistance, while Brookes et al. [107] and Jefferson et al. [108] showed that MLSS concentration is not a governing factor influencing the overall membrane fouling, and no consistent correlation was observed between MLSS and fouling intensity.

The influence of the relative hydrophobicity on system performance is also not fully comprehended. According to the findings by Deng et al. [40] and Huang et al. [109], high RH fosters the mitigation of fouling due to the weaker interactions of hydrophobic flocs with a hydrophilic membrane. In addition, lower RH values entail floc deterioration and the consequent increase of cake layer resistance [29], whereas higher RH values are associated with better flocculation [60]. Meanwhile, as specified by Meng et al. [36] and Tian et al. [64], higher RH of sludge causes the formation of a more dense cake layer on the membrane surface, resulting in a greater TMP rise.

There is a lack of data on the correlation between SVI and membrane fouling intensity. Chae et al. [110] stated that high SVI values corresponded to severe membrane fouling in an MBR system. Ng et al. [111] linked the increased SVI values to the higher ratio of non-flocculating components of the activated sludge but did not mention if this affected the fouling intensity. In contrast, according to Fan et al. [112] and Wu and Huang [113], this parameter is not a reliable indicator to predict the membrane fouling potential for MBR systems and has no effect on membrane filterability.

As found, COD is indirectly related to the fouling intensity. COD is linked to the concentration of soluble foulants which have a negative effect on membrane filterability [114]. In addition, COD in the effluent from aerobic and anaerobic biological systems is encountered in the form of soluble microbial products which are among the foulants in MBRs [115]. Meanwhile, Lesjean et al. [116] found no clear correlation between COD and the fouling intensity.

Hence, to gain a deeper understanding of the role of the mixed liquor characteristics in the filtration performance of the investigated system, it was decided to monitor these parameters and their variation over time in the separation chamber (Table 3) and to process the collected data statistically.

**Table 3.** Parameters of the mixed liquor in the separation chamber.

| Parameter | Value |
| --- | --- |
| MLSS, g/L | 5–6.5 |
| dMLSS/dt, (g/L)/day | −0.61–2.06 |
| DSVI, mL/g | 118–272 |
| dDSVI/dt, (mL/g)/day | −91–57 |
| RH, % | 20.5–61.5 |
| dRH/dt, %/day | −27–35 |
| $COD_{dis}$, $mgO_2$/L | 38–134 |
| dCOD/dt, $mgO_2$/L/day | −35–27.5 |

Since the operating conditions varied significantly throughout the whole filtration period (Table 2), which influenced both the activated sludge parameters and the fouling indicators, it was decided to split the whole data range into its characteristic phases and statistically analyze them separately from each other, excluding the data which covered the chemical cleanings. Hence, three basic periods were established: period A (days 3–34), period B (days 49–77) and period C (days 86–114).

PLS regression (also known as a projection of latent structures) was used as an advanced mathematical and statistical tool to model the relations between the X variables and the Y responses within every single period (Table 4).

**Table 4.** Model inputs.

| Period | Predictors | Responses |
| --- | --- | --- |
| A | MLSS, dMLSS/dt, DSVI dDSVI/dt, RH, dRH/dt, $COD_{dis}$, dCOD/dt | TMP, $P_n$, $dP_n$/dt |
| B | MLSS, dMLSS/dt, DSVI dDSVI/dt, $COD_{dis}$, dCOD/dt | TMP, $P_n$, $dP_n$/dt |
| C | MLSS, dMLSS/dt, DSVI dDSVI/dt, $COD_{dis}$, dCOD/dt | TMP, $P_n$, $dP_n$/dt |

The X- and Y-matrices were modelled simultaneously to find the latent variables in input X parameters that best predicted the latent variables in the corresponding Y responses (i.e., PCAs on the

X- and Y-data were performed with the subsequent acquisition of the relative scores). Then, the plotting of two sets of the scores (those related to X and Y) against each other was conducted, maximizing the covariance between X and Y [117].

The obtained model was validated by applying a random cross-validation in PLS. The number of PLS components (factors), was chosen according to the explained variance.

The results of the performed analyses of the data from the initial period of the system performance (Period A) are shown below (Figure 3).

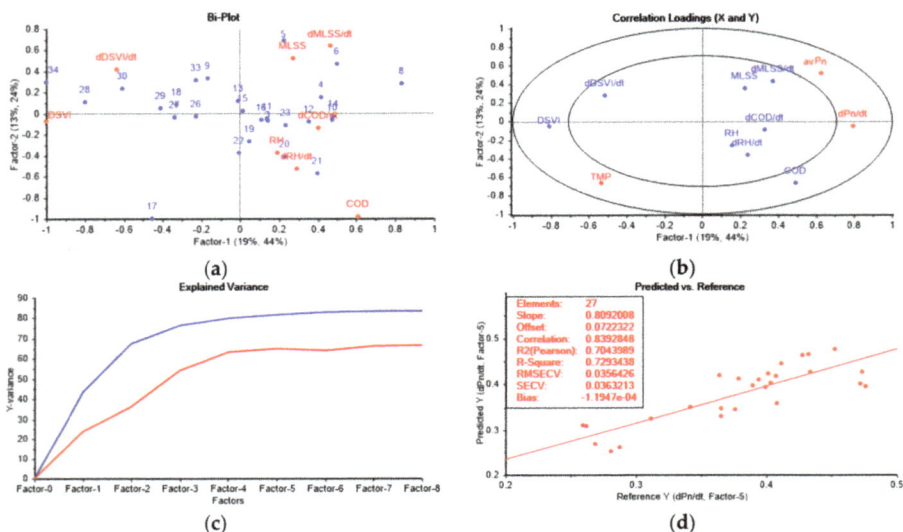

**Figure 3.** Results of PLS of the data from the period A of the filtration performance monitoring: (**a**) Bi-plot; (**b**) loadings plot; (**c**) explained variance plot; (**d**) fouling intensity prediction model.

The correlation loadings plot is computed by accounting for each variable for the displayed latent variables (factors). From the loadings plot, Factor-1 clearly describes DSVI, dDSVI/dt, TMP, COD, dMLSS/dt, permeability, $P_n$, and its slope, $dP_n/dt$, since the first three variables are located at the far left, and the rest at the far right along the Factor-1 axis. Factor-1 also accounts for dCOD/dt, while MLSS and dRH/dt mainly contribute to Factor-2. According to the PLS loadings plot, COD and DSVI explain more than 50% of the variance and are probably the most important variables. DSVI has a negative correlation with both permeability and permeability slope, but is positively linked to TMP. Particularly in this case, COD has a negative correlation with the variables DSVI, dDSVI/dt, MLSS and dMLSS/dt, and is negatively linked to the average normalized permeability (nP). Although the rest of the variables are located in the inner ellipse, which indicates up to 50% of the explained variance and thus does not contain enough structured variation to discriminate between the mixed liquor samples, it was decided to keep them to make the model more reliable.

The analysis of the scores and loadings plot and the bi-plot demonstrates that the samples from days 1–20 are mostly characterized by higher RH, dRH/dt, MLSS, dMLSS/dt, COD, and dCOD/dt, while the samples taken during the period 22–34 day have higher DSVI and dDSVI/dt values.

As demonstrated by the graph of explained variance (Figure 3c), it is preferable to use five components, since this number gives a lower residual variance.

According to the Figure 3d (the validation graph), the developed model is linear (R-squared = 0.73) and with a reasonable fit to the majority of data: Slope = 0.81, offset 0.07 and the dispersion of the validation samples around the regression line (Root Mean Square Error of Cross Validation–RMSEV) and the standard error of cross-validation (SECV) are approximately 0.036. Consequently, the model is reliable and can be used for future predictions for the defined number of factors under the operational conditions applied during the period A.

Relative hydrophobicity and its change required much more effort and time to be experimentally determined than other variables. In addition, RH and dRH/dt are characterized by relatively low-weighted regression coefficients: 0.02 and −0.086, respectively (Factor-2); and, 0.07 and 0.04, respectively (Factor-1) (i.e., these variables are not well explained by the components). Considering the above-mentioned aspects, it was decided to exclude RH and dRH/dt from further monitoring and analysis.

The second period, B, covers the filtration performance data collected between the first and the second chemical cleanings of the system. Obtained results of the PLS analysis are represented below (Figure 4).

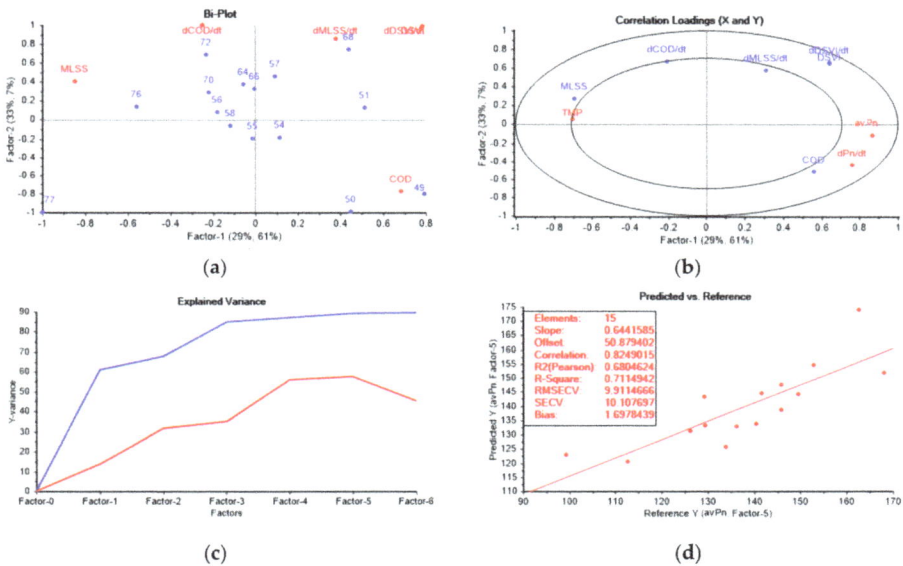

**Figure 4.** Results of PLS of the data from the period B of the filtration performance monitoring: (a) Bi-plot; (b) loadings plot; (c) explained variance plot; (d) fouling intensity prediction model.

According to the bi-plot (Figure 4b), the majority of the samples within period B are characterized by higher dCOD/dt values. Meanwhile, the samples taken on days 49–50 are characterized by higher COD values; on days 51, 57 and 68 by relatively high dMLSS/dt, DSVI, and dDSVI/dt values; on day 72 by comparatively high dCOD/dt values; and on days 76 and 77 by more significant MLSS values.

According to the correlation loadings plot, Factor-1 apparently describes TMP, MLSS, COD, average permeability ($avP_n$), $dP_n/dt$, DSVI and dDSVI/dt. Factor-2 is related to dCOD/dt and dMLSS/dt. All the variables were marked as significant according to the plot of correlation loadings, even though the MLSS variable gives slightly less than 50% of the explained variance. MLSS and dCOD/dt are positively linked to the TMP response, in contrast to dMLSS/dt, DSVI, dDSVI/dt, which have a negative correlation with TMP and the permeability slope ($dP_n/dt$). The COD variable has a high positive correlation with $dP_n/dt$ and is positively linked to the average permeability ($avP_n$).

Figure 4 demonstrates that the optimum number of factors is five, which provides more than 57% of the explained Y-variance.

An analysis of the validation plot shows that the developed model is linear, having R-squared = 0.71 and with a good fit to the majority of data (i.e., slope = 0.64). RMSEV and SECV are approximately 10, but it is essential to acknowledge that the mentioned errors have the same units as the reference Y (in this case, average normalized permeability, $avP_n$). R-squared (Pearson) is close to R-squared correlation (0.68 vs. 0.82), which indicates the reliability of the model. Consequently, a good prediction is attained with the developed model, which proves that the model is reliable and can be used during further stages when the operating conditions applied in the period B are replicated.

The output from the PLS modelling of the data acquired after the second CIP (the period C) is demonstrated below (Figure 5).

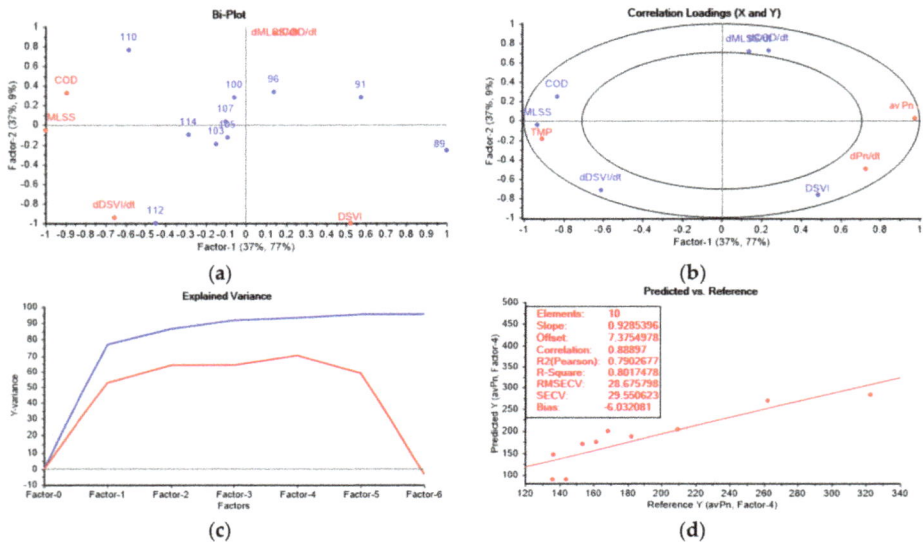

**Figure 5.** Results of PLS of the data from the period C of the filtration performance monitoring: (**a**) Bi-plot; (**b**) loadings plot; (**c**) explained variance plot; (**d**) fouling intensity prediction model.

The bi-plot shows that the samples from day 89 have a higher DSVI value, while dMLSS/dt and dCOD/dt are the most distinctive parameters for days 91 and 96. Days 100, 107 and 110 are characterized by higher COD content, whereas days 103, 105 and 114 have higher MLSS values. Day 112 is characterized by a higher dDSVI/dt.

From the correlation loadings plot (Figure 5b), COD, MLSS, TMP, dDSVI/dt, DSVI, $avP_n$ and $dP_n$/dt contribute to Factor-1, while Factor-2 describes dMLSS/dt and dCOD/dt. All the specified variables explain more than 50% of the variance and thus have high importance in relation to Factor-1 and Factor-2. MLSS and dDSVI/dt are positively linked to TMP and have a negative correlation with the permeability indicators, $avP_n$ and $dP_n$/dt. DSVI is positively correlated to $dP_n$/dt, while dMLSS/dt and dCOD/dt have a negative correlation with the permeability slope.

The explained variance plot indicates that the optimum number of factors is four, which provides more than 70% of explained Y-variance.

The points of the validation graph in Figure 5d have a linear trend (R-squared = 0.8), having a good fit to the majority of data (slope = 0.93). R-squared (Pearson) is close to R-squared correlation (0.79 vs. 0.89), which indicates the reliability of the model. Only the errors RMSEV and SECV are higher than in previous cases, but this can be explained by the higher values of the response function (average permeability) in this particular case.

Since the higher amount of data was available to be collected during the last period C (Table 5) in comparison to the previous modes, it was decided to apply the predict function to new data.

**Table 5.** Mixed liquor characteristics and fouling indicators during period VIII (new data).

| | TMP$_{av}$ [1], Bar | av dP$_n$/dt [2] | avP$_n$ [3], LMH/Bar | DSVI [4], mL/g | dDSVI/dt [5] | MLSS [6], g/L | dMLSS/dt [7] | COD$_f$ [8], mgO$_2$/L | dCOD/dt [9] |
|---|---|---|---|---|---|---|---|---|---|
| max. | 266.16 | 0.26 | 125.45 | 185.41 | 5.52 | 5.74 | 0.35 | 69.80 | 5.00 |
| min. | 232.30 | 0.23 | 112.98 | 142.60 | −7.79 | 5.32 | −0.17 | 45.40 | −3.83 |
| average | 249.26 | 0.24 | 120.66 | 166.56 | −1.96 | 5.48 | 0.02 | 55.52 | −0.44 |

Notes: [1] Average transmembrane pressure; [2] Average normalized permeability slope; [3] Average normalized permeability; [4] Diluted sludge volume index; [5] Diluted sludge volume index slope; [6] Mixed liquor suspended solids; [7] Mixed liquor suspended solids slope; [8] Chemical oxygen demand (filtered); [9] Chemical oxygen demand slope.

Full prediction with the identification of outliers was used. The following results were obtained (Figure 6).

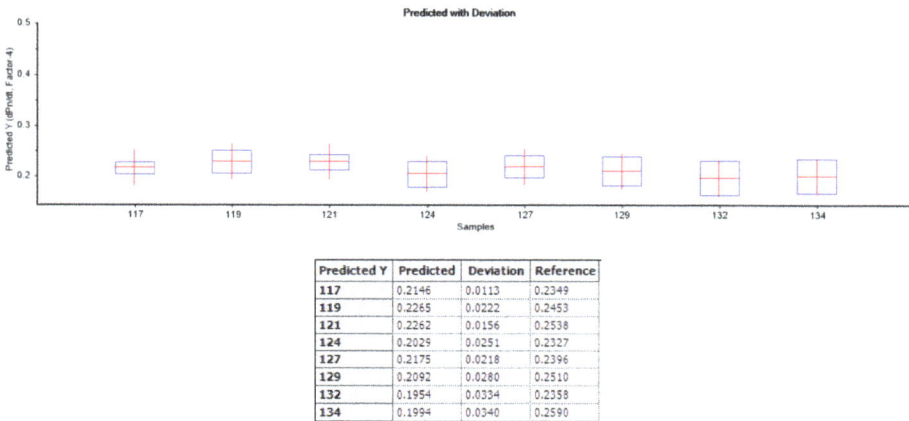

| Predicted Y | Predicted | Deviation | Reference |
|---|---|---|---|
| 117 | 0.2146 | 0.0113 | 0.2349 |
| 119 | 0.2265 | 0.0222 | 0.2453 |
| 121 | 0.2262 | 0.0156 | 0.2538 |
| 124 | 0.2029 | 0.0251 | 0.2327 |
| 127 | 0.2175 | 0.0218 | 0.2396 |
| 129 | 0.2092 | 0.0280 | 0.2510 |
| 132 | 0.1954 | 0.0334 | 0.2358 |
| 134 | 0.1994 | 0.0340 | 0.2590 |

**Figure 6.** Prediction results for the new data from Period C for four factors using the derived PLS model for the relevant period.

The deviation between the predicted and the reference values is in the range 0.01–0.034, which demonstrates the reliability of the applied model.

Consequently, a good prediction is attained by applying the developed model, which proves that the model is reliable and can be used during further stages under the operating conditions that were applied during period C.

In addition, MLR was performed using leverage correction. However, obtained results are unreliable since the same data was validated and used for the prediction, which provided overly optimistic results. The application of the test matrix in MLR would merely copy the PLS strategy but do so in a more difficult way. MLR is a simpler way of doing the calculations, but PLS is much more advanced due to the applied validation techniques.

SRT and permeate flux are among the key operating parameters controlling fouling intensity in MBR.

In order to estimate the influence of SRT on the performance of the current system, this parameter was included in the models as an additional variable. The acquired results are represented in Figure 7.

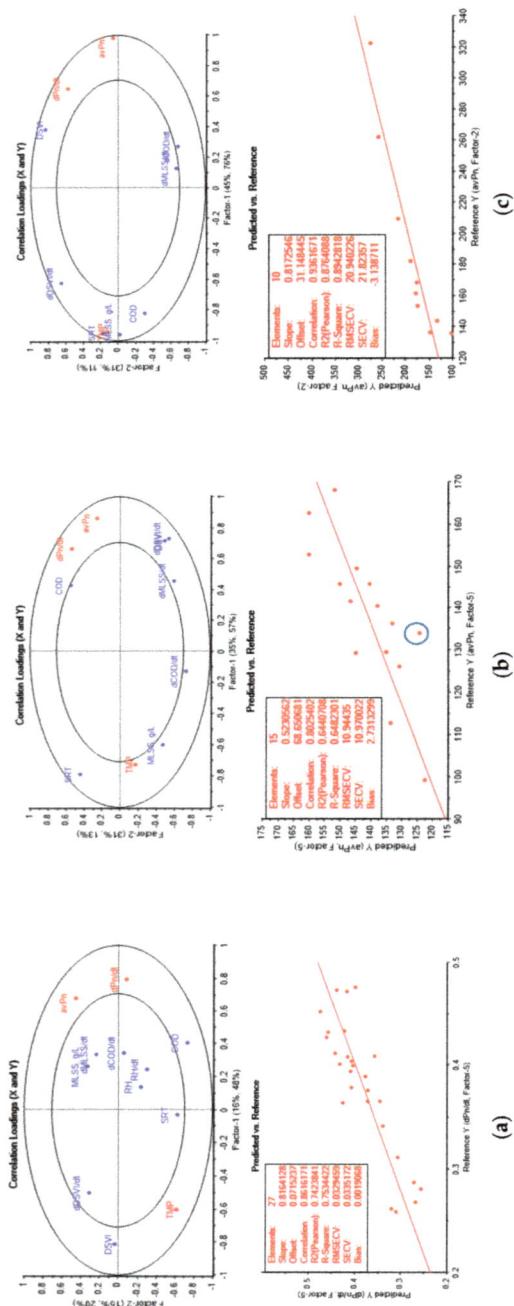

**Figure 7.** Results of PLSs of the data from all the periods of the filtration performance monitoring, including SRT: (**a**) period A; (**b**) period B; (**c**) period C.

According to the correlation loadings plot related to period A, SRT explains less than 50% of the variance and thus has relatively little influence. In this particular case, SRT exhibits an independent variation in relation to other variables, except for COD, which has a weak positive link with SRT. Meanwhile, SRT exhibits a slightly negative correlation with the normalized permeability and permeability slope for period A. Concerning the model enhancement, the introduction of the new variable did not entail any significant improvement: RMSECV was just 0.002 less than its value in the initial model, while the bias, on the contrary, showed an order of magnitude increase in absolute value.

The results related to period B demonstrate that SRT is an important variable which explains more than 50% of the variance in the dataset. It has a strong negative correlation with COD and the normalized permeability. In addition, SRT is positively correlated with MLSS along Factor-1. The negative correlation between SRT and COD during this period can be attributed to the higher treatment performance of the biomass, which becomes well-developed at SRT up to 40 days and thus is capable of a more efficient biodegradation of organic contaminants, particularly SMPs, causing the decrease of COD values [118,119]. Meanwhile, the increase in SRT promotes the development of higher MLSS concentrations [120], thus inducing membrane fouling.

The introduction of the new variable into the existing model decreased its linearity R-squared = 0.65 vs. R-squared = 0.71 (values in the new model vs. values characteristic for the basic model related to period B), with a slightly worse fit to the majority of data (slope = 0.52 vs. slope = 0.64), RMSEV 10.9 vs. 9.9, SECV 10.97 vs. 10.1, bias 2.73 vs. 1.7. In addition, the new model underestimated a sample from day 72 (marked with the blue circle).

The modelling of the dataset from period C demonstrates the importance of the SRT variable. SRT is highly positively correlated with MLSS and TMP, and is negatively linked to normalized permeability and its slope, hence indicating the fouling enhancement through the increase of MLSS at higher SRTs, which agrees with the previous findings by Le-Clech et al. [29], Van den Broeck et al. [120], Yigit et al. [121]. The positive link between SRT and COD along Factor-1 during this period can be attributed to the accumulation of small microbial by-products (SMP with the molecular weight (MW) < 1 kDa), which contribute to fouling through deflocculation at high SRTs (>31 days) [118,121,122]. However, further studies are required to confirm the presence of different groups of microorganisms at various SRTs in this system (for example, tightly and loosely bound EPS, small SMP, etc.), since the deep investigation of the biomass content was not in the scope of the current research.

The new model exhibits higher linearity (R-squared = 0.89 vs. R-squared = 0.80) and a slightly higher accuracy (RMSEV = 20.9 vs. RMSEV = 28.7; SECV = 21.8 vs. SECV = 29.6; and, bias = −3.4 vs. bias = −6.0) than the initial model.

It is noteworthy that the purpose of including SRT in modelling was not to improve the models for the relevant periods developed earlier in this work, since the inclusion of a new variable is undesirable as it could complicate the model (i.e., it is preferable to use as low a number of variables as possible) [123]. Besides, the introduction of the SRT variable to the model covering period C barely decreased the deviation in the prediction of the new dataset (Table 4; 0.016–0.0261 vs. 0.011–0.034), making the extension of the model size unreasonable for its further use in the system controller. The scope was to show the influence of SRT on the operational parameters and fouling intensity in the current system to achieve the highest possible fouling inhibition.

As discovered, SRT should be less than 31 days to avoid a severe membrane fouling. This can be called the critical SRT. The SRT that can be applied without a sharp decrease in permeability is 20 days for the current BF-MBR system. In the studied pilot plant, SRT was adjusted by changing the frequency of sludge removal and the volume of the removed sludge per batch.

Concerning the permeate flux, it can be decreased in order to minimize the filtration resistance if the biomass exhibits high fouling propensity. The current system worked at a constant permeate flux, which varied depending on the monitoring period (Table 2). In general, all the applied fluxes were below the critical flux value to avoid a severe membrane fouling [124–126]. The critical net flux was determined by the flux-step method, described by Miller et al. [127], and was in a range of 12–15 LMH.

In addition to the desludging option, the concentration of the mixed liquor in the separation and biological chambers was regulated by adjusting the RAS pumping intensity (i.e., pulse length and frequency). The introduction of the RAS line made it possible to build up the desired level of biomass in biological and separation chambers, and to adjust the endogenous decay of the biomass, thus providing sufficient COD and $NH_4^+$ removal.

To summarize, the monitored mixed liquor characteristics allowed the controlling of the fouling intensity by adjusting the operating conditions which helped to maintain the stability of the system performance and, hence, the permeate quality: BF-MBR installation assured 100% MLSS elimination and 67–90% treatment efficiency in terms of COD removal, keeping the TMP below 500 mbar.

## 4. Conclusions

The developed chemometric approach to the assessment of membrane fouling in membrane bioreactor advances the field of fouling monitoring and provides a statistical tool for its control in submerged membrane bioreactors.

The approach was based on PLS regression analysis and was used to detect membrane fouling patterns in the biofilm ceramic membrane bioreactor pilot system during 114 days of operation, varying membrane flux and solid retention time, and covering the periods of steady fouling and TMP jumps, followed by two chemical cleanings in the system.

The mixed liquor parameters MLSS, dMLSS/dt, DSVI, dDSVI/dt, COD, and dCOD/dt were found to be significant for estimation and prediction of fouling intensity, while relative hydrophobicity of mixed liquor and its slope seemed to play a secondary role. Normalized permeability and its slope were identified as the most reliable fouling indicators, while critical solid retention time was introduced as another quantitative parameter, influencing the intensity of membrane fouling.

The cross-validation of every model and the complete validation of the model, covering the last phase of the filtration, demonstrated low uncertainty of the predictions, and hence high reliability of the models, allowing further implementation of the developed fouling control strategies.

The models were used to adjust operational parameters of the pilot system according to the characteristics of biomass, keeping the system running below critical transmembrane pressure (500 mbar), with 67–90% removal of chemical oxygen demand and 100% retention of suspended solids, resulting in good recovery of membrane permeability after chemical cleanings, thus removing irreversible fouling.

Further work is foreseen in the validation of the developed approach in an operational environment of decentralized membrane bioreactors, where the sustainable operation is frequently a critical issue due to the lack of qualified supervision, and which raises the barrier to penetration of membrane bioreactors into cost-sensitive markets.

**Author Contributions:** O.K. and Z.M. conceived and designed the experiments; O.K. performed the experiments under supervision of Z.M. and analyzed the data; H.R. contributed reagents, materials and analysis tools, and contributed to the discussion of the article; O.K. wrote the paper with advice from Z.M.

**Funding:** This research was co-funded by the Erasmus+ Program of the European Union (project number: 561755-EPP-1-2015-1-NOEPPKA2-CBHE-JP).

**Acknowledgments:** Authors express their gratitude to Daniel Todt from Ecomotive AS for facilitating pilot plant studies; Yuliia Dzihora and Stella Saliu for their inputs as Erasmus+ exchange students; Knut Kvaal for his guidance on mathematical modelling using the Unscrambler software.

**Conflicts of Interest:** The authors declare no conflict of interest. The founding sponsors had no role in the design of the study; in the collection, analyses, or interpretation of data; in the writing of the manuscript, and in the decision to publish the results.

## References

1. World Economic Forum. The Global Risks Report 2018. Available online: https://www.weforum.org/reports/the-global-risks-report-2018 (accessed on 9 December 2017).

2. World Economic Forum. Grim Reaping. Available online: http://reports.weforum.org/global-risks-2018/grim-reaping/ (accessed on 10 December 2017).

3. United Nations World Water Assessment Programme (WWAP). *The United Nations World Water Development Report 2017, Wastewater: The Untapped Resource*; UNESCO: Paris, France, 2017.

4. Lautze, J.; Stander, E.; Drechsel, P.; da Silva, A.K.; Keraita, B. *Global Experiences in Water Reuse*; International Water Management Institute (IWMI): Colombo, Sri Lanka; CGIAR Research Program on Water, Land and Ecosystems (WLE): Montpellier, France, 2014; 31p.

5. European Commission. Water is too Precious to Waste. Available online: http://ec.europa.eu/environment/water/pdf/water_reuse_factsheet_en.pdf (accessed on 13 January 2018).

6. Hwang, H.; Forrester, A.; Lansey, K. Decentralized water reuse: Regional water supply system resilience benefits. *Procedia Eng.* **2014**, *70*, 853–856. [CrossRef]

7. Lesjean, B.; Leiknes, T.; Hochstrat, R.; Schories, R.; Gonzalez, G.; Gonzalez, A. MBR: Technology gets timely EU cash boost. *Filtr. Sep.* **2006**, *43*, 20–23. [CrossRef]

8. Hai, F.I.; Yamamoto, K.; Lee, C.-H. *Membrane Biological Reactors: Theory, Modeling, Design, Management and Applications to Wastewater Reuse*; IWA Publishing: London, UK, 2014; 504p, ISBN 9781780400655.

9. Geilvoet, S.P. The Delft Filtration Characterisation Method: Assessing Membrane Bioreactor Activated Sludge Filterability. Ph.D. Thesis, Delft University of Technology, Delft, The Netherlands, 12 February 2010.

10. Global $8.27 Bn Membrane Bioreactor Market, 2025. Available online: http://markets.businessinsider.com/news/stocks/global-8-27-bn-membrane-bioreactor-market-2025-1005680257 (accessed on 15 December 2017).

11. Böhm, L.; Drews, A.; Prieske, H.; Bérubé, P.R.; Kraume, M. The importance of fluid dynamics for MBR fouling mitigation. *Bioresour. Technol.* **2012**, *122*, 50–61. [CrossRef] [PubMed]

12. Ivanovic, I.; Leiknes, T.O. The biofilm membrane bioreactor (BF-MBR)—A review. *Desalination Water Treat.* **2012**, *37*, 288–295. [CrossRef]

13. Ivanovic, I.; Leiknes, T.O.; Ødegaard, H. Fouling control by reduction of submicron particles in a BF-MBR with an integrated flocculation zone in the membrane reactor. *Sep. Sci. Technol.* **2008**, *43*, 1871–1883. [CrossRef]

14. Yu, H.Y.; Xu, Z.K.; Lei, H.; Hu, M.X.; Yang, Q. Photoinduced graft polymerization of acrylamide on polypropylene microporous membranes for the improvement of antifouling characteristics in a submerged membrane-bioreactor. *Sep. Purif. Technol.* **2007**, *53*, 119–125. [CrossRef]

15. Brookes, A.; Jefferson, B.; Guglielmi, G.; Judd, S.J. Sustainable Flux Fouling in a Membrane Bioreactor: Impact of Flux and MLSS. *Sep. Sci. Technol.* **2006**, *41*, 1279–1291. [CrossRef]

16. Kraume, M.; Wedi, D.; Schaller, J.; Iversen, V.; Drews, A. Fouling in MBR: What use are lab investigations for full scale operation? *Desalination* **2009**, *236*, 94–103. [CrossRef]

17. Yusuf, Z.; Wahab, N.A.; Abusam, A. Neural Network-based Model Predictive Control with CPSOGSA for SMBR Filtration. *Int. J. Electr. Comput. Eng.* **2017**, *7*, 1538–1545. [CrossRef]

18. Song, W.; Li, Z.; Li, Y.; You, H.; Qi, P.; Liu, F.; Loy, D.A. Facile sol-gel coating process for anti-biofouling modification of poly (vinylidene fluoride) microfiltration membrane based on novel zwitterionic organosilica. *J. Membr. Sci.* **2018**, *550*, 266–277. [CrossRef]

19. Gkotsis, P.K.; Mitrakas, M.M.; Tolkou, A.K.; Zouboulis, A.I. Batch and continuous dosing of conventional and composite coagulation agents for fouling control in a pilot-scale MBR. *Chem. Eng. J.* **2016**, *311*, 255–264. [CrossRef]

20. Lee, J.C.; Kim, J.S.; Kang, I.J.; Cho, M.H.; Park, P.K.; Lee, C.H. Potential and limitations of alum or zeolite addition to improve the performance of a submerged membrane bioreactor. *Water Sci. Technol.* **2001**, *43*, 59–66. [CrossRef] [PubMed]

21. Zarei, A.; Moslemi, M.; Mirzaei, H. The Combination of KMnO4 Oxidation and Polymeric Flocculation for the Mitigation of Membrane Fouling in a Membrane Bioreactor. *Sep. Purif. Technol.* **2016**, *159*, 124–134. [CrossRef]

22. Drews, A. Membrane fouling in membrane bioreactors—Characterisation, contradictions, cause and cures. *J. Membr. Sci.* **2010**, *363*, 1–28. [CrossRef]

23. Zheng, Y.; Zhang, W.; Tang, B.; Ding, J.; Zheng, Y.; Zhang, Z. Membrane fouling mechanism of biofilm-membrane bioreactor (BF-MBR): Pore blocking model and membrane cleaning. *Bioresour. Technol.* **2018**, *250*, 398–405. [CrossRef] [PubMed]

24. Leiknes, T.; Ødegaard, H. The development of a biofilm membrane bioreactor. *Desalination* **2007**, *202*, 135–143. [CrossRef]

25. Jin, L.; Ong, S.L.; Ng, H.Y. Fouling control mechanism by suspended biofilm carriers addition in submerged ceramic membrane bioreactors. *J. Membr. Sci.* **2013**, *427*, 250–258. [CrossRef]

26. Arabi, S.; Nakhla, G. Impact of cation concentrations on fouling in membrane bioreactors. *J. Membr. Sci.* **2009**, *343*, 110–118. [CrossRef]

27. Chang, I.S.; Judd, S.J. Domestic wastewater treatment by a submerged MBR (membrane bio-reactor) with enhanced air sparging. *Water Sci. Technol.* **2003**, *47*, 149–154. [CrossRef] [PubMed]

28. Judd, S. *The MBR Book Principles and Applications of Membrane Bioreactors in Water and Wastewater Treatment*, 1st ed.; Elsevier Ltd.: London, UK, 2006; p. 325, ISBN-13 978-1-85-617481-7.

29. Le-Clech, P.; Chen, V.; Fane, T.A.G. Fouling in membrane bioreactors used in wastewater treatment. *J. Membr. Sci.* **2006**, *284*, 17–53. [CrossRef]

30. Mafirad, S.; Mehrnia, M.R.; Azami, H.; Sarrafzadeh, M.H. Effects of biofilm formation on membrane performance in submerged membrane bioreactors. *Biofouling* **2011**, *27*, 477–485. [CrossRef] [PubMed]

31. Meng, F.; Zhang, S.; Oh, Y.; Zhou, Z.; Shin, H.-S.; Chae, S.-R. Fouling in membrane bioreactors: An updated review. *Water Res.* **2017**, *114*, 151–180. [CrossRef] [PubMed]

32. Fallis, A. Experimental Methods in Wastewater Treatment. *J. Chem. Inf. Model.* **2013**, *53*. [CrossRef]

33. Ferreira, M.L. Filterability and Sludge Concentration in Membrane Bioreactors. Ph.D. Thesis, Delft University of Technology, Delft, The Netherlands, 15 September 2011.

34. Meng, F.; Shi, B.; Yang, F.; Zhang, H. Effect of hydraulic retention time on membrane fouling and biomass characteristics in submerged membrane bioreactors. *Bioprocess Biosyst. Eng.* **2007**, *30*, 359–367. [CrossRef] [PubMed]

35. Ji, L.; Zhou, J. Influence of aeration on microbial polymers and membrane fouling in submerged membrane bioreactors. *J. Membr. Sci.* **2006**, *276*, 168–177. [CrossRef]

36. Meng, F.; Zhang, H.; Yang, F.; Zhang, S.; Li, Y.; Zhang, X. Identification of activated sludge properties affecting membrane fouling in submerged membrane bioreactors. *Sep. Purif. Technol.* **2006**, *51*, 95–103. [CrossRef]

37. Vanysacker, L.; Boerjan, B.; Declerck, P.; Vankelecom, I.F.J. Biofouling ecology as a means to better understand membrane biofouling. *Appl. Microbiol. Biotechnol.* **2014**, *98*, 8047–8072. [CrossRef] [PubMed]

38. Wu, B.; Fane, A.G. Microbial relevant fouling in membrane bioreactors: Influencing factors, characterization, and fouling control. *Membranes* **2012**, *2*, 565–584. [CrossRef] [PubMed]

39. Zhou, Z.; Meng, F.; He, X.; Chae, S.-R.; An, Y.; Jia, X. Metaproteomic analysis of biocake proteins to understand membrane fouling in a submerged membrane bioreactor. *Environ. Sci. Technol.* **2015**, *49*, 1068–1077. [CrossRef] [PubMed]

40. Deng, L.; Guo, W.; Hao, H.; Farzana, M.; Zuthi, R.; Zhang, J.; Liang, S.; Li, J.; Wang, J.; Zhang, X. Membrane fouling reduction and improvement of sludge characteristics by bioflocculant addition in submerged membrane bioreactor. *Sep. Purif. Technol.* **2015**, *156*, 450–458. [CrossRef]

41. Lee, W.; Kang, S.; Shin, H. Sludge characteristics and their contribution to microfiltration in submerged membrane bioreactors. *J. Membr. Sci.* **2003**, *216*, 217–227. [CrossRef]

42. Meng, F.; Chae, S.R.; Drews, A.; Kraume, M.; Shin, H.S.; Yang, F. Recent advances in membrane bioreactors (MBRs): Membrane fouling and membrane material. *Water Res.* **2009**, *43*, 1489–1512. [CrossRef] [PubMed]

43. Meng, F.; Zhang, H.; Yang, F.; Li, Y.; Xiao, J.; Zhang, X. Effect of filamentous bacteria on membrane fouling in submerged membrane bioreactor. *J. Membr. Sci.* **2006**, *272*, 161–168. [CrossRef]

44. Krzeminski, P. Activated Sludge Filterability and Full-Scale Membrane Bioreactor Operation. Ph.D. Thesis, Delft University of Technology, Delft, The Netherlands, 22 January 2013.

45. Azami, H.; Sarrafzadeh, M.H.; Mehrnia, M.R. Influence of sludge rheological properties on the membrane fouling in submerged membrane bioreactor. *Desalin. Water Treat.* **2011**, *34*, 117–122. [CrossRef]

46. Chang, I.S.; Kim, S.N. Wastewater treatment using membrane filtration—Effect of biosolids concentration on cake resistance. *Process Biochem.* **2005**, *40*, 1307–1314. [CrossRef]

47. Wang, Z.; Chu, J.; Song, Y.; Cui, Y.; Zhang, H.; Zhao, X.; Li, Z.; Yao, J. Influence of operating conditions on the efficiency of domestic wastewater treatment in membrane bioreactors. *Desalination* **2009**, *245*, 73–81. [CrossRef]

48. Radjenović, J.; Matošić, M.; Mijatović, I. Membrane bioreactor (MBR) as an advanced wastewater treatment technology. In *Handbook of Environmental Chemistry*; Springer: Berlin/Heidelberg, Germany, 2008; Volume 5, pp. 37–101.

49. Iorhemen, O.T.; Hamza, R.A.; Tay, J.H. Membrane bioreactor (MBR) technology for wastewater treatment and reclamation: Membrane fouling. *Membranes* **2016**, *6*, 33. [CrossRef] [PubMed]

50. Reid, E.; Liu, X.; Judd, S.J. Sludge characteristics and membrane fouling in full-scale submerged membrane bioreactors. *Desalination* **2008**, *219*, 240–249. [CrossRef]

51. Fan, F.; Zhou, H. Interrelated Effects of Aeration and Mixed Liquor Fractions on Membrane Fouling for Submerged Membrane Bioreactor Processes in Wastewater Treatment. *Environ. Sci. Technol.* **2007**, *41*, 2523–2528. [CrossRef] [PubMed]

52. Hernandez Rojas, M.E.; Van Kaam, R.; Schetrite, S.; Albasi, C. Role and variations of supernatant compounds in submerged membrane bioreactor fouling. *Desalination* **2005**, *17*, 95–107. [CrossRef]

53. Salazar-Peláez, M.L.; Morgan-Sagastume, J.M.; Noyola, A. Influence of hydraulic retention time on UASB post-treatment with UF membranes. *Water Sci. Technol.* **2011**, *64*, 2299–2305. [CrossRef] [PubMed]

54. Chen, R.; Nie, Y.; Hu, Y.; Miao, R.; Utashiro, T.; Li, Q.; Xu, M.; Li, Y.Y. Fouling behaviour of soluble microbial products and extracellular polymeric substances in a submerged anaerobic membrane bioreactor treating low-strength wastewater at room temperature. *J. Membr. Sci.* **2017**, *531*, 1–9. [CrossRef]

55. Jiang, T. Characterization and Modelling of Soluble Microbial Products in Membrane Bioreactors. Ph.D. Thesis, Ghent University, Gent, Belgium, 2007.

56. Xie, W.M.; Ni, B.J.; Sheng, G.P.; Seviour, T.; Yu, H.Q. Quantification and kinetic characterization of soluble microbial products from municipal wastewater treatment plants. *Water Res.* **2016**, *88*, 703–710. [CrossRef] [PubMed]

57. Deng, L.; Guo, W.; Ngo, H.H.; Zhang, J.; Liang, S.; Xia, S.; Zhang, Z.; Li, J. A comparison study on membrane fouling in a sponge-submerged membrane bioreactor and a conventional membrane bioreactor. *Bioresour. Technol.* **2014**, *165*, 69–74. [CrossRef] [PubMed]

58. Jørgensen, M.K.; Nierychlo, M.; Nielsen, A.H.; Larsen, P.; Christensen, M.L.; Nielsen, P.H. Unified understanding of physico-chemical properties of activated sludge and fouling propensity. *Water Res.* **2017**, *120*, 117–132. [CrossRef] [PubMed]

59. Van den Broeck, R.; Krzeminski, P.; Van Dierdonck, J.; Gins, G.; Lousada-Ferreira, M.; Van Impe, J.F.M.; van der Graaf, J.H.J.M.; Smets, I.Y.; van Lier, J.B. Activated sludge characteristics affecting sludge filterability in municipal and industrial MBRs: Unraveling correlations using multi-component regression analysis. *J. Membr. Sci.* **2011**, *378*, 330–338. [CrossRef]

60. Liu, Y.; Fang, H.H.P. Influences of extracellular polymeric substances (EPS) on flocculation, settling, and dewatering of activated sludge. *Crit. Rev. Environ. Sci. Technol.* **2003**, *33*, 237–273. [CrossRef]

61. Tu, X.; Zhang, S.; Xu, L.; Zhang, M.; Zhu, J. Performance and fouling characteristics in a membrane sequence batch reactor (MSBR) system coupled with aerobic granular sludge. *Desalination* **2010**, *261*, 191–196. [CrossRef]

62. Jang, N.; Ren, X.; Choi, K.; Kim, I.S. Comparison of membrane biofouling in nitrification and denitrification for the membrane bioreactor (MBR). *Water Sci. Technol.* **2006**, *53*, 43–49. [CrossRef] [PubMed]

63. Nittami, T.; Tokunaga, H.; Satoh, A.; Takeda, M.; Matsumoto, K. Influence of surface hydrophilicity on polytetrafluoroethylene flat sheet membrane fouling in a submerged membrane bioreactor using two activated sludges with different characteristics. *J. Membr. Sci.* **2014**, *463*, 183–189. [CrossRef]

64. Tian, Y.; Chen, L.; Zhang, S.; Cao, C.; Zhang, S. Correlating membrane fouling with sludge characteristics in membrane bioreactors: An especial interest in EPS and sludge morphology analysis. *Bioresour. Technol.* **2011**, *102*, 8820–8827. [CrossRef] [PubMed]

65. Rahimi, Y.; Torabian, A.; Mehrdadi, N.; Habibi-Rezaie, M.; Pezeshk, H.; Nabi-Bidhendi, G.R. Optimizing aeration rates for minimizing membrane fouling and its effect on sludge characteristics in a moving bed membrane bioreactor. *J. Hazard. Mater.* **2011**, *186*, 1097–1102. [CrossRef] [PubMed]

66. Delrue, F.; Stricker, A.E.; Mietton-Peuchot, M.; Racault, Y. Relationships between mixed liquor properties, operating conditions and fouling on two full-scale MBR plants. *Desalination* **2011**, *272*, 9–19. [CrossRef]

67. Wang, Z.; Wu, Z.; Tang, S. Impact of temperature seasonal change on sludge characteristics and membrane fouling in a submerged membrane bioreactor. *Sep. Sci. Technol.* **2010**, *45*, 920–927. [CrossRef]

68. Li, X.F.; Zhang, L.N.; Du, G.C. Influence of sludge discharge on sludge settleability and membrane flux in a membrane bioreactor. *Environ. Technol.* **2010**, *31*, 1289–1294. [CrossRef] [PubMed]

69. Kim, D.S.; Kang, J.S.; Lee, Y.M. The Influence of Membrane Surface Properties on Fouling in a Membrane Bioreactor for Wastewater Treatment. *Sep. Sci. Technol.* **2005**, *39*, 833–854. [CrossRef]

70. Gitis, V.; Rothenberg, G. *Ceramic Membranes: New Opportunities and Practical Applications*; Wiley-VCH Verlag GmbH & Co. KGaA: Weinheim, Germany, 2016; p. 394, ISBN 978-3-527-33493-3.

71. Iversen, V. Comprehensive Assessment of Flux Enhancers in Membrane Bioreactors for Wastewater Treatment. Ph.D. Thesis, Technical University of Berlin, Berlin, Germany, 4 October 2010.

72. Yonekawa, H.; Tomita, Y.; Watanabe, Y. Behavior of micro-particles in monolith ceramic membrane filtration with pre-coagulation. *Water Sci. Technol.* **2004**, *50*, 317–325. [CrossRef] [PubMed]

73. Dickhout, J.M.; Moreno, J.; Biesheuvel, P.M.; Boels, L.; Lammertink, R.G.H.; de Vos, W.M. Produced water treatment by membranes: A review from a colloidal perspective. *J. Colloid Interface Sci.* **2017**, *487*, 523–534. [CrossRef] [PubMed]

74. Jin, L.; Ong, S.L.; Ng, H.Y. Comparison of fouling characteristics in different pore-sized submerged ceramic membrane bioreactors. *Water Res.* **2010**, *44*, 5907–5918. [CrossRef] [PubMed]

75. Meabe, E.; Lopetegui, J.; Ollo, J.; Lardies, S. Ceramic Membrane Bioreactor: Potential applications and challenges for the future. In Proceedings of the MBR Asia International Conference, Kuala Lumpur, Malaysia, 25–26 April 2011.

76. Shi, X.; Tal, G.; Hankins, N.P.; Gitis, V. Fouling and cleaning of ultrafiltration membranes: A review. *J. Water Process Eng.* **2014**, *1*, 121–138. [CrossRef]

77. Wang, Z.; Ma, J.; Tang, C.Y.; Kimura, K.; Wang, Q.; Han, X. Membrane cleaning in membrane bioreactors: A review. *J. Membr. Sci.* **2014**, *468*, 276–307. [CrossRef]

78. Chen, F.; Bi, X.; Ng, H.Y. Effects of bio-carriers on membrane fouling mitigation in moving bed membrane bioreactor. *J. Membr. Sci.* **2016**, *499*, 134–142. [CrossRef]

79. Lee, S.J.; Dilaver, M.; Park, P.K.; Kim, J.H. Comparative analysis of fouling characteristics of ceramic and polymeric microfiltration membranes using filtration models. *J. Membr. Sci.* **2013**, *432*, 97–105. [CrossRef]

80. Larrea, A.; Rambor, A.; Fabiyi, M. Ten years of industrial and municipal membrane bioreactor (MBR) systems—Lessons from the field. *Water Sci. Technol.* **2014**, *70*, 279–288. [CrossRef] [PubMed]

81. Bérubé, P.R.; Hall, E.R.; Sutton, P.M. Parameters Governing Permeate Flux in an Anaerobic Membrane Bioreactor Treating Low-Strength Municipal Wastewaters: A Literature Review. *Water Environ. Res.* **2006**, *78*, 887–896. [CrossRef] [PubMed]

82. Lin, H.; Gao, W.; Meng, F.; Liao, B.-Q.; Leung, K.-T.; Zhao, L.; Chen, J.; Hong, H. Membrane Bioreactors for Industrial Wastewater Treatment: A Critical Review. *Crit. Rev. Environ. Sci. Technol.* **2012**, *42*, 677–740. [CrossRef]

83. Park, S.H.; Park, Y.G.; Lim, J.-L.; Kim, S. Evaluation of ceramic membrane applications for water treatment plants with a life cycle cost analysis. *Desalin. Water Treat.* **2015**, *54*, 973–979. [CrossRef]

84. Çiçek, N.; Franco, H.P.; Suidan, M.T.; Urbain, V.; Manem, J. Characterization and Comparison of a Membrane Bioreactor and a Conventional Activated-Sludge System in the Treatment of Wastewater Containing High-Molecular-Weight Compounds. *Water Environ. Res.* **1999**, *71*, 64–70. [CrossRef]

85. Shang, R.; Verliefde, A.R.D.; Hu, J.; Heijman, S.G.J.; Rietveld, L.C. The impact of EfOM, NOM and cations on phosphate rejection by tight ceramic ultrafiltration. *Sep. Purif. Technol.* **2014**, *132*, 289–294. [CrossRef]

86. Jeong, Y.; Kim, Y.; Jin, Y.; Hong, S.; Park, C. Comparison of filtration and treatment performance between polymeric and ceramic membranes in anaerobic membrane bioreactor treatment of domestic wastewater. *Sep. Purif. Technol.* **2018**, *199*, 182–188. [CrossRef]

87. Hofs, B.; Ogier, J.; Vries, D.; Beerendonk, E.F.; Cornelissen, E.R. Comparison of ceramic and polymeric membrane permeability and fouling using surface water. *Sep. Purif. Technol.* **2011**, *79*, 365–374. [CrossRef]

88. Jeong, Y.; Cho, K.; Kwon, E.E.; Tsang, Y.F.; Rinklebe, J.; Park, C. Evaluating the feasibility of pyrophyllite-based ceramic membranes for treating domestic wastewater in anaerobic ceramic membrane bioreactors. *Chem. Eng. J.* **2017**, *328*, 567–573. [CrossRef]

89. Jeong, Y.; Lee, S.; Hong, S.; Park, C. Preparation, characterization and application of low-cost pyrophyllite-alumina composite ceramic membranes for treating low-strength domestic wastewater. *J. Membr. Sci.* **2017**, *536*, 108–115. [CrossRef]

90. Kaniganti, C.M.; Emani, S.; Thorat, P.; Uppaluri, R. Microfiltration of Synthetic Bacteria Solution Using Low Cost Ceramic Membranes. *Sep. Sci. Technol.* **2015**, *50*, 121–135. [CrossRef]

91. Li, L.; Chen, M.; Dong, Y.; Dong, X.; Cerneaux, S.; Hampshire, S.; Caoa, J.; Zhua, L.; Zhua, Z.; Liu, J. A low-cost alumina-mullite composite hollow fiber ceramic membrane fabricated via phase-inversion and sintering method. *J. Eur. Ceram. Soc.* **2016**, *36*, 2057–2066. [CrossRef]

92. Lorente-ayza, M.; Pérez-fernández, O.; Alcalá, R.; Sánchez, E.; Mestre, S.; Coronas, J.; Menéndez, M. Comparison of porosity assessment techniques for low-cost ceramic membranes. *Boletín de La Sociedad Española de Cerámica Y Vidrio* **2017**, *56*, 29–38. [CrossRef]

93. Tewari, P.K.; Singh, R.K.; Batra, V.S.; Balakrishnan, M. Membrane bioreactor (MBR) for wastewater treatment: Filtration performance evaluation of low cost polymeric and ceramic membranes. *Sep. Purif. Technol.* **2010**, *71*, 200–204. [CrossRef]

94. Chemometric Analysis for Spectroscopy. Available online: http://www.camo.com/downloads/resources/application_notes/Chemometric%20Analysis%20for%20Spectroscopy.pdf (accessed on 19 July 2018).

95. Singh, K.P.; Malik, A.; Mohan, D.; Sinha, S.; Singh, V.K. Chemometric data analysis of pollutants in wastewater—A case study. *Anal. Chim. Acta* **2005**, *532*, 15–25. [CrossRef]

96. Torgersen, G.; Rød, J.K.; Kvaal, K.; Bjerkholt, J.T.; Lindholm, O.G. Evaluating flood exposure for properties in Urban areas using a multivariate modelling technique. *Water* **2017**, *9*, 318. [CrossRef]

97. Philippe, N.; Racault, Y.; Stricker, A.E.; Spérandio, M.; Vanrolleghem, P.A. Modelling the long-term evolution of permeability in full-scale municipal MBRs: A multivariate statistical modelling approach. *Procedia Eng.* **2012**, *44*, 574–580. [CrossRef]

98. Kaneko, H.; Funatsu, K. Visualization of Models Predicting Transmembrane Pressure Jump for Membrane Bioreactor. *Ind. Eng. Chem. Res.* **2012**, *51*, 9679–9686. [CrossRef]

99. De Temmerman, L.; Naessens, W.; Maere, T.; Marsili-Libelli, S.; Villez, K.; Nopens, I.; Temmink, H.; Nopens, I. Detecting membrane fouling occurrences in a full-scale membrane bioreactor with principal component analysis. In Proceedings of the 11th IWA Conference on Instrumentation Control and Automation (ICA), Narbonne, France, 18–20 September 2013.

100. Ji, J.; Qiu, J.; Wong, F.; Li, Y. Enhancement of filterability in MBR achieved by improvement of supernatant and floc characteristics via filter aids addition. *Water Res.* **2008**, *42*, 3611–3622. [CrossRef] [PubMed]

101. Wu, J.; Huang, X. Use of ozonation to mitigate fouling in a long-term membrane bioreactor. *Bioresour. Technol.* **2010**, *101*, 6019–6027. [CrossRef] [PubMed]

102. Todt, D.; Heistad, A.; Jenssen, P.D. Load and distribution of organic matter and nutrients in a separated household wastewater stream. *Environ. Technol.* **2015**, *36*, 1584–1593. [CrossRef] [PubMed]

103. Ying, Z.; Ping, G. Effect of powdered activated carbon dosage on retarding membrane fouling in MBR. *Sep. Purif. Technol.* **2006**, *52*, 154–160. [CrossRef]

104. Judd, S. The status of membrane bioreactor technology. *Trends Biotechnol.* **2008**, *26*, 109–116. [CrossRef] [PubMed]

105. Rosenberg, M.; Gutnick, D.; Rosenberg, E. Adherence of bacteria to hydrocarbons: A simple method for measuring cell-surface hydrophobicity. *FEMS Microbiol. Lett.* **1980**, *9*, 29–33. [CrossRef]

106. Effect of MLSS on Flux—MLSS Paradox. Available online: http://onlinembr.info/membrane-process/effect-of-mlss-on-flux-mlss-paradox/ (accessed on 22 January 2018).

107. Brookes, A.; Judd, S.; Reid, E.; Germain, E.; Smith, S.; Alvarez-Vazquez, H.; Le-Clech, P.; Stephenson, T.; Turra, E.; Jefferson, B. Biomass characterisation in membrane bioreactors. In Proceedings of the International Membrane Science and Technology Conference (IMSTEC), Sydney, Australia, 10–14 November 2003.

108. Jefferson, B.; Brookes, A.; Le-Clech, P.; Judd, S.J. Methods for understanding organic fouling in MBRs. *Water Sci. Technol.* **2004**, *49*, 237–244. [CrossRef] [PubMed]

109. Huang, X.; Wu, J. Improvement of membrane filterability of the mixed liquor in a membrane bioreactor by ozonation. *J. Membr. Sci.* **2008**, *318*, 210–216. [CrossRef]

110. Chae, S.R.; Ahn, Y.T.; Kang, S.T.; Shin, H.S. Mitigated membrane fouling in a vertical submerged membrane bioreactor (VSMBR). *J. Membr. Sci.* **2006**, *280*, 572–581. [CrossRef]

111. Ng, H.Y.; Hermanowicz, S.W. Membrane bioreactor operation at short solids retention times: Performance and biomass characteristics. *Water Res.* **2005**, *39*, 981–992. [CrossRef] [PubMed]

112. Fan, F.; Zhou, H.; Husain, H. Identification of wastewater sludge characteristics to predict critical flux for membrane bioreactor processes. *Water Res.* **2006**, *40*, 205–212. [CrossRef] [PubMed]

113. Wu, J.; Huang, X. Effect of mixed liquor properties on fouling propensity in membrane bioreactors. *J. Membr. Sci.* **2009**, *342*, 88–96. [CrossRef]

114. Lee, W.-N.; Chang, I.-S.; Hwang, B.-K.; Park, P.-K.; Lee, C.-H.; Huang, X. Changes in biofilm architecture with addition of membrane fouling reducer in a membrane bioreactor. *Process Biochem.* **2007**, *42*, 655–661. [CrossRef]

115. Kunacheva, C.; Stuckey, D. Analytical methods for soluble microbial products (SMP) and extracellular polymers (ECP) in wastewater treatment systems: A review. *Water Res.* **2014**, *61*, 1–18. [CrossRef] [PubMed]

116. Lesjean, B.; Rosenberger, S.; Laabs, C.; Jekel, M.; Gnirss, R.; Amy, G. Correlation between membrane fouling and soluble/colloidal organic substances in membrane bioreactors for municipal wastewater treatment. *Water Sci. Technol.* **2005**, *51*, 1–8. [CrossRef] [PubMed]

117. CAMO. *The Unscrambler, Tutorials CAMO Process AS 2006*; CAMO Software AS: Oslo, Norway, 2006.

118. Deng, L.; Guo, W.; Hao Ngo, H.; Zhang, H.; Wang, J.; Li, J.; Xia, S.; Wu, Y. Biofouling and control approaches in membrane bioreactors. *Bioresour. Technol.* **2016**, *221*, 656–665. [CrossRef] [PubMed]

119. Isma Aida, M.I.; Idris, A.; Omar, R.; Razreena Putri, A.R. Effects of SRT and HRT on Treatment Performance of MBR and Membrane Fouling. *Int. J. Chem. Mol. Nucl. Mater. Metall. Eng.* **2014**, *8*, 488–492.

120. Van den Broeck, R.; Van Dierdonck, J.; Nijskens, P.; Dotremont, C.; Krzeminski, P.; van der Graaf, J.H.J.M.; van Lier, J.B.; Van Impe, J.F.M.; Smets, I.Y. The influence of solids retention time on activated sludge bioflocculation and membrane fouling in a membrane bioreactor (MBR). *J. Membr. Sci.* **2012**, *401–402*, 48–55. [CrossRef]

121. Yigit, N.O.; Harman, I.; Civelekoglu, G.; Koseoglu, H.; Cicek, N.; Kitis, M. Membrane fouling in a pilot-scale submerged membrane bioreactor operated under various conditions. *Desalination* **2008**, *231*, 124–132. [CrossRef]

122. Malamis, S.; Andreadakis, A. Fractionation of proteins and carbohydrates of extracellular polymeric substances in a membrane bioreactor system. *Bioresour. Technol.* **2009**, *100*, 3350–3357. [CrossRef] [PubMed]

123. Sivchenko, N.; Kvaal, K.; Ratnaweera, H. Evaluation of image texture recognition techniques in application to wastewater coagulation. *Cogent Eng.* **2016**, *3*. [CrossRef]

124. Jiang, T.; Kennedy, M.D.; Guinzbourg, B.F.; Vanrolleghem, P.A.; Schippers, J.C. Optimising the operation of a MBR pilot plant by quantitative analysis of the membrane fouling mechanism. *Water Sci. Technol.* **2005**, *51*, 19–25. [CrossRef] [PubMed]

125. Le Clech, P.; Jefferson, B.; Chang, I.S.; Judd, S.J. Critical flux determination by the flux-step method in a submerged membrane bioreactor. *J. Membr. Sci.* **2003**, *227*, 81–93. [CrossRef]

126. Ognier, S.; Wisniewski, C.; Grasmick, A. Membrane bioreactor fouling in sub-critical filtration conditions: A local critical flux concept. *J. Membr. Sci.* **2004**, *229*, 171–177. [CrossRef]

127. Miller, D.J.; Kasemset, S.; Paul, D.R.; Freeman, B.D. Comparison of membrane fouling at constant flux and constant transmembrane pressure conditions. *J. Membr. Sci.* **2014**, *454*, 505–515. [CrossRef]

*water*

MDPI

Article

# Quantitative Analysis of Membrane Fouling Mechanisms Involved in Microfiltration of Humic Acid–Protein Mixtures at Different Solution Conditions

Chunyi Sun [1], Na Zhang [1], Fazhan Li [2], Guoyi Ke [3], Lianfa Song [4], Xiaoqian Liu [1] and Shuang Liang [1,*]

[1] Shandong Provincial Key Laboratory of Water Pollution Control and Resource Reuse, School of Environmental Science and Engineering, Shandong University, Jinan 250100, China; scy3735@163.com (C.S.); 15650576825@163.com (N.Z.); shl0701@163.com (X.L.)

[2] School of Environmental and Municipal Engineering, North China University of Water Resources and Electric Power, Zhengzhou 450046, China; lifazhan@ncwu.edu.cn

[3] Department of mathematics and Physical Sciences, Louisiana State University of Alexandria, 8100 Hwy 71 S, Alexandria, LA 71302-9119, USA; gke@lsua.edu

[4] Department of Civil and Environmental Engineering, Texas Tech University, 10th and Akron, Lubbock, TX 79409-1023, USA; lianfa.song@ttu.edu

\* Correspondence: liangshuang@sdu.edu.cn; Tel.: +86-531-88361712

Received: 23 August 2018; Accepted: 19 September 2018; Published: 22 September 2018

**Abstract:** A systematical quantitative understanding of different mechanisms, though of fundamental importance for better fouling control, is still unavailable for the microfiltration (MF) of humic acid (HA) and protein mixtures. Based on extended Derjaguin–Landau–Verwey–Overbeek (xDLVO) theory, the major fouling mechanisms, i.e., Lifshitz–van der Waals (LW), electrostatic (EL), and acid–base (AB) interactions, were for the first time quantitatively analyzed for model HA–bovine serum albumin (BSA) mixtures at different solution conditions. Results indicated that the pH, ionic strength, and calcium ion concentration of the solution significantly affected the physicochemical properties and the interaction energy between the polyethersulfone (PES) membrane and HA–BSA mixtures. The free energy of cohesion of the HA–BSA mixtures was minimum at pH = 3.0, ionic strength = 100 mM, and $c(Ca^{2+})$ = 1.0 mM. The AB interaction energy was a key contributor to the total interaction energy when the separation distance between the membrane surface and HA–BSA mixtures was less than 3 nm, while the influence of EL interaction energy was of less importance to the total interaction energy. The attractive interaction energies of membrane–foulant and foulant–foulant increased at low pH, high ionic strength, and calcium ion concentration, thus aggravating membrane fouling, which was supported by the fouling experimental results. The obtained findings would provide valuable insights for the quantitative understanding of membrane fouling mechanisms of mixed organics during MF.

**Keywords:** microfiltration; xDLVO theory; HA–BSA mixtures; interaction energy; membrane fouling; solution conditions

## 1. Introduction

Microfiltration (MF) is increasingly applied in water treatment due to the continuously decreasing cost and progressively more efficient performance of MF membranes [1,2]. However, membrane fouling, mainly caused by natural organic matter (NOM), still remains the primary impediment for the widespread application of MF technology [3]. Humic acid (HA) was identified as the main component of NOM [4], and is considered to be one of the main culprits causing membrane fouling [5]. Furthermore, protein is also ubiquitous in natural water and more hydrophilic than HA. Some investigations found that hydrophilic matter could result in more serious membrane fouling than hydrophobic matter [6,7]. Thus, protein can also contribute to membrane fouling in spite of its lower content in natural water [8]. Early literature demonstrated the fouling mechanisms of individual HA or protein in low-pressure membranes, as well as the impact of solution conditions on fouling propensity [9,10]. However, natural water does not contain only one kind of foulant, but a mixture of organic foulants. Therefore, it is necessary to explore more complicated fouling mechanisms of mixed organics during MF.

The mechanisms involved in HA–protein mixture fouling were studied to certain extent. Madaeni et al. [11] found that the co-existence of bovine serum albumin (BSA) resulted in higher HA rejection and lower flux during ultrafiltration (UF), which was mainly attributed to pore blocking and cake deposition. Salehi et al. [12] studied the adsorption behavior of HA to a UF membrane in the presence of protein and proposed that intermolecular electrostatic interactions played an important role. Myat et al. [13] evaluated the importance of interactions between HA and BSA in membrane fouling and put forward that electrostatic interactions, hydrophobic interactions, and hydrogen-bonding interactions were the dominant types of interactions. These previous studies provided a sound starting point for understanding the complex fouling mechanisms of HA–BSA mixtures. However, to date, the relative contribution to membrane fouling of each individual interaction, at the quantitative level, still remains unknown. The lack of such useful information greatly hinders the development of more precise fouling control strategies.

The extended Derjaguin–Landau–Verwey–Overbeek (xDLVO) theory is widely acknowledged as the most notable approach for the quantitative analysis of major membrane fouling mechanisms [14,15]. However, its application mainly focuses on the case of single foulant, and it is rarely reported for more complicated organic mixture fouling. Recently, Lin et al. [16] utilized xDLVO theory to elucidate the relative roles of different fouling mechanisms involved in the UF of HA and fulvic acid (FA) mixtures, and demonstrated the feasibility of using xDLVO theory for predicting UF membrane fouling. Nevertheless, this study only focused on the interactions between membrane and foulants in the initial stage, neglecting those between foulants and foulants in the subsequent stage. Ding et al. [17] adopted xDLVO theory to explore the influence of varying proportions of BSA and sodium alginate (SA) mixtures on MF membrane fouling. It was reported that the physicochemical interactions between foulants and membrane and between foulant molecules are very complex, which are greatly impacted by solution conditions (i.e., ionic strength and divalent cations). However, to date, the xDLVO theory is yet to be applied for the quantitative analysis of membrane fouling mechanisms involved in MF of humic acid–protein mixtures.

In this work, the major mechanisms (i.e., Lifshitz–van der Waals (LW), electrostatic (EL), and acid–base (AB) interactions) underlying HA–protein mixture fouling was, for the first time, quantitatively analyzed in a systematical manner. The xDLVO theory was applied for both the initial membrane–foulant interactions and the subsequent foulant–foulant interactions. Furthermore, the effects of all classically investigated solution conditions (i.e., pH, ionic strength, and calcium ion concentration) on fouling mechanisms were quantitatively evaluated. The acquired results would further extend xDLVO application in elucidating organic mixture fouling, and would particularly provide valuable quantitative insight into humic acid–protein mixture fouling.

## 2. Materials and Methods

### 2.1. Microfiltration Membrane and Model Foulants

A polyethersulfone (PES) membrane with a pore size of 0.22 μm (Haichengshijie filtering equipments Co. Ltd., Beijing, China) was adopted in this study. Prior to use, the PES membrane was immersed in deionized (DI) water for 24 h to remove impurities or additives. Powdered humic acid

(HA, Sigma Aldrich, Saint Louis, MO, USA) was chosen as the representative of humic substances in natural organic matter. The stock solution (1 g·L$^{-1}$) was prepared by dissolving pre-weighed amounts of powdered HA in DI water, followed by filtration through a 0.45-μm nylon membrane to remove insoluble substances. Commercially available bovine serum albumin (BSA; Roche, Mannhein, Germany) was used as the protein-like substance in natural organic matter. According to the manufacturer, BSA has a molecular weight of 68 kDa and a molecular size of 14 nm × 4 nm × 4 nm. Humic substances in NOM accounted for 50–90% in natural water [18], and the concentrations usually ranged from 2 to 10 mg·L$^{-1}$ [19]. To simulate natural water components, the applied raw concentrations of HA and BSA in the solution were set at 4:1, respectively, leading to a total concentration of 10 mg·L$^{-1}$.

The solution ionic strength (NaCl) was amended to 10 mM and the pH was adjusted to 3.0, 4.7, 7.0, and 9.0 using small amounts of either 0.1 M HCl or NaOH. At the same time, the mixed solutions with different NaCl concentrations of 10, 20, 50, and 100 mM were used to investigate the influence of ionic strength when the solution pH was 7.0. The mixed solutions with different Ca$^{2+}$ concentrations of 0, 0.2, 0.5, and 1.0 mM were prepared to study the effect of divalent cations at a constant pH of 7.0 and an NaCl concentration of 10 mM. In addition, the baseline solution conditions were set as pH = 7.0, ionic strength = 10 mM, and c(Ca$^{2+}$) = 0 mM.

## 2.2. xDLVO Theory

According to Van Oss [20], the total interfacial interaction energy for aqueous systems comprises Lifshitz–van der Waals (LW), electrostatic (EL), and acid–base (AB) interactions, which can be written as follows:

$$U_{mlf}^{TOT} = U_{mlf}^{LW} + U_{mlf}^{EL} + U_{mlf}^{AB}, \tag{1}$$

where $U_{mlf}^{TOT}$ is the total interaction energy between a membrane surface and a foulant immersed in water, and $U_{mlf}^{LW}$, $U_{mlf}^{EL}$, and $U_{mlf}^{AB}$ are the LW, EL, and AB interaction energy terms, respectively. The subscripts $m$, $l$, and $f$ correspond to the membrane, bulk liquid (e.g., water in this study), and foulants (HA and BSA mixtures in this study), respectively.

### 2.2.1. Surface Thermodynamic Parameters

In order to calculate interfacial interaction energies, the surface tension parameters ($\gamma_s^{LW}$, $\gamma_s^+$, and $\gamma_s^-$) of the membrane and HA–BSA mixtures were obtained using contact angle measurements, performing three probe liquids with well-known surface tension parameters and employing the extended Young equation, which can be given as follows [21,22]:

$$(1 + \cos\theta)\gamma_l^{TOT} = 2\left(\sqrt{\gamma_s^{LW}\gamma_l^{LW}} + \sqrt{\gamma_s^+\gamma_l^-} + \sqrt{\gamma_s^-\gamma_l^+}\right), \tag{2}$$

where $\theta$ is the contact angle and $\gamma^{TOT} (= \gamma^{LW} + \gamma^{AB})$ is the total surface tension. $\gamma^{LW}$ is the Lifshitz–van der Waals component and $\gamma^{AB} (= 2\sqrt{\gamma^+\gamma^-})$ is the acid–base component. $\gamma^+$ and $\gamma^-$ are the electron-acceptor and electron-donor components, respectively. The subscripts $s$ and $l$ represent solid surface and liquid, respectively.

The interfacial free energy per unit area between membrane and foulant contact in aqueous solution could be determined using the surface tension parameters calculated above. It is assumed that contact occurs at the minimum equilibrium cut-off distance, $d_0$, which represents a value of 0.158 nm (±0.009 nm) [23]. The LW and AB free energies per unit are expressed as follows:

$$\Delta G_{d_0}^{LW} = 2\left(\sqrt{\gamma_l^{LW}} - \sqrt{\gamma_m^{LW}}\right)\left(\sqrt{\gamma_f^{LW}} - \sqrt{\gamma_l^{LW}}\right); \tag{3}$$

$$\Delta G_{d_0}^{AB} = 2\sqrt{\gamma_l^+}\left(\sqrt{\gamma_m^-} + \sqrt{\gamma_f^-} - \sqrt{\gamma_l^-}\right) + 2\sqrt{\gamma_l^-}\left(\sqrt{\gamma_m^+} + \sqrt{\gamma_f^+} - \sqrt{\gamma_l^+}\right) - 2\left(\sqrt{\gamma_m^+\gamma_f^-} + \sqrt{\gamma_m^-\gamma_f^+}\right). \tag{4}$$

### 2.2.2. Interfacial Interaction Energy

As the separation distance ($d$) between two surfaces increases, the LW and AB interaction energy components are gradually reduced by the interaction energy following the specific attenuation form. In order to obtain the actual interaction energies, Derjaguin's technique was applied to calculate interaction energies between a spherical foulant and an infinite planar surface. The LW, AB, and EL interaction energy components can be given by [24]

$$U_{mlf}^{LW}(d) = -\frac{A_H a_f}{6d},$$ (5)

$$U_{mlf}^{AB}(d) = 2\pi a_f \lambda \Delta G_{d_0}^{AB} \exp\left(\frac{d_0 - d}{\lambda}\right),$$ (6)

$$U_{mlf}^{EL}(d) = \pi \varepsilon_r \varepsilon_0 a_f \left(2\zeta_m \zeta_f \ln\left(\frac{1 + \exp(-\kappa d)}{1 - \exp(-\kappa d)}\right) + (\zeta_m^2 + \zeta_f^2) \ln(1 - \exp(-2\kappa d))\right),$$ (7)

where $A_H$ ($= -12\pi d_0^2 \Delta G_{d_0}^{LW}$) is the Hamaker constant, $a_f$ is the radius of the spherical foulant, $d$ is the separation distance between foulant and membrane, $\lambda$ ($=0.6$ nm) is the characteristic decay length of AB interaction in water, $\varepsilon_0$ ($=8.854 \times 10^{-12}$ C·V$^{-1}$·m$^{-1}$) is the dielectric permittivity in vacuum, $\varepsilon_r$ ($=78.5$) is the relative permittivity of water, $\zeta_m$ and $\zeta_f$ are the zeta potentials of membrane and foulant, respectively, and $\kappa$ is the inverse Debye screening length, which is determined by [25]

$$\kappa = \sqrt{\frac{e^2 \Sigma n_i z_i^2}{\varepsilon_r \varepsilon_0 kT}},$$ (8)

where $e$ ($=1.6 \times 10^{-19}$ C) is the electron charge, $n_i$ is the number concentration of ion $i$ in the bulk solution, $z_i$ is the valence of ion $i$, $k$ ($=1.38 \times 10^{-23}$ J·K$^{-1}$) is the Boltzmann's constant, and $T$ is the absolute temperature. The background electrolyte concentration in this study was 0.01 M NaCl.

Likewise, interaction energies between two spherical foulant particles could also be calculated using Derjaguin's technique.

$$U_{flf}^{LW}(d) = \frac{-A_H a_1 a_2}{6d(a_1 + a_2)},$$ (9)

$$U_{flf}^{AB}(d) = \frac{2\pi a_1 a_2}{a_1 + a_2} \lambda \Delta G_{d_0}^{AB} \exp(\frac{d_0 - d}{\lambda}),$$ (10)

$$U_{flf}^{EL}(d) = \pi \varepsilon_0 \varepsilon_r \frac{a_1 a_2}{a_1 + a_2} \zeta_f^2 \ln(1 + e^{-\kappa d}),$$ (11)

where $a_1$ and $a_2$ are the radii of the interacting foulant particles.

### 2.3. Analytical Methods

Contact angles of the PES membrane and HA–BSA mixtures at different solution conditions were measured using the sessile drop method with a goniometer (JC2000C Contact Angle Meter, Shanghai Zhongchen Experiment Equipments Co. Ltd., Shanghai, China). The three kinds of probe liquids selected for contact angle measurements were DI water, glycerol, and diiodomethane. These probe liquids were chosen on the premise that two of them must be polar (DI water and glycerol) and the other must be non-polar (diiodomethane) [23]. At least seven measurements at different locations were averaged to obtain a reliable value for each sample.

The zeta potential of the PES membrane was determined using a zeta potential analyzer (SurPASS, Anton Paar, Graz, Austria). The zeta potential of foulants was measured using a zetasizer (3000HSa, Malvern Instruments, Malvern, UK). Dynamic light scattering (DLS; BI-200SM/BI-9000, Brookhaven, Holtsville, NY, USA) was used to measure the hydraulic diameters of foulant molecules at different solution conditions in order to calculate the interfacial interaction energies. Each data value is the average of three measurements. All the measurements in the study were performed at $20.0 \pm 1.0\,^{\circ}$C. The morphology of the fouled membrane surface was observed with scanning electron microscopy (SEM; JSM-7600F, JEOL, Tokyo, Japan) under baseline solution conditions. In order to ensure the reliability of experimental results, the same side of the HA–BSA fouled membrane was chosen for SEM image analysis.

### 2.4. Fouling Experiments

The dead-end MF experiments were conducted at constant pressure mode at room temperature ($20 \pm 1.0\,^{\circ}$C). The experimental system is shown in Figure S1 of the Supplementary Materials. The stirred cell had an inner diameter of 8 cm, providing an effective filtration area of 50.26 cm$^2$, and it was equipped with a built-in rotor. The stirring speed was set at 180 rpm [26] throughout the whole filtration process to prevent concentration polarization by means of applying shear stress on the membrane surface. Before each filtration of feed solution, DI water was filtered through the PES membrane with an operating pressure of 20 kPa for about 20 min to stabilize the filtration system, and the initial flux $J_0$ was measured. Then, the mixed solution was introduced into the stirred cell, and permeate flux was measured by a balance connected to a computer. Filtration was stopped when permeate flux leveled off (approximately 5 L) and no change happened within 30 min.

To analyze fouling behaviors in different filtration stages, the entire filtration process was separated into initial and final stages, with correspondence to adhesion stage and cohesion stage, respectively. Following the initial stage, the final stage started (i.e., where the two stages were separated) when the ratio of filtration time $t$ to cumulative permeate volume $V$ was proportional to $V$, as expressed in the cake filtration model [27]:

$$t/V = aV + b, \tag{12}$$

where $a$ and $b$ are model parameters.

Fouling potential ($K$) was adopted as a parameter indicating the severity of membrane fouling. It is defined as the reduction in relative flux caused by a unit mass of HA–BSA mixtures that is brought in contact with the membrane surface. Fouling potentials of initial and final stages were determined according to the following equation:

$$K = \frac{\Delta(J/J_0)}{C_0 \times \Delta V}, \tag{13}$$

where $C_0$ is the concentration of feed solution.

## 3. Results and Discussion

### 3.1. Physicochemical Properties of the PES Membrane and Foulants

The average contact angles, as well as the zeta potentials of the clean membranes before usage and the fouled membranes fully covered with HA–BSA mixtures after usage, were systemically measured, and are listed in Table 1. Despite the contact angles of the PES membrane being reported in many references [28,29], the variation in PES membrane contact angle in different solution conditions remains unknown. It was found that the water contact angles ($\theta_W$) of the PES membrane decreased with increasing pH, suggesting that water molecules are energetically favorable for contacting with the membrane. The reduced $\theta_W$ probably resulted from the more intensive hydrogen bonding between water molecules and the membrane surface at higher pH. A similar trend was also observed by Meng et al. measuring the $\theta_W$ of polyamide (PA) and polypropylene (PP) membranes at different solution pH [30]. On the contrary, the $\theta_W$ of the PES membrane increased with increasing ionic strength and calcium ion concentration. Though the contact angles of HA or BSA alone with solution conditions were measured in previous work [30,31], no information was available regarding HA–BSA mixtures. As shown in Table 1, it is interesting to note that the $\theta_W$ of the HA–BSA mixtures exhibited a similar trend with solution conditions to that of the PES membrane. The glycerol contact angles ($\theta_G$) increased with decreasing pH or increasing ionic strength and calcium ion concentration, but not as significantly as $\theta_W$. In contrast, no significant trends regarding the variation in diiodomethane contact angle ($\theta_D$) with solution conditions were observed regardless of membrane and foulant.

**Table 1.** Contact angles and zeta potentials of the polyethersulfone (PES) membrane and humic acid (HA)–bovine serum albumin (BSA) mixtures at different solution conditions.

| Solution Conditions | PES Membrane | | | | HA–BSA | | | |
|---|---|---|---|---|---|---|---|---|
| | $\theta_W$ (°) | $\theta_G$ (°) | $\theta_D$ (°) | Zeta (mV) | $\theta_W$ (°) | $\theta_G$ (°) | $\theta_D$ (°) | Zeta (mV) |
| pH = 3.0 | 37.7 ± 1.9 [a] | 44.4 ± 2.1 | 40.0 ± 2.5 | −16.8 ± 2.1 | 90.2 ± 2.1 | 72.9 ± 3.1 | 30.4 ± 1.0 | −16.5 ± 2.5 |
| pH = 4.7 | 32.0 ± 2.2 | 43.4 ± 2.0 | 37.7 ± 1.8 | −21.8 ± 1.8 | 84.6 ± 3.0 | 72.0 ± 0.9 | 41.9 ± 2.9 | −28.4 ± 2.7 |
| pH = 7.0 | 27.5 ± 1.6 | 42.8 ± 1.8 | 36.9 ± 1.5 | −37.8 ± 2.3 | 69.1 ± 1.2 | 71.9 ± 2.7 | 39.0 ± 2.1 | −32.2 ± 1.9 |
| pH = 9.0 | 23.0 ± 2.5 | 41.8 ± 1.9 | 40.5 ± 2.2 | −51.8 ± 1.5 | 62.6 ± 1.5 | 70.1 ± 2.7 | 35.8 ± 1.6 | −43.0 ± 2.1 |
| IS = 10 mM | 27.5 ± 1.6 | 42.8 ± 1.8 | 36.9 ± 1.5 | −37.8 ± 2.3 | 69.1 ± 1.2 | 71.9 ± 2.7 | 39.0 ± 2.1 | −32.2 ± 1.9 |
| IS = 20 mM | 34.1 ± 1.4 | 46.4 ± 1.1 | 36.4 ± 1.8 | −31.3 ± 1.9 | 70.4 ± 2.1 | 72.3 ± 1.3 | 39.8 ± 1.6 | −26.3 ± 1.2 |
| IS = 50 mM | 41.8 ± 2.3 | 50.6 ± 1.5 | 36.8 ± 1.5 | −23.1 ± 1.3 | 71.9 ± 2.3 | 72.8 ± 2.2 | 47.9 ± 1.5 | −19.2 ± 2.4 |
| IS = 100 mM | 53.7 ± 2.1 | 51.5 ± 2.9 | 38.5 ± 2.3 | −13.8 ± 2.7 | 76.3 ± 2.0 | 73.5 ± 2.1 | 50.2 ± 2.3 | −10.9 ± 2.6 |
| c(Ca²⁺) = 0 mM | 27.5 ± 1.6 | 42.8 ± 1.8 | 36.9 ± 1.5 | −37.8 ± 2.3 | 69.1 ± 1.2 | 71.9 ± 2.7 | 39.0 ± 2.1 | −32.2 ± 1.9 |
| c(Ca²⁺) = 0.2 mM | 53.4 ± 2.6 | 55.3 ± 2.0 | 29.3 ± 1.3 | −34.9 ± 2.4 | 76.8 ± 2.8 | 72.8 ± 0.7 | 36.5 ± 1.0 | −29.3 ± 2.3 |
| c(Ca²⁺) = 0.5 mM | 54.5 ± 0.9 | 55.3 ± 2.2 | 30.2 ± 0.5 | −32.4 ± 1.4 | 81.1 ± 2.3 | 73.8 ± 1.2 | 43.0 ± 0.8 | −22.2 ± 1.8 |
| c(Ca²⁺) = 1.0 mM | 64.7 ± 3.5 | 58.3 ± 2.9 | 32.2 ± 2.2 | −30.7 ± 1.2 | 85.3 ± 2.1 | 74.1 ± 1.9 | 40.6 ± 2.7 | −16.9 ± 2.6 |

[a] Sample mean ± standard deviation, number of measurements: $n = 7$ (contact angle); $n = 3$ (zeta potential). IS: ionic strength; $\theta_W$: water contact angle; $\theta_G$: glycerol contact angle; $\theta_D$: diiodomethane contact angle.

The variations in zeta potential of the PES membrane and HA–BSA mixtures with solution conditions were also measured, as shown in Table 1. Zeta potentials of the membrane and foulants decreased remarkably (i.e., more negatively charged) with the increase in solution pH. This can be attributed to the intensified deprotonation of –COOH groups at higher solution pH. The increase in ionic strength lowered the absolute value of zeta potentials (i.e., less negatively charged). This phenomenon may be related with the shielding effect or double-layer compression by the increase in the number of counter ions with the increasing ionic strength [32]. Zeta potentials increased with the addition of calcium ions, probably due to the preferential adsorption of divalent cations to the negatively charged membrane/foulant surface.

### 3.2. Surface Tension Parameters of the PES Membrane and Foulants

The calculated surface tension parameters and free energy of cohesion of the PES membrane and HA–BSA mixtures at different solution conditions are summarized in Table 2. Both the PES membrane and HA–BSA mixtures possessed high electron-donor components ($\gamma^-$) and relatively low electron-acceptor components ($\gamma^+$). This is consistent with previous studies, reporting that polymeric membranes and organic matter typically showed high electron-donor properties [33,34]. The $\gamma^-$ values of the PES membrane and HA–BSA mixtures increased with increasing pH, while they decreased at a high ionic strength and calcium concentration. It can be noted that the trend of $\gamma^-$ values with solution conditions was similar to that of the absolute value of zeta potentials. There seems to be a strong relationship between electron-donor components and negative surface charge. This may be ascribed to the deprotonation of surface groups, the enhancement of negative charge at high pH, and low ionic strength and calcium ion concentration. In contrast, no significant trend of $\gamma^+$ values with solution conditions was observed. In addition, compared with LW surface tension ($\gamma^{LW}$), acid–base surface tension ($\gamma^{AB}$) was found to be much lower, which can be attributed to the smaller $\gamma^+$ values.

$\Delta G_{sls}$ represents the interaction energy per unit area when two surfaces with the same material composition come into contact with each other [35,36]. It can be used as an indicator of hydrophobic/hydrophilic properties with negative and positive values indicating hydrophobic and hydrophilic surfaces, respectively. As shown in Table 2, the positive $\Delta G_{sls}$ of the PES membrane increased with increasing solution pH, suggesting that higher pH enhanced membrane hydrophilicity. In particular, the PES membrane changed from hydrophilic to hydrophobic when the ionic strength and calcium ion concentration reached 100 mM and 0.2 mM, respectively. This implies that the increasing ionic strength and the presence of calcium ions would make the HA–BSA mixtures deposit/adsorb onto the membrane surface easier. It can also be seen from Table 2 that the calculated $\Delta G_{sls}$ of the HA–BSA mixtures at pH = 3.0, ionic strength = 100 mM, and $c(Ca^{2+})$ = 1.0 mM was minimum, indicating that the HA–BSA mixtures was more hydrophobic and thermodynamically unstable. In addition, the $\Delta G_{sls}$ of the HA–BSA mixtures under all the conditions tested was negative, suggesting that the HA–BSA mixtures was hydrophobic. However, previous studies found that BSA was hydrophilic and HA was more hydrophobic in natural water [37]. These results probably appeared because the high proportion of hydrophobic HA played a decisive role in the mixed solution.

**Table 2.** Surface tension parameters ($\gamma$) and interfacial free energy of cohesion, $\Delta G_{sls}$ (mJ/m$^2$) of the PES membrane and HA–BSA mixtures at different solution conditions. Surface tension components: $\gamma^+$, electron acceptor; $\gamma^-$, electron donor; $\gamma^{AB}$, acid–base; $\gamma^{LW}$, Lifshitz–van der Waals; $\gamma^{TOT}$, total.

| Solution Conditions | PES Membrane | | | | | | HA–BSA | | | | | |
|---|---|---|---|---|---|---|---|---|---|---|---|---|
| | $\gamma^+$ | $\gamma^-$ | $\gamma^{AB}$ | $\gamma^{LW}$ | $\gamma^{TOT}$ | $\Delta G_{sls}$ | $\gamma^+$ | $\gamma^-$ | $\gamma^{AB}$ | $\gamma^{LW}$ | $\gamma^{TOT}$ | $\Delta G_{sls}$ |
| pH = 3.0 | 0.52 | 34.53 | 8.51 | 34.47 | 42.98 | 11.42 | 0.01 | 0.87 | 0.21 | 44.06 | 44.27 | −89.16 |
| pH = 4.7 | 0.37 | 39.32 | 7.60 | 35.37 | 42.97 | 18.42 | 0.06 | 3.56 | 0.95 | 38.64 | 39.59 | −65.58 |
| pH = 7.0 | 0.29 | 42.93 | 7.06 | 35.67 | 42.73 | 23.71 | 0.24 | 19.47 | 4.31 | 40.11 | 44.42 | −17.16 |
| pH = 9.0 | 0.37 | 46.18 | 8.27 | 34.27 | 42.54 | 28.21 | 0.45 | 27.31 | 6.98 | 41.66 | 48.64 | −3.29 |
| IS = 10 mM | 0.29 | 42.93 | 7.06 | 35.67 | 42.73 | 23.71 | 0.24 | 19.47 | 4.31 | 40.11 | 44.42 | −17.16 |
| IS = 20 mM | 0.19 | 39.38 | 5.54 | 35.86 | 41.40 | 19.20 | 0.20 | 18.06 | 3.82 | 39.71 | 43.53 | −20.12 |
| IS = 50 mM | 0.13 | 34.37 | 4.25 | 35.71 | 39.96 | 11.77 | 0.03 | 17.05 | 1.53 | 35.44 | 36.97 | −21.33 |
| IS = 100 mM | 0.45 | 21.79 | 6.28 | 35.06 | 41.34 | −9.92 | 0.001 | 12.24 | 0.09 | 34.16 | 34.25 | −34.02 |
| $c(Ca^{2+})$ = 0 mM | 0.29 | 42.93 | 7.06 | 35.67 | 42.73 | 23.71 | 0.24 | 19.47 | 4.31 | 40.11 | 44.42 | −17.16 |
| $c(Ca^{2+})$ = 0.2 mM | 0.05 | 24.05 | 2.09 | 38.33 | 40.42 | −7.45 | 0.10 | 10.41 | 2.04 | 41.32 | 43.36 | −40.71 |
| $c(Ca^{2+})$ = 0.5 mM | 0.07 | 22.86 | 2.52 | 38.04 | 40.56 | −9.63 | 0.002 | 7.07 | 0.21 | 38.07 | 38.28 | −52.37 |
| $c(Ca^{2+})$ = 1.0 mM | 0.18 | 13.99 | 3.14 | 37.37 | 40.51 | −28.40 | 0.06 | 3.12 | 0.83 | 39.31 | 40.14 | −68.23 |

### 3.3. Interfacial Interaction Energies of Membrane–Foulant and Foulant–Foulant Combinations

Based on the above surface tension parameters, the interfacial interaction energies of membrane–foulant and foulant–foulant combinations at different solution conditions were calculated and are shown in Table 3. According to the xDLVO theory, a positive value of interaction energy implies repulsive interaction that hinders membrane fouling, while a negative value indicates attractive interaction that aggravates membrane fouling [16,38]. The greater absolute value of interfacial interaction energy signifies a stronger repulsive/attractive interaction between two surfaces.

According to the corresponding measurements, the change in membrane adhesive features before and after usage can be determined, corresponding to the clean membrane–foulant and fouled membrane–foulant interactions, respectively. Usually, membrane fouling behavior during the initial filtration period can be reasonably expected to be determined by the clean membrane–foulant interaction. As shown in Table 3, the PES–foulant combination had a negative LW interaction energy regardless of the variation in solution conditions, indicating that the LW component accelerates membrane fouling in the initial stage. The EL interaction energy was positive in the solution conditions studied, suggesting that the EL component can prevent initial membrane fouling. However, the AB interaction energy ($U_{mlf}^{AB}$) of the PES–foulant combination was negative at low pH, high ionic strength, and calcium ion concentration, indicating that the AB interaction can accelerate initial fouling. This is attributed to the more hydrophobic nature of the PES membrane at these solution conditions, which is indicated by $\Delta G_{sls}$ shown in Table 2. Furthermore, compared with the LW and EL interaction energies, the absolute value of AB interaction energy was much higher. Therefore, the AB interaction energy plays an important role in determining both the sign and absolute value of the overall interaction energies between membrane and foulants.

**Table 3.** The calculated interfacial interaction energies (kT) of PES–foulant and foulant-foulant combinations at different solution conditions. LW, AB, and EL represent the Lifshitz–van der Waals, acid–base, and electrostatic contributions to energy, respectively, while $m$, $l$, and $f$ represent the contact involving membrane, liquid, and foulant, respectively; $d_0$ is the minimum equilibrium cut-off distance.

| Solution Conditions | PES–Foulant | | | | Foulant–Foulant | | | |
|---|---|---|---|---|---|---|---|---|
| | $U_{mlf}^{LW}(d_0)$ | $U_{mlf}^{AB}(d_0)$ | $U_{mlf}^{EL}(d_0)$ | $U_{mlf}^{TOT}(d_0)$ | $U_{flf}^{LW}(d_0)$ | $U_{flf}^{AB}(d_0)$ | $U_{flf}^{EL}(d_0)$ | $U_{flf}^{TOT}(d_0)$ |
| pH = 3.0 | −34.56 | −765.35 | 12.04 | −787.87 | −28.49 | −1136.38 | 1.50 | −1163.37 |
| pH = 4.7 | −28.68 | −455.13 | 25.26 | −458.55 | −17.61 | −848.66 | 4.44 | −861.83 |
| pH = 7.0 | −31.91 | 221.98 | 51.71 | 241.78 | −20.37 | −162.22 | 5.74 | −176.85 |
| pH = 9.0 | −30.88 | 470.49 | 93.88 | 533.49 | −23.53 | 43.14 | 10.17 | 29.78 |
| IS = 10 mM | −31.91 | 221.98 | 51.71 | 241.78 | −20.37 | −162.22 | 5.74 | −176.85 |
| IS = 20 mM | −31.62 | 108.90 | 34.41 | 111.69 | −19.59 | −206.47 | 3.75 | −222.31 |
| IS = 50 mM | −24.26 | −21.50 | 18.00 | −27.76 | −12.13 | −251.71 | 1.93 | −261.91 |
| IS = 100 mM | −21.32 | −488.64 | 5.84 | −504.12 | −10.18 | −436.26 | 0.60 | −445.84 |
| c(Ca$^{2+}$) = 0 mM | −31.91 | 221.98 | 51.71 | 241.78 | −20.37 | −162.22 | 5.74 | −176.85 |
| c(Ca$^{2+}$) = 0.2 mM | −66.18 | −879.55 | 72.03 | −873.70 | −37.93 | −803.19 | 4.72 | −836.40 |
| c(Ca$^{2+}$) = 0.5 mM | −53.92 | −1186.69 | 45.60 | −1195.01 | −27.63 | −1113.58 | 4.51 | −1136.70 |
| c(Ca$^{2+}$) = 1.0 mM | −56.37 | −1999.22 | 26.05 | −2029.54 | −31.37 | −1468.17 | 1.56 | −1497.98 |

As the entire membrane surface is covered with HA–BSA mixtures, the following membrane fouling behavior would be controlled by foulant–foulant interaction. It can also be seen from Table 3 that the trend of absolute value with solution conditions of all foulant–foulant interaction energies was similar as that of membrane–foulant energies. The LW interaction energy was negative, suggesting that LW interaction can boost the attachment of the approaching HA–BSA mixtures to the deposited mixtures. The EL interaction energy was positive, and thus, resisted the HA–BSA mixtures. The AB interaction energy, because of its much larger absolute value, also plays a critical role in determining the value of total interaction energy between foulants and foulants.

To further elucidate the role of different mechanisms in membrane fouling, the variation in interaction energy components with separation distance under baseline solution conditions is shown in Figure 1. It can be found from Figure 1a that the HA–BSA mixtures was subject to repulsive interactions (AB component) with a decrease in the distance between membrane and foulants (d < 3 nm). When the foulants approached the membrane surface, the total interaction energy became attractive, resulting in the adsorption of foulants. In addition, Figure 1b depicts that the AB and LW interaction energies between foulants and foulants exhibited entirely attractive interactions. The effect of EL interaction energy was of less importance to the total interaction energy, which agrees with certain studies [15,39]. These results clearly showed that the AB interaction energy was a main contributor to the total interaction energy when the separation distance between the membrane surface and HA–BSA mixtures was less than 3 nm.

Variations in total interaction energy were significantly influenced by different solution conditions. Figure 2 displays the variation in interaction energies of membrane–foulant and foulant–foulant with separation distance at different solution conditions. It can be found from Figure 2a that the energy barrier decreased with the decrease in pH from 9.0 to 7.0, and then disappeared with the further decrease in pH to 4.7 and 3.0. The energy barrier means that the foulants should have sufficient kinetic energy to overcome the barrier to arrive at the membrane surface [40]. Under alkaline conditions, the interaction energy between membrane and foulants exhibited entirely repulsive interactions because of the deprotonation of the HA–BSA mixtures, resulting in the increase of the energy barrier and reducing the adsorption of foulants. In addition, Hoek et al. [41] and Chen et al. [28] reported that surface morphology significantly influenced the fouling behavior of the membrane, and found that the great influence of roughness on the membrane surface was to reduce the primary energy barrier's height, thus rendering rough surfaces more favorable for foulant deposition. As seen from Figure 2b, the interaction energy of foulant–foulant exhibited similar variation with solution pH as that of membrane–foulant. Figure 2c–f show that ionic strength and calcium ion concentration had significant influences on the total interaction energy. It was obvious that the attractive interaction energy increased substantially with the increases in ionic strength and calcium ion concentration. When the ionic strength and calcium ion concentration were 50 mM and 0.2 mM, respectively, the interaction energies were entirely attractive. The reduction in repulsive interaction can be ascribed to the charge neutralization effect with the addition of electrolytes. Thus, it can be concluded that the HA–BSA mixtures was subject to greater attractive interactions with the PES membrane at low pH, and high ionic strength and calcium ion concentration.

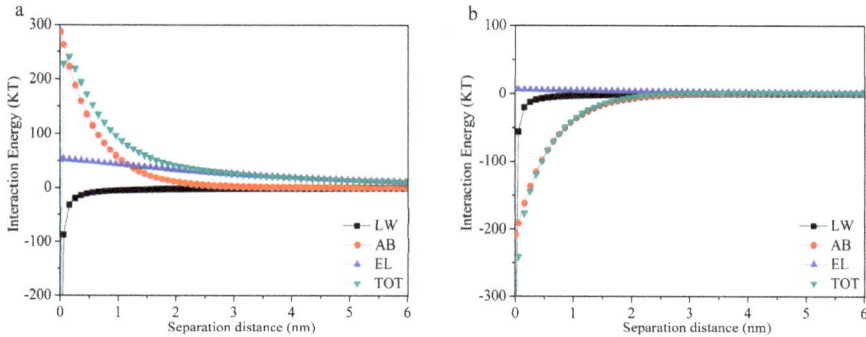

**Figure 1.** Variation in interaction energy components with separation distance under baseline solution conditions (pH = 7.0, ionic strength (IS) = 10 mM, c(Ca$^{2+}$) = 0 mM): (**a**) membrane–foulant; (**b**) foulant–foulant.

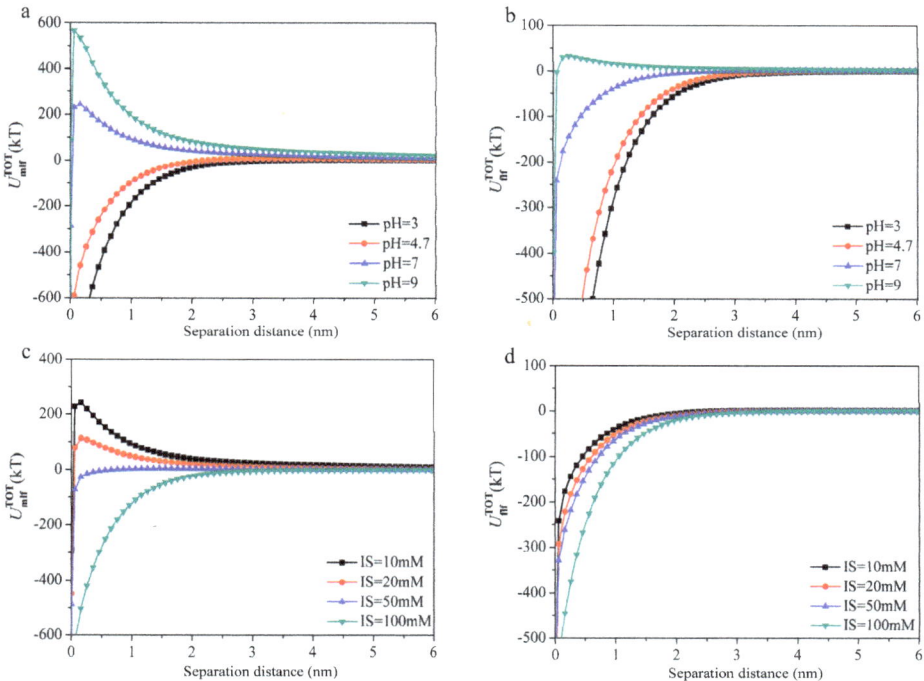

**Figure 2.** *Cont.*

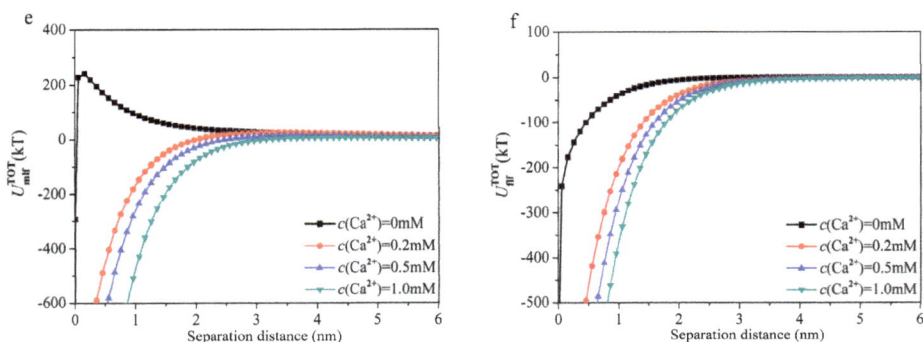

**Figure 2.** Variations in interaction energies of membrane–foulant and foulant–foulant with separation distance at different solution conditions: (**a,b**) pH; (**c,d**) IS; (**e,f**) $Ca^{2+}$ concentration.

According to the above trends of total interaction energies with solution conditions, some qualitative predictions about MF membrane fouling caused by HA–BSA mixtures can be yielded. The membrane fouling potential of HA–BSA mixtures would follow the order of pH = 3.0 > pH = 4.7 > pH = 7.0 > pH = 9.0; IS = 100 mM > IS = 50 mM > IS = 20 mM > IS = 10 mM; and $c(Ca^{2+})$ = 1.0 mM > $c(Ca^{2+})$ = 0.5 mM > $c(Ca^{2+})$ = 0.2 mM > $c(Ca^{2+})$ = 0 mM.

### 3.4. Experimental Verification

In order to verify the above theoretical predictions, the normalized flux reduction curves of the HA–BSA mixtures at different solution conditions are depicted in Figure 3. It can be observed that great flux decline was found at the beginning of filtration, and then, the speed of flux decline gradually decreased. This observation is consistent with previous reports on the behavior of membrane fouling during different filtration periods [42,43]. It can be seen from Figure 3a that when filtrating the HA–BSA mixtures at different solution pH, fouling propensity was in the order of pH = 3.0 > pH = 4.7 > pH = 7.0 > pH = 9.0. Moreover, the SEM image also shows that the HA–BSA-fouled membrane with baseline solution conditions displayed a homogeneous porous surface structure (Figure S2a). The porosity of the membrane surface decreased significantly when the solution pH decreased to 3.0 (Figure S2b). This may be attributed to the greater attractive interactions enhancing the HA–BSA mixture's attachment at low pH.

Based on the xDLVO predictions, filtrating HA–BSA mixtures with an increase in ionic strength was expected to have greater fouling propensity. In fact, this fouling trend can be reflected in Figure 3b. When the ionic strength increased to 100 mM, the surface porosity of the PES membrane also decreased (Figure S2c). As expected (Figure 3c), the flux decline rate of the PES membrane was faster with the increase in calcium ion concentration. Following the addition of calcium ions, it would not only neutralize the negative charge on the membrane and foulant surface, but it would bridge between membrane surface and foulant molecules, resulting in the formation of a cross-linked chelate in the fouling layer [44,45]. A further addition of calcium ions can be evidenced by the fact that there were no pores on the membrane surface, and the deposited layer structure seemed to be much more compact (Figure S2d). The corresponding fouling experimental results were consistent with the above theoretical predictions.

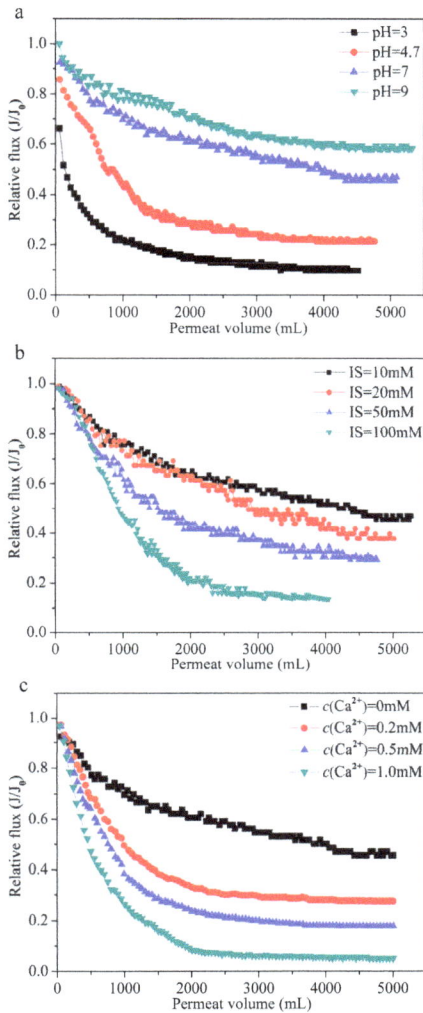

**Figure 3.** Normalized flux reduction curves of the humic acid (HA)–bovine serum albumin (BSA) mixtures during microfiltration (MF) at different solution conditions: (**a**) pH; (**b**) IS; (**c**) $Ca^{2+}$ concentration.

Figure 4 presents the correlation analysis between interaction energies and fouling potentials at initial and final stages with different solution conditions. An obvious negative linear relationship was observed between fouling potential and interaction energy under various solution conditions for different filtration stages. The same correlations between fouling behaviors and adhesive and cohesive interaction energies were observed in previous studies [46,47]. It can be seen from Figure 4 that the attractive interaction energy in the initial and final stages increased with decreasing pH or increasing ionic strength and calcium ion concentration; thus membrane fouling was aggravated. These results would further extend the xDLVO application in elucidating the mechanisms of MF membrane fouling by HA–BSA mixtures at different solution conditions.

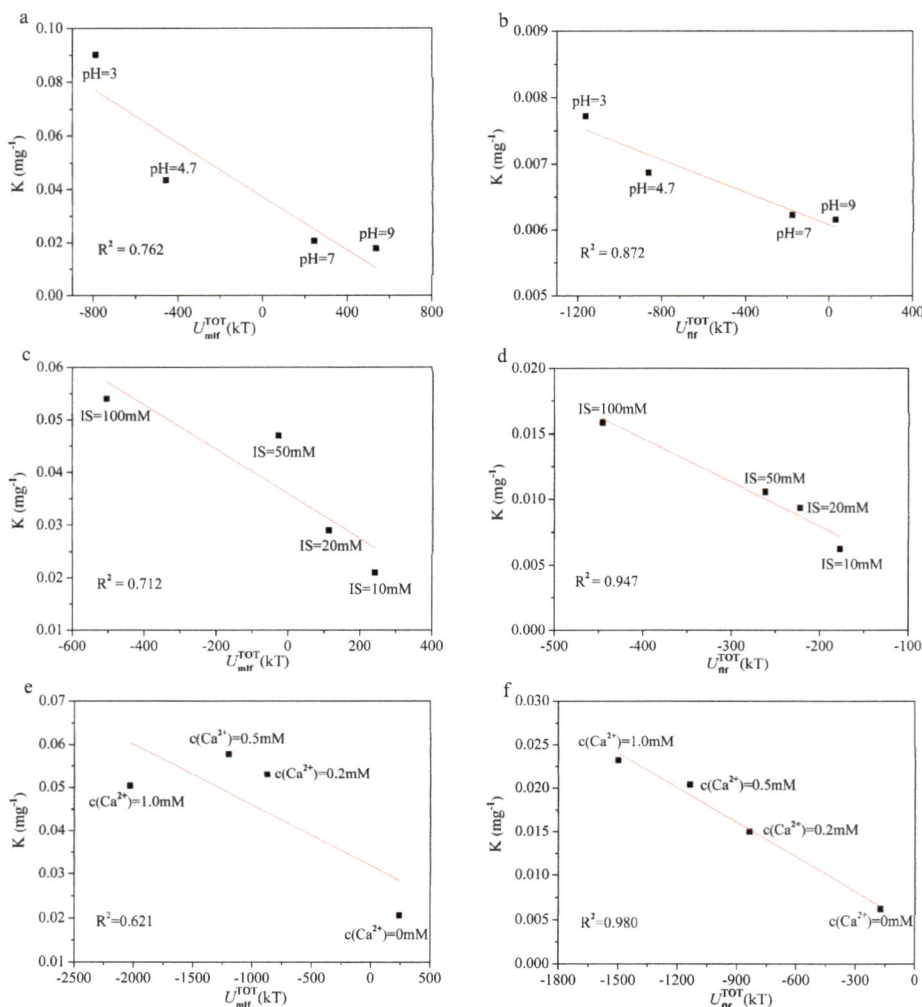

**Figure 4.** The correlation analysis between interaction energies and fouling potentials at initial and final stages with different solution conditions: (**a,b**) pH; (**c,d**) IS; (**e,f**) $Ca^{2+}$ concentration.

## 4. Conclusions

The xDLVO theory was used to quantitatively analyze the MF membrane fouling mechanisms of HA–protein mixtures at different solution conditions. Measured physicochemical properties revealed that the contact angles, as well as the zeta potentials of membrane and foulants, varied with solution conditions, which was due to the deprotonation of functional groups and the electrostatic shielding effect. Both the PES membrane and HA–BSA mixtures exhibited high electron-donor components ($\gamma^-$) and relatively low electron-acceptor components ($\gamma^+$), and the minimum free energy of cohesion of the HA–BSA mixtures was exhibited at pH = 3.0, ionic strength = 100 mM, and c($Ca^{2+}$) = 1.0 mM, indicating the mixed solution was unstable and hydrophobic. The calculated interaction parameters showed that the AB interaction energy played an important role in the total interaction energy when the separation distance was less than 3 nm, while the contribution of EL interaction energy to the total interaction energy was of less importance. The attractive interaction energies of membrane–foulant and foulant–foulant were substantially increased with decreasing pH or increasing ionic strength and calcium ion concentration, thus aggravating membrane fouling. Fouling experiments showed that MF membrane fouling by the HA–BSA mixtures was more serious at low pH, and high ionic strength and calcium ion concentration, which was consistent with theoretical predictions. In addition, a strong negative linear relationship between fouling potential and corresponding interaction energy in both stages was observed. This study would provide valuable quantitative information for a more detailed understanding of membrane fouling mechanisms involved in the MF of humic acid–protein mixtures.

**Supplementary Materials:** The following are available online at http://www.mdpi.com/2073-4441/10/10/1306/s1. Figure S1: Schematic diagram of experimental system for constant pressure dead-end microfiltration; Figure S2: SEM characterization results at different solution conditions: (a) pH = 7.0, IS = 10 mM, $c$ (Ca$^{2+}$) = 0 mM; (b) pH = 3.0, IS = 10 mM, $c$ (Ca$^{2+}$) = 0 mM; (c) pH = 7.0, IS = 100 mM, $c$ (Ca$^{2+}$) = 0 mM; (d) pH = 7.0, IS = 10 mM, $c$ (Ca$^{2+}$) = 1.0 mM.

**Author Contributions:** The experimental work was conducted by S.L., C.S., N.Z., and X.L. The manuscript was written by C.S. and S.L. Data analysis was performed by C.S., N.Z., F.L., G.K., L.S., X.L., and S.L. All authors approved the final version of the article, including the authorship list.

**Funding:** This research was funded by the China Major Science and Technology Program for Water Pollution Control and Treatment (No. 2017ZX07101003), the National Natural Science Foundation of China (50908133), the Fundamental Research Funds of Shandong University (2017JC024), the Henan Province Science and Technology Major Project (161100310700), and the APC was funded by the China Major Science and Technology Program for Water Pollution Control and Treatment (No. 2017ZX07101003).

**Acknowledgments:** Assistance from Meiqi Yang and Yuxuan Ren for their cooperation in completion of laboratory experiments are acknowledged gratefully.

**Conflicts of Interest:** The authors declare no conflicts of interest.

## References

1. Guo, H.; Wyart, Y.; Perot, J.; Nauleau, F.; Moulin, P. Low-pressure membrane integrity tests for drinking water treatment: A review. *Water Res.* **2010**, *44*, 41–57. [CrossRef] [PubMed]
2. Zhang, X.; Fan, L.; Roddick, F.A. Influence of the characteristics of soluble algal organic matter released from Microcystis aeruginosa on the fouling of a ceramic microfiltration membrane. *J. Membr. Sci.* **2013**, *425*, 23–29. [CrossRef]
3. Hashino, M.; Hirami, K.; Katagiri, T.; Kubota, N.; Ohmukai, Y.; Ishigami, T.; Maruyama, T.; Matsuyama, H. Effects of three natural organic matter types on cellulose acetate butyrate microfiltration membrane fouling. *J. Membr. Sci.* **2011**, *379*, 233–238. [CrossRef]
4. Lv, X.; Gao, B.; Sun, Y.; Shi, X.; Xu, H.; Wu, J. Effects of humic acid and solution chemistry on the retention and transport of cerium dioxide nanoparticles in saturated porous media. *Water Air Soil Pollut.* **2014**, *225*, 2167. [CrossRef]
5. Dong, H.; Gao, B.; Yue, Q.; Sun, S.; Wang, Y.; Li, Q. Floc properties and membrane fouling of polyferric silicate chloride and polyferric chloride: The role of polysilicic acid. *Environ. Sci. Pollut. Res.* **2015**, *22*, 4566–4574. [CrossRef] [PubMed]
6. Zhou, S.; Shao, Y.; Gao, N.; Li, L.; Deng, J.; Tan, C.; Zhu, M. Influence of hydrophobic/hydrophilic fractions of extracellular organic matters of Microcystis aeruginosa on ultrafiltration membrane fouling. *Sci. Total Environ.* **2014**, *470*, 201–207. [CrossRef] [PubMed]
7. Gray, S.R.; Ritchie, C.; Tran, T.; Bolto, B. Effect of NOM characteristics and membrane type on microfiltration performance. *Water Res.* **2007**, *41*, 3833–3841. [CrossRef] [PubMed]
8. Zheng, X.; Zietzschmann, F.; Plume, S.; Paar, H.; Ernst, M.; Wang, Z.; Jekel, M. Understanding and Control of Biopolymer Fouling in Ultrafiltration of Different Water Types. *Water* **2017**, *9*, 298. [CrossRef]
9. Shao, J.; Hou, J.; Song, H. Comparison of humic acid rejection and flux decline during filtration with negatively charged and uncharged ultrafiltration membranes. *Water Res.* **2011**, *45*, 473–482. [CrossRef] [PubMed]

10. Lim, Y.P.; Mohammad, A.W. Effect of solution chemistry on flux decline during high concentration protein ultrafiltration through a hydrophilic membrane. *Chem. Eng. J.* **2010**, *159*, 91–97. [CrossRef]

11. Madaeni, S.; Sedeh, S.N.; De Nobili, M. Ultrafiltration of humic substances in the presence of protein and metal ions. *Transp. Porous Media* **2006**, *65*, 469–484. [CrossRef]

12. Salehi, E.; Madaeni, S. Adsorption of humic acid onto ultrafiltration membranes in the presence of protein and metal ions. *Desalination* **2010**, *263*, 139–145. [CrossRef]

13. Myat, D.T.; Stewart, M.B.; Mergen, M.; Zhao, O.; Orbell, J.D.; Gray, S. Experimental and computational investigations of the interactions between model organic compounds and subsequent membrane fouling. *Water Res.* **2014**, *48*, 108–118. [CrossRef] [PubMed]

14. Shen, L.G.; Lei, Q.; Chen, J.-R.; Hong, H.-C.; He, Y.-M.; Lin, H.-J. Membrane fouling in a submerged membrane bioreactor: Impacts of floc size. *Chem. Eng. J.* **2015**, *269*, 328–334. [CrossRef]

15. Lin, T.; Lu, Z.; Chen, W. Interaction mechanisms of humic acid combined with calcium ions on membrane fouling at different conditions in an ultrafiltration system. *Desalination* **2015**, *357*, 26–35. [CrossRef]

16. Lin, T.; Lu, Z.; Chen, W. Interaction mechanisms and predictions on membrane fouling in an ultrafiltration system, using the XDLVO approach. *J. Membr. Sci.* **2014**, *461*, 49–58. [CrossRef]

17. Ding, Y.; Tian, Y.; Li, Z.; Wang, H.; Chen, L. Interaction energy evaluation of the role of solution chemistry and organic foulant composition on polysaccharide fouling of microfiltration membrane bioreactors. *Chem. Eng. Sci.* **2013**, *104*, 1028–1035. [CrossRef]

18. Nishijima, W.; Speitle, G.E., Jr. Fate of biodegradable dissolved organic carbon produced by ozonation on biological activated carbon. *Chemosphere* **2004**, *56*, 113–119. [CrossRef] [PubMed]

19. Chang, H.; Qu, F.; Liu, B.; Yu, H.; Li, K.; Shao, S.; Li, G.; Liang, H. Hydraulic irreversibility of ultrafiltration membrane fouling by humic acid: Effects of membrane properties and backwash water composition. *J. Membr. Sci.* **2015**, *493*, 723–733. [CrossRef]

20. Van Oss, C. Acid—Base interfacial interactions in aqueous media. *Colloids Surf. A Physicochem. Eng. Asp.* **1993**, *78*, 1–49. [CrossRef]

21. Subramani, A.; Huang, X.; Hoek, E.M. Direct observation of bacterial deposition onto clean and organic-fouled polyamide membranes. *J. Colloid Interface Sci.* **2009**, *336*, 13–20. [CrossRef] [PubMed]

22. Wang, Z.; Chen, Z.; Yang, L.; Tan, F.; Wang, Y.; Li, Q.; Chang, Y.-I.; Zhong, C.-J.; He, N. Effect of surface physicochemical properties on the flocculation behavior of Bacillus licheniformis. *RSC Adv.* **2017**, *7*, 16049–16056. [CrossRef]

23. Brant, J.A.; Childress, A.E. Assessing short-range membrane–colloid interactions using surface energetics. *J. Membr. Sci.* **2002**, *203*, 257–273. [CrossRef]

24. Hoek, E.M.; Agarwal, G.K. Extended DLVO interactions between spherical particles and rough surfaces. *J. Colloid Interface Sci.* **2006**, *298*, 50–58. [CrossRef] [PubMed]

25. Lee, S.; Kim, S.; Cho, J.; Hoek, E.M. Natural organic matter fouling due to foulant–membrane physicochemical interactions. *Desalination* **2007**, *202*, 377–384. [CrossRef]

26. Liang, S.; Zhao, Y.; Liu, C.; Song, L. Effect of solution chemistry on the fouling potential of dissolved organic matter in membrane bioreactor systems. *J. Membr. Sci.* **2008**, *310*, 503–511. [CrossRef]

27. Ye, Y.; Le Clech, P.; Chen, V.; Fane, A.G.; Jefferson, B. Fouling mechanisms of alginate solutions as model extracellular polymeric substances. *Desalination* **2005**, *175*, 7–20. [CrossRef]

28. Chen, L.; Tian, Y.; Cao, C.Q.; Zhang, J.; Li, Z.N. Interaction energy evaluation of soluble microbial products (SMP) on different membrane surfaces: Role of the reconstructed membrane topology. *Water Res.* **2012**, *46*, 2693–2704. [CrossRef] [PubMed]

29. Shen, L.; Wang, X.; Li, R.; Yu, H.; Hong, H.; Lin, H.; Chen, J.; Liao, B.-Q. Physicochemical correlations between membrane surface hydrophilicity and adhesive fouling in membrane bioreactors. *J. Colloid Interface Sci.* **2017**, *505*, 900–909. [CrossRef] [PubMed]

30. Meng, X.; Tang, W.; Wang, L.; Wang, X.; Huang, D.; Chen, H.; Zhang, N. Mechanism analysis of membrane fouling behavior by humic acid using atomic force microscopy: Effect of solution pH and hydrophilicity of PVDF ultrafiltration membrane interface. *J. Membr. Sci.* **2015**, *487*, 180–188. [CrossRef]

31. Wang, X.; Zhou, M.; Meng, X.; Wang, L.; Huang, D. Effect of protein on PVDF ultrafiltration membrane fouling behavior under different pH conditions: Interface adhesion force and XDLVO theory analysis. *Front. Environ. Sci. Eng.* **2016**, *10*, 1–11. [CrossRef]

32. Mo, H.; Tay, K.G.; Ng, H.Y. Fouling of reverse osmosis membrane by protein (BSA): Effects of pH, calcium, magnesium, ionic strength and temperature. *J. Membr. Sci.* **2008**, *315*, 28–35. [CrossRef]

33. Wang, Q.; Wang, Z.; Zhu, C.; Mei, X.; Wu, Z. Assessment of SMP fouling by foulant–membrane interaction energy analysis. *J. Membr. Sci.* **2013**, *446*, 154–163. [CrossRef]

34. Hong, H.; Peng, W.; Zhang, M.; Chen, J.; He, Y.; Wang, F.; Weng, X.; Yu, H.; Lin, H. Thermodynamic analysis of membrane fouling in a submerged membrane bioreactor and its implications. *Biores. Technol.* **2013**, *146*, 7–14. [CrossRef] [PubMed]

35. Ding, Y.; Tian, Y.; Li, Z.; Wang, H.; Chen, L. Microfiltration (MF) membrane fouling potential evaluation of protein with different ion strengths and divalent cations based on extended DLVO theory. *Desalination* **2013**, *331*, 62–68. [CrossRef]

36. Bayoudh, S.; Othmane, A.; Mora, L.; Ouada, H.B. Assessing bacterial adhesion using DLVO and XDLVO theories and the jet impingement technique. *Colloids Surf. Biointerfaces* **2009**, *73*, 1–9. [CrossRef] [PubMed]

37. Yamamura, H.; Okimoto, K.; Kimura, K.; Watanabe, Y. Hydrophilic fraction of natural organic matter causing irreversible fouling of microfiltration and ultrafiltration membranes. *Water Res.* **2014**, *54*, 123–136. [CrossRef] [PubMed]

38. Bower, M.J.D.; Bank, T.L.; Giese, R.F.; Oss, C.J.V. Nanoscale forces of interaction between glass in aqueous and non-aqueous media: A theoretical and empirical study. *Colloids Surf. A Physicochem. Eng. Asp.* **2010**, *362*, 90–96. [CrossRef]

39. Subramani, A.; Hoek, E.M. Direct observation of initial microbial deposition onto reverse osmosis and nanofiltration membranes. *J. Membr. Sci.* **2008**, *319*, 111–125. [CrossRef]

40. Redman, J.A.; Walker, S.L.; Elimelech, M. Bacterial adhesion and transport in porous media: Role of the secondary energy minimum. *Environ. Sci. Technol.* **2004**, *38*, 1777–1785. [CrossRef] [PubMed]

41. Hoek, E.M.; Bhattacharjee, S.; Elimelech, M. Effect of Membrane Surface Roughness on Colloid−Membrane DLVO Interactions. *Langmuir* **2003**, *19*, 4836–4847. [CrossRef]

42. Wang, L.; Miao, R.; Wang, X.; Lv, Y.; Meng, X.; Yang, Y.; Huang, D.; Feng, L.; Liu, Z.; Ju, K. Fouling behavior of typical organic foulants in polyvinylidene fluoride ultrafiltration membranes: Characterization from microforces. *Environ. Sci. Technol.* **2013**, *47*, 3708–3714. [CrossRef] [PubMed]

43. Li, Q.; Xu, Z.; Pinnau, I. Fouling of reverse osmosis membranes by biopolymers in wastewater secondary effluent: Role of membrane surface properties and initial permeate flux. *J. Membr. Sci.* **2007**, *290*, 173–181. [CrossRef]

44. Yang, Q.; Liu, Y.; Li, Y. Control of protein (BSA) fouling in RO system by antiscalants. *J. Membr. Sci.* **2010**, *364*, 372–379. [CrossRef]

45. Li, Q.; Elimelech, M. Organic fouling and chemical cleaning of nanofiltration membranes: Measurements and mechanisms. *Environ. Sci. Technol.* **2004**, *38*, 4683–4693. [CrossRef] [PubMed]

46. Kim, S.; Hoek, E.M. Interactions controlling biopolymer fouling of reverse osmosis membranes. *Desalination* **2007**, *202*, 333–342. [CrossRef]

47. Jin, X.; Huang, X.; Hoek, E.M. Role of specific ion interactions in seawater RO membrane fouling by alginic acid. *Environ. Sci. Technol.* **2009**, *43*, 3580–3587. [CrossRef] [PubMed]

*water*

MDPI

Article

# The Effect of Ca and Mg Ions on the Filtration Profile of Sodium Alginate Solution in a Polyethersulfone-2-(methacryloyloxy) Ethyl Phosphorylchloline Membrane

Nasrul Arahman [1,*], Suffriandy Satria [2], Fachrul Razi [1] and M. Roil Bilad [3]

1   Department of Chemical Engineering, Universitas Syiah Kuala, Jl. Syeh A Rauf, 7, Darussalam, Banda Aceh 23111, Indonesia; fachrurrazi@che.unsyiah.ac.id
2   Graduate School of Chemical Engineering, Universitas Syiah Kuala, Banda Aceh 23111, Indonesia; satria.suffriandy@gmail.com
3   Department of Chemical Engineering, UniversitiTeknologi Petronas, Bandar Seri Iskandar, Perak 32610, Malaysia; mroil.bilad@utp.edu.my
*   Correspondence: nasrular@unsyiah.ac.id; Tel.: +62-8136-0927-917

Received: 14 August 2018; Accepted: 4 September 2018; Published: 7 September 2018

**Abstract:** The efforts to improve the stability of membrane filtration in applications for wastewater treatment or the purification of drinking water still dominate the research in the field of membrane technology. Various factors that cause membrane fouling have been explored to find the solution for improving the stability of the filtration and prolong membrane lifetime. The present work explains the filtration performance of a hollow fiber membrane that is fabricated from polyethersulfone-2-(methacryloyloxy) ethyl phosphorylchloline while using a sodium alginate (SA) feed solution. The filtration process is designed in a pressure driven cross-flow module using a single piece hollow fiber membrane in a flow of outside-inside We investigate the effect of Ca and Mg ions in SA solution on the relative permeability, membrane resistance, cake resistance, and cake formation on the membrane surface. Furthermore, the performance of membrane filtration is predicted while using mathematical models that were developed based on Darcy's law. Results show that the presence of Ca ions in SA solution has the most prominent effect on the formation of a cake layer. The formed cake layer has a significant effect in lowering relative permeability. The developed models have a good fit with the experimental data for pure water filtration with $R^2$ values between 0.9200 and 0.9999. When treating SA solutions, the developed models fit well with experimental with the best model (Model I) shows $R^2$ of 0.9998, 0.9999, and 0.9994 for SA, SA + Ca, and SA + Mg feeds, respectively.

**Keywords:** membrane filtration; membrane fouling; cake resistance; membrane resistance; sodium alginate

## 1. Introduction

In the last few decades, the use of membrane technology for water and wastewater treatment has undergone rapid development. Membrane material developments have also been comprehensively and continuously conducted [1–4]. In comparison with conventional processes, membrane technology offers some advantages. Membrane filtration is highly flexible and it can easily be applied under a required specification. It can be combined with other processes (i.e., membrane bioreactor). The process operation can be performed continuously and automatically, with high selectivity. Membrane is able to separate small particles, such as microbes, bacteria, viruses, and ions [5]. Additionally, the technology can easily be scaled-up in a low foot-print, and the separation properties can be easily fine-tuned with solvent treatment of the membrane material, as described elsewhere [6]. Membrane processes also offer sustainable solutions for various applications, such as waste valorization [7], desalination [8], organocatalysis [9], and extraction [10].

Hollow fiber membranes are the most frequently used in the water processing industry. They are in form of pipe-shape with a diameter 340 μm–1 mm [11,12], where the flow of permeate can be adjusted from inside to outside or vice versa [6]. They have the most prominent advantage over flat membranes in terms its much higher packing density. However, besides those advantages, membrane fouling remains the major obstacle that limits the process productivity via blockage of pores by foulant. Membrane fouling is a serious problem that decreases the hydraulic performance. Therefore, extensive researches have been done for understanding the mechanisms and finding methods for membrane fouling control.

The studies of membrane modification to minimize the fouling and investigation of the factors affecting the foulant deposition and build-up constitute one of the main focuses of membrane research. Some of the common foulant materials include inorganics (clay, silica, salt, and oil), living creatures (microorganisms) [13], and synthetic or natural organic substances [14]. Fouling decreases flux and indirectly affect membrane lifetime because of the use of chemical agent for maintenance and intensive cleanings. Fouling management also dictates process operation and increases the operational costs due to the need for more energy and chemicals for physical and chemical cleanings. Both are required for membrane performance recovery and for regularly maintaining the system productivity.

Based on removability of the foulant, fouling is categorized into reversible and irreversible [15]. Foulant is formed inside the pore or resides on top of the membrane surface as a result of the concentration of polarization, cake layer development, adsorption of foulant materials, and the diffusion of the substance on the membrane pores [16,17]. The nature of foulant can be detected while using different tools, such as Fourier Transform Infrared spectroscopy (FTIR) for chemical bonding identification, Scanning Electron Microscopy (SEM), or Atomic Force Microscopy (AFM) for obtaining surface morphology.

In addition to actual autopsy of foulant from a fouled membrane, many other approaches have also been developed to comprehend the nature of membrane fouling, including the use of mathematical model. Mathematical model help one to understand the mechanism of membrane fouling and formation of fouling layer. By analyzing the pattern of the data obtained from experimental results, one can envisage models that explain such a pattern and later can relate it with the actual physical phenomena [18].

The study of the fouling phenomenon while using a mathematical model for the commercial microfiltration polypropylene hollow-fiber membrane for dead-end $TiO_2$ catalyst separation has been conducted elsewhere [19]. In the study, the model was driven in a constant-flux system, as such the pressure develops overtime to compensate membrane fouling. The developed model can explain the cake layer formation that corresponds to cake resistance ($R_c$) [19]. In another study, a model has been developed to predict flux decline of polypropylene and polyethersulfone membranes for filtration of yeast suspension in a cross-flow system. The developed model offers a good predictions of flux trend as function of time [20].

This study investigates the membrane fouling of sodium alginate (SA) solution in the presence of Ca and Mg ions. The nature of the fouling is monitored by measuring water permeability, membrane resistance, and the rejection coefficient. Such performances were later modelled using mathematical formula. SA is one of the natural organic substances (NOM) that is often found in industrial wastewater and seawater [21]. In membrane filtration, this substance may generate adverse fouling effects [15]. Therefore, a comprehensive study is required to understand the phenomenological nature, the type, and filtration condition that may affect the membrane fouling involving SA [22]. A mathematical model was developed by studying the effect of the appearance of Ca and Mg ions. The available models in literature do not include the presence of Ca and Mg ions. The focus was on cake-formation by assessing the membrane resistance ($R_m$), cake resistance ($R_c$), and flux decrease ($J_p$). The models were validated by fitting it with experimental data to predict the $R_m$, $R_c$, and $J_p$.

## 2. Theory

A transport model can be developed based on Darcy's law by considering both mass and momentum balances under constant temperature and pressure. The flow of feed solution passing the membrane pores through the walls of the hollow fiber membrane is considered to be laminar, where the axial flow is ignored.

For fouling investigation, the filtration is driven by pressure ang the transport is from outside to inside. The separation is dominant on the surface, so the cake layer is assumed only on the outer wall of the membrane with minimum extent of pore blocking. We assume that the formed cake is homogeneous and clogged up, so the thickness of the cake is uniform surrounding the outer membrane surface. From such assumption, mathematical models to predict the relative permeability $(J_p/J_o)$ profile as function of $R_m$, $R_c$, radius of cake (rc), and $\delta_c$ have been developed. Illustration on the filtration flow and formation of cake layer on the membrane surface are shown in Figure 1.

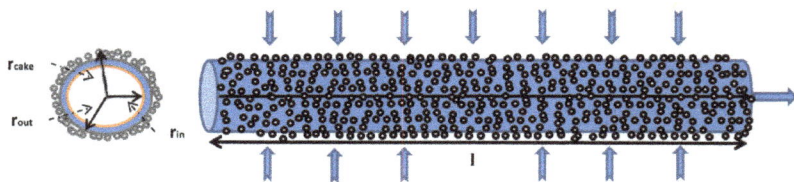

**Figure 1.** Illustration of the outside-inside flow in which cake layer is formed on the outer surface of the hollow fiber membrane. The $r_{in}$ is inner radius of hollow fiber membrane, $r_{out}$ outer radius, $r_{cake}$ radius of cake, l length of the fiber.

## 3. Method

### 3.1. Material

The polymer, the solvent and membrane modifying agent were polyethersulfone (PES-Ultrason E6020 P, BASF Co, Ludwigshafen, Germany) and dimethyl formamide (DMF, WAKO Pure Chemical Industries, Ltd., Osaka, Japan) and 2-(methacryloyloxy)ethyl phosphorylchloline (MPC, Sigma Aldrich, Steinheim, Germany), respectively. The samples of Ca and Mag ions were formulated from $CaCl_2$ and $Mg_2SO_4$ compounds (both from WAKO Pure Chemical Industries, Ltd., Osaka, Japan). NaOH and $H_2SO_4$ were used to control the pH of the sample water (Sigma Aldrich, Steinheim, Germany). All of the chemicals were used as received without further purification.

### 3.2. Membrane Preparation and Characterization

A lab-made hydrophilic hollow fiber ultrafiltration membrane was used. It was made and conditioned to have good properties by adding 2-(methacryloyloxy) ethyl phosphorylchloline (MPC) in the polymer solution. The polymer solution (PES 20 wt% and MPC 2 wt% in DMF solvent) was pumped with a flow rate of 0.06 ms$^{-1}$ through the spinneret to the coagulation bath with an air gap distance of 5 cm. The formed membrane was tied on a spindle cylinder, with a rotation speed of 0.22 ms$^{-1}$.

The morphology of the resulting membrane was analyzed using SEM (FE-SEM, JSM-7500F, JEOL Ltd., Tokyo, Japan). One piece of hollow fiber membrane was freeze dried for one night in a tube (FD-1000, Eyela, Tokyo, Japan). The membrane was then put and fractured under liquid nitrogen, followed by the coating procedure with Pt/Pd sputtering before SEM analysis. The degree of hydrophilicity of the membrane was analyzed with a water contact angle meter (Kyowa Interface Science Co., Drop Master 300, CA-A, Saitama, Japan). The sample was fixed on a metal plate. Around 0.5 µL of water was poured on the membrane surface, and the contact angle between the water and the membrane surface was then recorded. The data for the water contact angle were based on the results of ten measurements.

### 3.3. Ultrafiltration Process

The membrane filtration profile was obtained while using a cross-flow filtration module operating in pressure different outside (PDO) system, in which water penetrates the membrane from out to inside of the fiber. A lab-scale PDO filtration unit is shown in Figure 2. The ultrafiltration set-up consisted of a peristaltic pump (Watson Marlow, Sci. 323, Rommerskirchen, Germany), two pressure gauges to regulate the pressure, a cross-flow module of out-in flow type, one strand of hollow fiber membrane, a feedback tank, and a permeate tank.

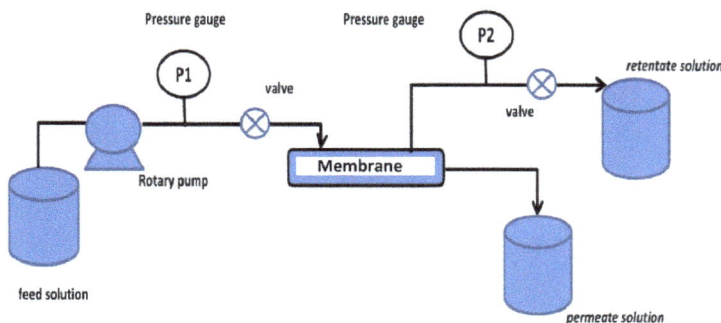

**Figure 2.** Lab-scale of cross-flow filtration for single piece of hollow fiber membrane. The filtrations were run at pressure of 1 atm, and the permeate was collected as an overflow and weighted every 10 min. The set-up was used for filtration of different feeds: (1) 50 ppm sodium alginate (SA) solution, (2) 50 ppm SA + 0.125 mM of Ca ion, and (3) 50 ppm SA + 0.125 mM of Mg ion.

In order to examine the filtration profile and to monitor membrane fouling, three models of feed solution were configured: (1) SA solution of 50 ppm, (2) SA solution of 50 ppm with 0.125 mM of Ca ion, and (3) SA solution of 50 ppm with 0.125 mM of Mg ion. The concentration range was taken as the upper values that lead cake layer fouling, as reported elsewhere by [23]. All the feeds were set at pH of 7. The filtration was started by draining the feed solution with a peristaltic pump in which with feed flow from the outer to inner side of the fiber a gauge pressure of 1 atm. The permeate was allowed to accumulate in the lumen (inner part of the hollow fiber), and after it was full, the permeate overflowed into the collection tank. The volume and weight of the permeate were recorded every 10 min, and the filtration lasted for one hour.

The water permeability of the membrane at the first 10 min ($J_{10}$) and at the next $n$ min ($J_n$) were calculated using Equations (1) and (2). With the same procedure, the water permeability was also calculated for the permeate of the feeds containing SA, SA + Ca, and SA + Mg.

$$J_{10} \times \left( \frac{1}{m^2 \times hr \times atm} \right) = \frac{V_{10}}{AtP} \tag{1}$$

$$J_n \times \left( \frac{1}{m^2 \times hr \times atm} \right) = \frac{V_n}{AtP} \tag{2}$$

where $V_{10}$ and $V_n$ are the permeate volume over a filtration time of 10 min and the next 10 min. A, t, and P are the membrane surface area, filtration time, and membrane pressure, respectively. The relative permeability ($R_p$) over time is presented relative to the first 10 min value, as in Equation (3).

$$R_p = \frac{J_n}{J_{10}} \tag{3}$$

*3.4. Cake Layer Model*

Membrane fouling for filtration of SA solution, as revealed later from the SEM image, is largely in the form of cake layer on the outer membrane surface. The cake-formation model, in this case, is assumed to follow the mass transfer of fluid in porous media and the formation of a cake layer as a part of the fouling phenomenon obeys the Darcy's law, as in Equation (4) [24].

$$\frac{dV}{dt} = V_p = -\frac{k}{\mu} A \frac{dP}{dr} \tag{4}$$

Differentiation of Equation (4) results in Equation (5).

$$\frac{dP}{dr} = \frac{V_p}{Ak} \mu \tag{5}$$

Under this condition, there is a change in the radius of the membrane (r) as a result of the increasing thickness of cake and the decreasing filtration flux. The membrane surface area is assumed perfectly cylindrical, as formula shown in Equation (6).

$$A(t) = 2\pi r(t) l \tag{6}$$

If the general direction of permeate flow is perpendicular to the center of the membrane, and the axial rate is negligible, with the initial limit before filtration $r_i \leq r \leq r_o$, and during filtration until the forming of cake $r_o \leq r \leq r_c(t)$, Equation (5) can be expanded, as follows:

$$P = \frac{V_{p(t)}}{2\pi l k_m} \mu \left[ \ln \frac{r_0}{r_i} \right] + \frac{V_{p(t)}}{2\pi l k_c} \mu \left[ \ln \frac{r_c(t)}{r_0} \right] \tag{7}$$

By linking the result of the integration of Equation (7) with the membrane resistance ($R_m$) in Equation (8) and cake resistance ($R_c$) in Equation (9). [25], Equation (7) can be formulated as Equation (10).

$$R_m = \frac{1}{k_m} r_0 \ln \frac{r_0}{r_i} \tag{8}$$

$$R_c(t) = \frac{1}{k_c} r_0 \ln \frac{r_c(t)}{r_0} \tag{9}$$

$$P = \frac{V_{p(t)}}{2\pi r_0 l} \mu \left( \frac{1}{k_m} r_0 \left[ \ln \frac{r_0}{r_i} \right] + \frac{1}{k_c} r_0 \left[ \ln \frac{r_c(t)}{r_0} \right] \right) \tag{10}$$

when $\frac{V_{p(t)}}{2\pi r_0 l}$ is the permeate flux change against time at constant pressure (P). Equation (10) can be simplified into Equations (11) and (12), as also proposed in literature [26].

$$P = J_{p(t)} \mu (R_m + R_c(t)) \tag{11}$$

$$J_{p(t)} = \frac{P}{\mu (R_m + R_c(t))} \tag{12}$$

By assuming that the formation of the cake layer on the membrane surface equals to the particles approaching the outer surface, in which $R_c(t)$ [19] is the radius of cake against time; the equation can be written, as follow:

$$V_p C_b dt - 2\pi h r_c dr_c (1 - \varepsilon_c)\rho_s + 2\pi h r_c dr_c C_b = 0 \tag{13}$$

$$\frac{dr_c}{dt} = \frac{V_p C_b}{2\pi h r_c ((1 - \varepsilon_c)\rho_s - C_b)} \tag{14}$$

The mass balance of the component transfer is shown in Equation (15) [27]:

$$J_c + D\frac{dc}{dx} = J_{cp} \tag{15}$$

By assuming that there is no reverse diffusion from the bulk toward the direction of inflow implying that:

$$D\frac{dc}{dx} = 0 \tag{16}$$

Equation (15) can be simplified into:

$$J_c = J_{cp} \tag{17}$$

When the inflow and the outflow rate are written as formula shown in Equations (18) and (19);

$$J_c = V_0 dC_b \tag{18}$$

$$J_{cp} = J_0 A\, C_b dt \tag{19}$$

in which $V_0$ is the feed volume, Equation (17) can then be expanded to Equation (20).

$$V_0 dC_b = J_0 A\, C_b dt \tag{20}$$

here, $J_0$ and A are the initial flux of filtration and surface area of the membrane, respectively. $J_0 A = V_P$ is the permeate flow rate:

$$V_0 dC_b = V_p C_b dt \tag{21}$$

$$\frac{dC_b}{dt} = \frac{V_p}{V_0} C_b \tag{22}$$

At $t = 0$, $C_b = C_0$, the change in bulk concentration against time can be written as

$$C_b(t) = C_0 \exp\left(-\frac{V_p}{V_0}t\right) \tag{23}$$

By substituting Equation (23) into Equation (20) and integrated the resulting equation in respect to time with the boundary condition of $t = 0$ and $r_c = r_0$, it results in Equation (24).

$$r_c(t) = \sqrt{r^2 + \frac{V_0}{\pi l}\ln\frac{(1-\varepsilon)\rho_s - C_0\exp\left(-\frac{V_p}{V_0}t\right)}{(1-\varepsilon)\rho_s - C_0}} \tag{24}$$

The $R_c$ value then can be written by substituting Equation (24) into Equation (25), resulting Equation (26), as follows:

$$R_c(t) = \frac{1}{k_c} r_0 \ln\frac{r_c(t)}{r_0} \tag{25}$$

$$R_c(t) = \frac{1}{k_c} r_0 \ln\frac{\sqrt{r^2 + \frac{V_0}{\pi l}\ln\frac{(1-\varepsilon)\rho_s - C_0\exp\left(-\frac{V_p}{V_0}t\right)}{(1-\varepsilon)\rho_s - C_0}}}{r_0} \tag{26}$$

*3.5. Model Validation*

Equation (26) is used to examine $R_c$ by fitting the model with experimental data via the minimum sum squares of errors (SSE) methods [28]. This way, the smallest deviation between model and experimental data is obtained to predict the unknown variables:

$$SSE = \sum_{i=1}^{n} \left( X_{experiment} - X_{Model} \right)^2 \tag{27}$$

The obtained value is then used to determine the other independent variables, and is validated with the $R^2$ value as reference (Equation (28)).

$$R^2 = \sum_{i=1}^{n} \frac{\left( X_{experiment} - \ddot{X}_{experiment} \right)^2}{\left( X_{Model} - \ddot{X}_{experiment} \right)^2} \tag{28}$$

Here, $X_{experiment}$ is the result of experimental data; $\ddot{X}_{experiment}$ the average of the results of the experimental data with respect to the filtration time; and, $X_{Model}$ the result calculated based on the model. The result of the $R^2$ calculation is better when the value is close to 1, which indicates that the model is in good agreement with the experimental result [29].

## 4. Results and Discussion

*4.1. Membrane Characteristic*

The results of the SEM characterization are shown in Figure 3. They show morphology of membrane comprising of skin layers inside and outside of the fiber. Macrovoid structures are seen underneath the skins and sponge-like structure in the middle of the membrane. Such structures are common for the membrane that was prepared by the dry-wet spinning method. The water contact angle of membrane surface is 62.8 ± 1.4°, showing a good hydrophilic trait.

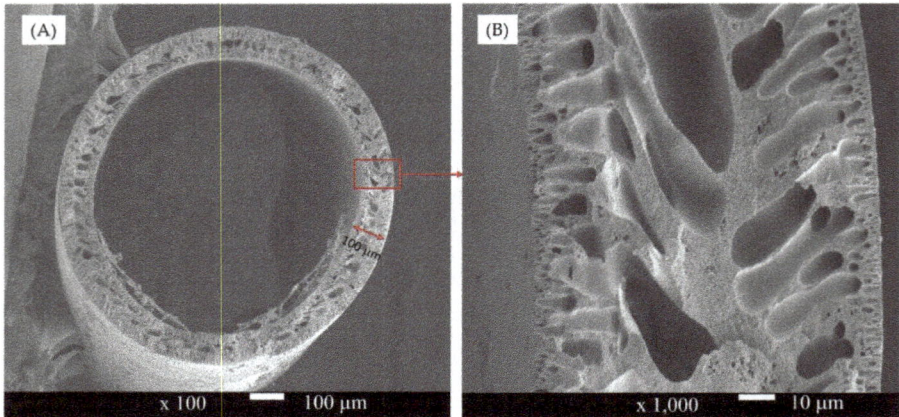

**Figure 3.** Scanning Electron Microscopy (SEM) image of hollow fiber membrane of the whole cross-section (**A**), and enlarged cross-section (**B**).

*4.2. Pure Water Filtration*

Figure 4 shows decreasing water permeability profile for 1 h filtration plotted against the total permeate volume. Water permeability gradually decreases until reaching a constant value.

The decrease in water permeability in this case is not caused by the membrane fouling, but as a result of pressure compaction, which alters the membrane structure until a certain shape and flux are constant.

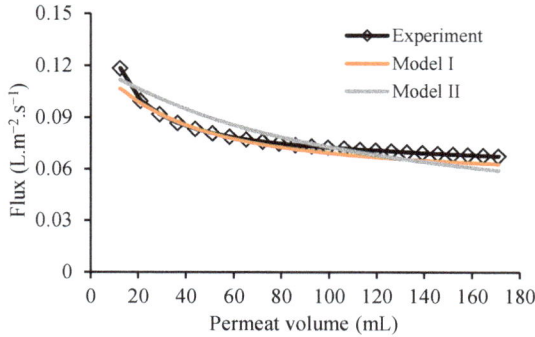

**Figure 4.** The profile of pure water flux: comparison between experimental data and models.

Figure 4 also shows the predicted water permeability according the developed mass transfer model. The model is used to predict the tendency of water permeability in identifying weather the flux has reached a threshold value. The prediction can also be used to evaluate the membrane lifetime [20].

In this study, the profile of water permeability is predicted while using the mathematical models, as in Equations (29) and (30).

$$J_t = J_0 \exp^{\left(-\frac{t}{A+Bt}\right)}, \text{ (model I)} \tag{29}$$

$$J_t = J_0(1 + (t/\tau))^{-0.5}, \text{ (model II)} \tag{30}$$

$J_t$ is water permeability at a certain time, $J_0$ the water permeability at the beginning of filtration, t the filtration time, A and B constants, and $\tau$ constant time. The calculation results of models I and II are shown in Table 1. The values of $R^2$ for model I is 0.95, and for model II is 0.61. This means that model I is more accurate than model II with a good correlation against the experiment result.

**Table 1.** Result of data processing of pure water filtration with Model I and Model II.

| Membrane Type | Model I | | | | Model II | | |
|---|---|---|---|---|---|---|---|
| | Constant | | Standard Error | | Constant | Standard Error | |
| | A | B | SSE | $R^2$ | T | SSE | $R^2$ |
| CA | 4920.28 | 1.24 | 0.00038 | 0.95 | 4772.56 | 0.000898 | 0.61 |

*4.3. Filtration of Sodium Alginate Solution*

The profile of water permeability for filtration of SA solution test is calculated while using Equation (3), and also predicted using Model I and Model II. Since the feed contains SA, it is assumed that the cake resistance causes the decrease in water permeability. Hence, we use a water permeability reduction model by the substitution of Equations (26) and (12) to form Equation (31) (Model III). Thus, the $J_p(t)$ value can be compared with the result of the experiment, Model I and Model II.

$$J_p(t) = \frac{P}{\mu\left(R_m + \frac{1}{k_c}r_0 \ln\frac{\sqrt{r^2 + \frac{V_0}{\pi l}\ln\frac{(1-\varepsilon)\rho_s - C_0\exp\left(-\frac{V_p}{V_0}t\right)}{(1-\varepsilon)\rho_s - C_0}}}{r_0}\right)} \tag{31}$$

The tendency of the relative permeability of the SA solution based on the experimental results, and prediction through Model I, II, and III are described in Figure 5A–C, respectively. The correlation of the experimental data and the calculation results with Model I, Model II, and Model III were concluded based on the $R^2$ value (Table 2). From Table 2, it can be observed that the validation result of the Models I is quite good with the $R^2$ values between 0.9994 and 0.9999, which means that Model I have a good fit with the experimental results.

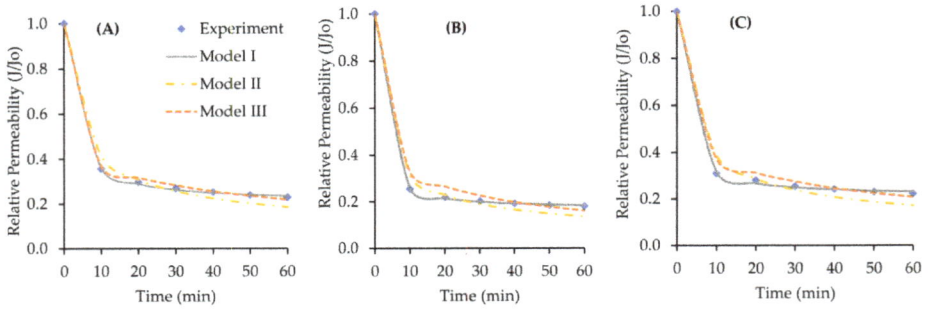

**Figure 5.** The decrease in permeability for (**A**) the SA; (**B**) the SA + Ca; and, (**C**) the SA + Mg sample.

**Table 2.** Validation model and experimental results of relative permeability shown in Figure 5.

| Sample Solution | Relative Permeability Model Validation ($R^2$) | | |
|---|---|---|---|
| | Model I | Model II | Model III |
| SA | 0.9998 | 0.9261 | 0.9943 |
| SA + Ca | 0.9999 | 0.9438 | 0.9837 |
| SA + Mg | 0.9994 | 0.9200 | 0.9898 |

Overall, the results described in Figure 5 indicate a drastic decrease in the relative permeability from the first data point. The reduction in the relative permeability gradually continues until the end of the filtration. The phenomenon of the permeability loss in the filtration of SA solution is quite different than the filtration process using deionized water. For the latter, the decrease in the permeability is below 50% of the initial value (Figure 4).

From experimental data and the models, it can be observed that the metal ions affect the filtration performances (Figure 5B,C). Metal ions interact with SA in the feed solution. When accumulate on the membrane surface, it leads to polarization concentration. The rates of permeability decline using the SA + Ca and SA + Mg are greater than of the SA solution. Moreover, when counting the ratio of water permeability reduction, it is clear that the decrease in the relative permeability of the membrane filtration while using SA + Ca solution is higher than the SA + Mg. As described in Figure 6, the change in relative permeability after 10 min with the SA + Ca is 74.75% higher than the rests. Extending filtration beyond one hour also shows the further reduction of relative permeability for the SA + Ca feed.

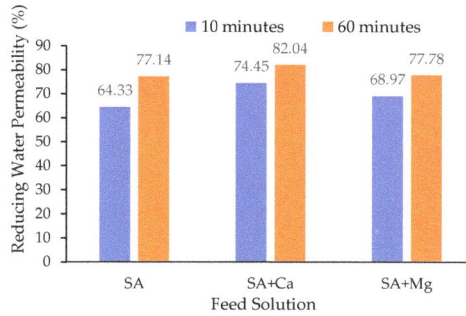

**Figure 6.** Reducing water permeability when treating several types of feeds.

In summary, the presence of $Ca^{2+}$ and $Mg^{2+}$ ions in the feeds accelerates membrane fouling. They decrease SA charge, and influence the adhesion force between the SA particles and the membrane surface, which turns to accelerate the membrane fouling [30]. The other reason is due to the nature of the Ca ion, which has a special property to the carboxylic group to form an aggregate with a larger molecular size [31].

*4.4. Membrane Resistance ($R_m$)*

Membrane resistance ($R_m$) is the intrinsic properties of the membrane in hindering the permeating water [32]. Its value is determined by filtering pure water and by fitting to Equation (32):

$$\frac{dt}{dV} = \frac{\mu \alpha c_s}{A^2(-\Delta p)}V + \frac{\mu}{A(-\Delta p)}R_m = K_p V + B \tag{32}$$

Integration of Equation (32), results:

$$\frac{t}{V} = \frac{K_p V}{2} + B \tag{33}$$

By plotting $tV^{-1}$ (filtration time on the permeate volume) versus V (volume of permeate), as described in Figure 7, the value of $K_p^{-2}$ (as slope) and B (as intercept to y-axis of linear equation) are obtained. This way, the Rm value is provided in Table 3. The Rm-value is then used to calculate the equation model for further determination of the permeation reduction and the resistance of the membrane due to formation of cake layer.

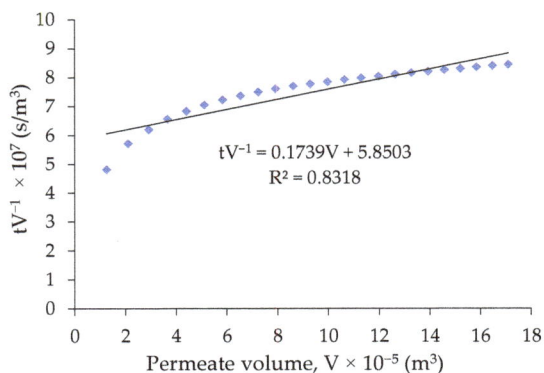

$$tV^{-1} = 0.1739V + 5.8503$$
$$R^2 = 0.8318$$

**Figure 7.** Relation between t/V and V.

**Table 3.** Result of calculation from the values of membrane resistance ($R_m$), $\varepsilon_c$ permeability of cake ($K_c$), and $R^2$ for every membrane and sample treatment.

| $R_m$ (m$^{-1}$) | Feed | $\varepsilon_c$ | $K_c$ (m$^{-2}$) | $R_c$ Validation ($R^2$) |
|---|---|---|---|---|
| | SA | 0.7123 | $1.66 \times 10^{-20}$ | 0.8622 |
| $5.75 \times 10^{12}$ | SA + Ca | 0.5372 | $4.94 \times 10^{-21}$ | 0.6701 |
| | SA + Mg | 0.5500 | $9.42 \times 10^{-21}$ | 0.7165 |

*4.5. Cake Resistance Model*

The cake resistance is additional persistence of the membrane in allowing mass transport due to the formation of the cake layer on its surface. This condition leads to the concentration of polarization, and consequently, causes a reduction in the permeability [33,34]. The $R_c$ value depends on the type of particle contained in the feed solution. For the experimental data, the $R_c$ might be determined while using Equation (34).

$$R_c(t) = \frac{P}{\mu J_{p(t)}} - R_m \tag{34}$$

in which t, P, and $J_p$ are the filtration time (hour), membrane pressure (atm), and permeation, respectively (l·m$^{-2}$·h$^{-1}$). Furthermore, $\mu$ and Rm are the solution viscosity (Pa.s) and membrane resistance (m$^{-1}$), respectively. The $R_c$ value is also predicted while using the developed model (Equation (19)), and the results were validated using Equation (28).

The permeability of cake ($K_c$), and cake porosity ($\varepsilon$) can be obtained from Equation (19) by considering the SSE on $R_c$ of the experimental data. The cake resistances based on the filtration results of SA, SA + Ca, and SA + Mg are depicted in Figure 8A–C, respectively. It is shown that the resistance of the cake increases with respect of the filtration time. The maximum resistance was obtained for filtration system while using the SA + Mg feed (Figure 8B).

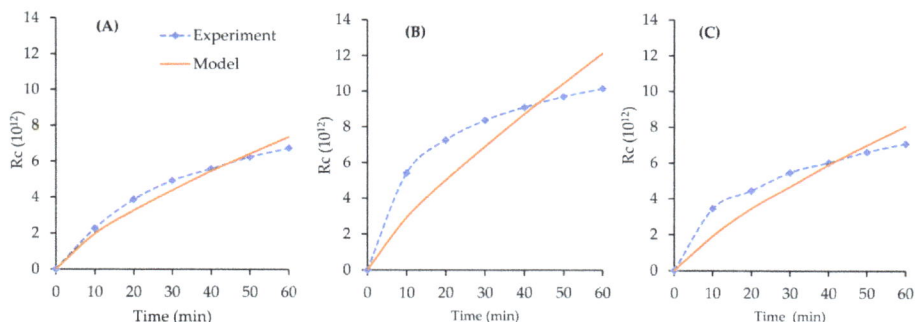

**Figure 8.** The increase in the cake resistance for (**A**) the SA; (**B**) the SA + Ca; and, (**C**) the SA + Mg feeds.

The increase in the $R_c$ provides additional persistence of the filtration process besides the effect of Rm, which directly and significantly reduces the water permeation. The increasing cake resistance based on the mathematics model for the filtration system using SA, SA + Ca, and SA + Mg as the feeds is also shown in Figure 8. The best correlation between the experimental data and the results of the model development is in the filtration system while using the SA feed (Figure 8B), with the $R^2$ of 0.8622 (Table 3).

### 4.6. Cake Layer Build-Up

Membrane fouling occurs via several mechanisms, e.g., the forming of a cake layer on the surface, concentration of polarization, and pore blocking (entrapment of foulant in membrane pores). In this study, we ascribe the fouling as build-up of cake layer on the outer surface of the hollow fiber membrane. The cake resistance was previously discussed and obtained while using Equation (34). Based on the $R_c$ data, the increase in the cake radius can be determined using Equation (9). The radius of the cake layer is considered to be the thickness of the cake formed on the outer surface of the membrane. The build-ups of cake layer thickness on the membrane surface for the filtration of the three feeds are shown in Figure 9. The presence of Ca ions in the SA solution has the greatest effect on the formation of cake layer.

The increase in the cake layer thickness on the membrane surface in this work instantly affects the decrease in the water permeability. The greatest cake layer thickness has the lowest water permeability (Figure 6). This is because an increase in the thickness of the cake (δ) leads to an increase in the membrane resistance, thereby inhibiting the mass transport and thus decreasing the permeate volume. Regarding this permeation performance, some researcher has modified the membrane in order to minimize the formation of cake layer with various hydrophilic polymer [4,35].

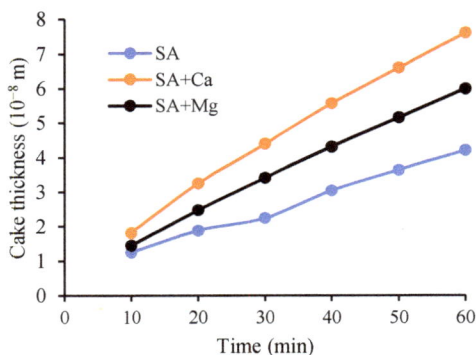

**Figure 9.** The improvement in the cake layer thickness versus the filtration time.

## 5. Conclusions

The fouling mechanisms of PES-MPC membrane for filtration of SA, SA + Ca, and SA + Mg solutions were studied and modelled. The presence of CA and Mg ions significantly lowers hydraulic performance. The developed models fit well with experimental with the best model (Model I) shows $R^2$ of 0.9998, 0.9999, and 0.9994 for SA, SA + Ca, and SA + Mg feeds, respectively. The effect of Ca was greater than the Mg ions in compacting the cake layer. When filtering the SA + Ca feed, the relative permeability decreased by up to 82% after 60 min filtration. As revealed by the developed model for filtration of this feed, the rate of cake resistance increment against time is directly proportional to the increase in the cake thickness reaching the highest δc of $7.59 \times 10^{-7}$ m. Due to severe effect of Ca and Mg ions on membrane fouling, their presences in a feed solution must be strongly considered when designing membrane filtration system. An unconventional fouling control system maybe required to achieve good system performance.

**Author Contributions:** N.A. proposed the research topic and experimental concept, prepared and editing the manuscript. S.S. was responsible for laboratory experiment. F.R. and M.R.B. contributed in providing research materials and advising the research data.

**Funding:** The research is funded by the Competency-Based Research Grant, the Ministry of Research, Technology and Higher Education of Indonesia (Grant: 02/UN11.2/PP/SP3/2018).

**Acknowledgments:** We express our gratitude to the Ministry of Research, Technology and Higher Education of Indonesia and the institutions that gave technical support, such as Syiah Kuala University of Banda Aceh, Indonesia, and Kobe University, Japan.

**Conflicts of Interest:** The authors declare no conflict of interest.

## References

1. Arahman, N.; Mulyati, S.; Rahmah, M.; Takagi, R. The removal of fluoride from water based on applied current and membrane types in electrodialyis. *J. Fluor. Chem.* **2016**, *191*, 97–102. [CrossRef]
2. Sheng, A.L.K.; Bilad, M.R.; Osman, N.B.; Arahman, N. Sequencing batch membrane photobioreactor for real secondary effluent polishing using native microalgae: Process performance and full-scale projection. *J. Clean. Prod.* **2017**, *168*, 708–715. [CrossRef]
3. Arahman, N.; Mulyati, S.; Rahmah, M.; Takagi, R.; Matsuyama, H. Removal pro file of sulfate ion from mix ion solution with diff erent type and configuration of anion exchange membrane in elctrodialysis. *J. Water Process Eng.* **2017**, *20*, 173–179. [CrossRef]
4. Plisko, T.V.; Liubimova, A.S.; Bildyukevich, A.V.; Penkova, A.V.; Dmitrenko, M.E.; Mikhailovskii, V.Y.; Melnikova, G.B.; Semenov, K.N.; Doroshkevich, N.V.; Kuzminova, A.I. Fabrication and characterization of polyamide-fullerenol thin film nanocomposite hollow fiber membranes with enhanced antifouling performance. *J. Membr. Sci.* **2018**, *551*, 20–36. [CrossRef]

5.  Shahkaramipour, N.; Tran, T.N.; Ramanan, S.; Lin, H. Membranes with surface-enhanced antifouling properties for water purification. *Membranes* **2017**, *7*, 13. [CrossRef] [PubMed]
6.  Razali, M.; Didaskalou, C.; Kim, J.F.; Babaei, M.; Drioli, E.; Lee, Y.M.; Szekely, G. Exploring and exploiting the effect of solvent treatment in membrane separations. *ACS Appl. Mater. Interface* **2017**, *9*, 11279–11289. [CrossRef] [PubMed]
7.  Didaskalou, C.; Buyuktiryaki, S.; Kecili, R.; Fonte, C.P.; Szekely, G. Valorisation of agricultural waste with adsorption/nanofiltration hybrid process: From materials to sustainable process design. *Green Chem.* **2017**, *19*, 3116–3125. [CrossRef]
8.  Li, G.; Law, W.C.; Chan, K.C. Floating, highly efficient, and scalable graphene membranes for seawater desalination using solar energy. *Green Chem.* **2018**, *20*, 3689–3695. [CrossRef]
9.  Didaskalou, C.; Kupai, J.; Cseri, L.; Barabas, J.; Vass, E.; Holtz, T.; Szekely, G. Membrane-grafted asymmetric organocatalyst for an integrated synthesis-separation platform. *ACS Catal.* **2018**, *8*, 7430–7438. [CrossRef]
10. Kumar, A.; Thakur, A.; Panesar, P.S. Lactic acid extraction using environmentally benign green emulsion ionic liquid membrane. *J. Clean. Prod.* **2018**, *181*, 574–583. [CrossRef]
11. Lee, M.J.; Hamid, M.R.A.; Lee, J.; Kim, J.S.; Lee, Y.M.; Jeong, H.K. Ultrathin zeolitic-imidazolate framework ZIF-8 membranes on polymeric hollow fibers for propylene/propane separation. *J. Membr. Sci.* **2018**, *559*, 28–34. [CrossRef]
12. Huang, A.; Feng, B. Synthesis of novel graphene oxide-polyimide hollow fiber membranes for seawater desalination. *J. Membr. Sci.* **2018**, *548*, 59–65. [CrossRef]
13. Rana, D.; Matsuura, T. Surface modifications for antifouling membranes. *Chem. Rev.* **2010**, *110*, 2448–2471. [CrossRef] [PubMed]
14. Motsa, M.M.; Mamba, B.B.; D'Haese, A.; Hoek, E.M.V.; Verliefde, A.R.D. Organic fouling in forward osmosis membranes: The role of feed solution chemistry and membrane structural properties. *J. Membr. Sci.* **2014**, *460*, 99–109. [CrossRef]
15. Charfi, A.; Yang, Y.; Harmand, J.; Ben Amar, N.; Heran, M.; Grasmick, A. Soluble microbial products and suspended solids influence in membrane fouling dynamics and interest of punctual relaxation and/or backwashing. *J. Membr. Sci.* **2015**, *475*, 156–166. [CrossRef]
16. Vargas, A.; Moreno-Andrade, I.; Buitrón, G. Controlled backwashing in a membrane sequencing batch reactor used for toxic wastewater treatment. *J. Membr. Sci.* **2008**, *320*, 185–190. [CrossRef]
17. Suwal, S.; Doyen, A.; Bazinet, L. Characterization of protein, peptide and amino acid fouling on ion-exchange and filtration membranes: Review of current and recently developed methods. *J. Membr. Sci.* **2015**, *496*, 267–283. [CrossRef]
18. Teychene, B.; Collet, G.; Gallard, H. Modeling of combined particles and natural organic matter fouling of ultra filtration membrane. *J. Membr. Sci.* **2016**, *505*, 185–193. [CrossRef]
19. Popović, S.; Dittrich, M.; Cakl, J. Modelling of fouling of outside-in hollow-fiber membranes by TiO2particles. *Sep. Purif. Technol.* **2015**, *156*, 28–35. [CrossRef]
20. Mallubhotla, H.; Belfort, G. Semiempirical Modeling of cross-flow microfiltration with periodic reverse filtration. *Ind. Eng. Chem. Res* **1996**, *5*, 2920–2928. [CrossRef]
21. Hashino, M.; Katagiri, T.; Kubota, N.; Ohmukai, Y.; Maruyama, T. Effect of membrane surface morphology on membrane fouling with sodium alginate. *J. Membr. Sci.* **2011**, *366*, 258–265. [CrossRef]
22. Xu, W.T.; Zhao, Z.P.; Liu, M.; Chen, K.C. Morphological and hydrophobic modifications of PVDF flat membrane with silane coupling agent grafting via plasma flow for VMD of ethanol-water mixture. *J. Membr. Sci.* **2015**, *491*, 110–120. [CrossRef]
23. Charfi, A.; Jang, H.; Kim, J. Membrane fouling by sodium alginate in high salinity conditions to simulate biofouling during seawater desalination. *Bioresour. Technol.* **2017**, *240*, 106–114. [CrossRef] [PubMed]
24. Rathore, A.S.; Kumar, V.; Arora, A.; Lute, S.; Brorson, K.; Shukla, A. Mechanistic modeling of viral filtration. *J. Membr. Sci.* **2014**, *458*, 96–103. [CrossRef]
25. Kin, J.; Dole, P. Filtration model for hollow fi ber membranes with compressible cake formation. *Desalination* **2009**, *240*, 2–6.
26. Zhang, H.; Gao, J.; Jiang, T.; Gao, D.; Zhang, S.; Li, H.; Yang, F. Bioresource technology a novel approach to evaluate the permeability of cake layer during cross-flow filtration in the flocculants added membrane bioreactors. *Bioresour. Technol.* **2011**, *102*, 11121–11131. [CrossRef] [PubMed]

27. Mulder, M. *Basic Principles of Membrane Technology*, 4th ed.; Kluwer Academic Publisher: Dordrecht, The Netherlands, 1991.

28. Badrnezhad, R.; Mirza, B. Journal of industrial and engineering chemistry modeling and optimization of cross-flow ultrafiltration using hybrid neural network-genetic algorithm approach. *J. Ind. Eng. Chem.* **2014**, *20*, 528–543. [CrossRef]

29. Netcher, A.C.; Duranceau, S.J. Modeling the improvement of ultra fi ltration membrane mass transfer when using biofiltration pretreatment in surface water applications. *Water Res.* **2016**, *90*, 258–264. [CrossRef] [PubMed]

30. Chang, H.; Liang, H.; Qu, F.; Shao, S.; Yu, H.; Liu, B. Role of backwash water composition in alleviating ultra filtration membrane fouling by sodium alginate and the effectiveness of salt backwashing. *J. Membr. Sci.* **2016**, *499*, 429–441. [CrossRef]

31. Hao, Y.; Moriya, A.; Maruyama, T.; Ohmukai, Y.; Matsuyama, H. Effect of metal ions on humic acid fouling of hollow fiber ultrafiltration membrane. *J. Membr. Sci.* **2011**, *376*, 247–253. [CrossRef]

32. Sarkar, D.; Chakraborty, D.; Naskar, M.; Bhattacharjee, C. Characterization and modeling of radial flow membrane (RFM) module in ultra filtration. *Desalination* **2014**, *354*, 76–86. [CrossRef]

33. Paipuri, M.; Kim, S.H.; Hassan, O.; Hilal, N.; Morgan, K. Numerical modelling of concentration polarisation and cake formation in membrane filtration processes. *Desalination* **2015**, *365*, 151–159. [CrossRef]

34. Garcia-Ivars, J.; Iborra-Clar, M.-I.; Alcaina-Miranda, M.-I.; Van der Bruggen, B. Comparison between Hydrophilic and Hydrophobic metal nanoparticles on the phase separation phenomena during formation of asymmetric polyethersulphone membranes. *J. Membr. Sci.* **2015**, *493*, 709–722. [CrossRef]

35. Arahman, N.; Mulyati, S.; Lubis, M.R.; Razi, F.; Takagi, R.; Matsuyama, H. Modification of polyethersulfone hollow fiber memrbanes with different polymeric additives. *Membr. Water Treat.* **2016**, *4*, 355–365. [CrossRef]

*water*

MDPI

*Article*

# Influence of the Backwash Cleaning Water Temperature on the Membrane Performance in a Pilot SMBR Unit

**Loukas Lintzos [1],\*, Kostas Chatzikonstantinou [1], Nikolaos Tzamtzis [1] and Simos Malamis [2]**

[1]   Department of Chemical Science, School of Chemical Engineering, National Technical University of Athens,
     9 Iroon Polytechniou St., 15780, Athens, Greece; coolwate@otenet.gr (K.C.); nipitz@central.ntua.gr (N.T.)
[2]   Department of Water Resources and Environmental Engineering, School of Civil Engineering, National
     Technical University of Athens, 5 Iroon Polytechniou St., 15780, Athens, Greece; malamis.simos@gmail.com
*    Correspondence: loukaslintzos07@gmail.com; Tel.: +30-210-772-4470

Received: 5 January 2018; Accepted: 22 February 2018; Published: 25 February 2018

**Abstract:** In this work, different backwash (BW) schemes were applied on identical hollow fiber (HF) membranes in a membrane bioreactor (MBR) treating municipal wastewater. The effect of BW duration (1 min, 3 min and 8 min) and water temperature (8 °C, 18 °C, 28 °C and 38 °C) on membrane fouling were investigated. Specifically, the transmembrane pressure (TMP) drop and the membrane permeability increase caused by the BW was investigated. Furthermore, the time required for the membrane to return to the state just before each BW experiment, was also examined. It was found that membranes presented better operating performance, as the BW temperature and the backwash duration were increased. Specifically, for 1 min backwash duration at the BW temperatures of 8 °C, 18 °C, 28 °C and 38 °C, TMP decreased by 7.1%, 8.7%, 11.2% and 14.2% respectively. For 8 min BW duration at 8 °C, 18 °C, 28 °C and 38 °C, TMP values decreased by 12%, 17.5%, 23.7% and 30.2% respectively. Increased BW water temperature and duration also improved the membrane permeability. Using higher BW water temperatures, more hours were required to return the membranes to the condition just before cleaning. The selected BW water temperatures did not adversely affect the permeate quality.

**Keywords:** Membrane Bioreactor; hollow fiber membranes; TMP; backwash duration; temperature; membrane fouling

---

## 1. Introduction

Membrane bioreactor (MBR) systems are now a mature technology for wastewater treatment. The importance of membrane technology, is growing in the field of environmental protection. In 2008, a 22.4% compound annual growth rate (CAGR) was predicted for the world MBR market for the period 2008–2018 [1,2]. The global MBR market was worth $838.2 million in 2011 and is expected to witness positive growth and revenue sales through 2018 [3]. According to the recent report from BCC Research, the global market for MBRs was $425.7 million in 2014 and is projected to approach $777.7 million by 2019, registering a CAGR of 12.8% in the period 2014–2019 [4].

Although MBR technology has many advantages over the conventional activated sludge process, membrane fouling and the increased energy consumption are a major barrier to further use of this technology [5,6]. Membrane fouling is a severe problem and affects operating cost due to the frequent membrane cleaning and the increased aeration demands [7,8]. The degree of fouling in submerged membrane systems is a complex function of feed characteristics, membrane properties but more importantly of biomass characteristics and operating conditions [9,10]. To address the problem of membrane fouling, several measures are undertaken, including, wastewater pretreatment, hydraulic and chemical cleaning of the membranes, membrane modification and operation under conservative fluxes [11].

One of the most common hydraulic membrane cleaning methods, especially in hollow fiber (HF) submerged MBR is backwash (BW) cleaning. Membranes require periodical backwash to control fouling [12]. During the BW cleaning procedure, the filtration process is reversed so that permeate is sent back through the membranes. BW cleaning is particularly effective in the removal of accumulated particles over the membrane surface, which mainly constitute the reversible fouling, dislodging loosely adherent aggregates sludge from the membrane's surface [13,14]. BW is applied in both ultrafiltration (UF) and microfiltration (MF) hollow fiber and also in specific flat sheet membranes. The above procedures are performed in immersed MBR systems. After the BW process, the membrane recovers part of its initial permeability. However, over time, an irreversible loss of membranes productivity is observed, using the above cleaning procedures particularly when the system operates at high membrane flux [15]. The BW frequency, the duration and the BW to filtration flux ratio, are key operating parameters for the design of an effective backwashing procedure [16]. The optimization of backwashing is necessary for energy and permeate consumption [17]. The optimal duration of a backwash cycle is related to the effective removal of reversible foulants from the membrane's surface [11]. In the literature there are several studies which investigate the effect of BW on the mitigation of membrane fouling. The fouling deposition and removal during filtration with short periodical backwash by direct observation coupled with hydraulic resistance measurement in hollow fiber membranes was investigated [18]. Other work [19] stress that the intermittent backwashing and relaxation are mandatory in the MBR for its effective operation; also examined different relaxation and BW scenarios and found that the provision of relaxation or backwashing at small intervals prolonged the MBR operation by reducing fouling rates. There is a lack of experimental data on the influence of water temperature on the efficiency of BW cleaning. Moreover, the influence of BW water temperature is considered important in patent No WO2015198080 A1 in which it claims that, water's or permeate water's temperature is foreseen as adjustable [20].

Research works have also focused on the effect of the duration of BW. The effect of BW duration, on the membrane fouling of a pilot pressurized hollow fiber membrane module applied for the pre-treatment of seawater was investigated [21]. The BW duration is one of the most significant parameters in order to minimize the specific energy consumption. It was shown by other work [22] that were able to prolong the time constant TMP operation by 50% when BW together with air scouring during BW was implemented compared to simple dead end filtration without BW. The effect of filtration duration on membrane fouling of real seawater filtration while other operation parameters were kept constant was investigated [18]. There are also works in which demonstrating the importance of backwash duration in MBR systems. It was pointed that backwashing conditions (i.e., duration, interval, strength) considerably affected the fouling rate in membrane bioreactor systems [23]. In addition were presented various combinations in duration of filtration and backwash in MBR systems in order to achieve better results with respect to membrane fouling and the lowest fouling rates were found for 15 s in each 5 min of filtration [24]. In this work, the impact of BW water temperature and duration on TMP and permeability was investigated in a MBR treating municipal wastewater. Specifically, four different temperatures and three different backwash cleaning time periods were investigated.

## 2. Materials and Methods

### 2.1. MBR Operation

The MBR consists of a 60 L aeration tank fed with synthetic wastewater which simulates municipal wastewater (Figure 1). To test the impact of BW water temperature and duration, three identical HF type Khong membranes were used. The HF membrane were immersed in the aeration tank of the pilot MBR which was fed with synthetic municipal wastewater. The HF membrane modules characteristics are given in Table 1. The MBR was operated continuously with repeating cycles of filtration followed by relaxation. During the normal MBR operation, suction was applied to each HF module using a peristaltic, suction pump (0.975 L/h–9.750 L/h). There were two effluent lines which allow to conduct two parallel experimental cycles. At each effluent line a glycerin pressure indicator and an analog vacuum pressure transducer (−1/0 bar) in series with an analogue flow meter (FM) were installed. The filtration (8 min) and relaxation (2 min) time of the suction pump were adjusted according to the manufacturer for the protection of HF membrane elements. The operating conditions of the MBR process are listed in Table 2.

**Table 1.** HF Membrane Characteristics.

| Membrane Type | Filtration Type | Membrane Material | Pore Size (μm) | Membrane Area (m$^2$) | Frame Dimensions (mm) | Critical Flux (L/m$^2$·h) |
|---|---|---|---|---|---|---|
| HF | UF | R-PVDF | 0.1 | 0.05 | 24 × 22 | 25 |

**Table 2.** Experimental conditions.

| Operating Parameter | Value |
|---|---|
| Working Time/Cycle (min) | 8 |
| Relaxing Time/Cycle (min) | 2 |
| Gross Flux (L/m$^2$·h) | 24 |
| pH | 7–8 |
| Aeration type | Coarse bubble |
| Max TMP (mbar) | 220 |
| MLSS (mg/L) | 7450–11250 |
| Backwash period/frequency | see experimental procedure |
| Backwash recommended flow (L/m$^2$·h) | 30 |
| Max Backwash Pressure (mbar) | <50 |

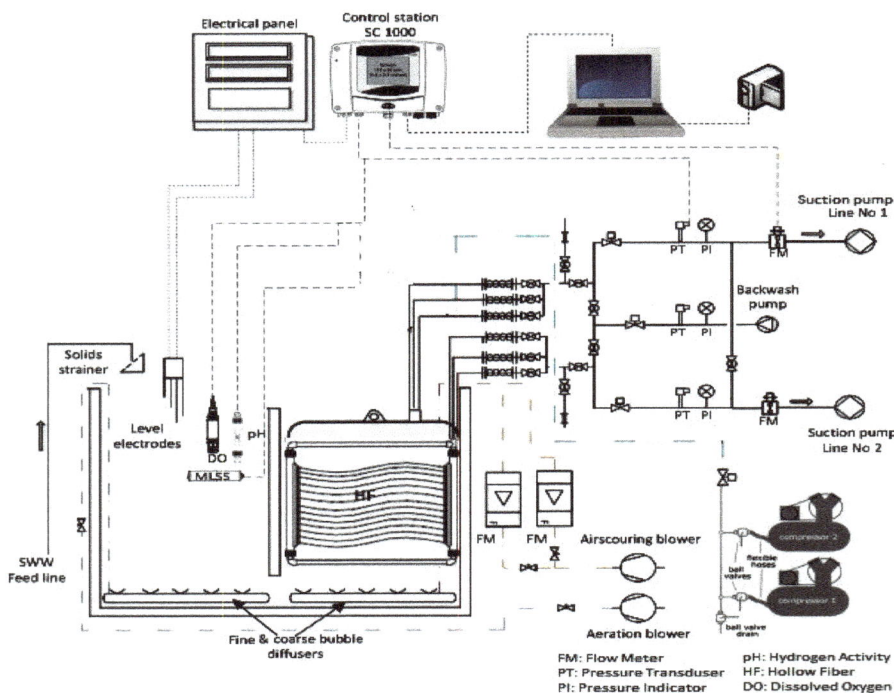

**Figure 1.** MBR pilot plant and backwashing layout.

The BW procedure was applied each time the membranes presented significant fouling, by monitoring the TMP value which was chosen not to exceed 210 mbar. The range of BW water temperature was in the permitted limits of manufacturers and did not exceed 40 °C. In the backwash line the following equipment was placed in series: a flow control solenoid valve, a glycerin pressure indicator and an analog pressure transducer (0/1 bar). The backwash pump was connected to a volume calibrated water container which was equipped with a thermometer, a regulated thermostat and an immersed heater. The backwash pump was manually calibrated to provide the recommended flow of backwash cleaning water. As seen in Figure 1, the BW was partly performed in a separate line compared to the permeate collection line. The dead volume of water in the BW tubes is considered to be negligible compared to the total amount of BW water. The MBR pilot unit was continuously working, for a period of 12 months. The air scouring amount was maintained constant but lower than manufactures suggest for faster membrane fouling simulation. More extensive description of the MBR pilot plant and backwashing equipment also presented elsewhere [25].

### 2.2. Backwashing Experimental Procedure

A consistent number of backwashing experimental cycles was performed to confirm the accuracy and repeatability of the working method. In each backwashing experimental cycle, an HF membrane (Khong) was used, working continuously at an operational scheme of 8 min filtration and 2 min relaxation period. Each cycle was carried out once the TMP had reached close to 210 mbar so that the membrane was considerably fouled. At the first period, three experimental cycles were implemented: the backwash water temperature was set at 8 °C and three different backwashing durations of 1 min, 3 min and 8 min were applied. At the second period, three BW cycles were carried out: the backwash water temperature was set at 18 °C which was near the ambient mixed liquor temperature. The above water temperature was tested for three different backwashing durations of 1 min, 3 min and 8 min. The same procedure repeated using backwash water temperature of 28 °C and 38 °C for BW duration of 1 min, 3 min and 8 min. In all cases, each backwash experimental cycle was repeated three times, in each selected membrane for repeatability reasons.

The TMP and permeability experimental data are mean values of each backwash scheme. All the values are normalized to a standard temperature of 20 °C. An adequate number of treated effluent samples were collected for physicochemical analyses after an hour of each cleaning procedure. For all BW cleaning cycles, tap water instead of the effluent was chosen, so as to assure potential external factors as internal clogging from solids or biofouling from microorganisms.

## 3. Results and Discussion

In Figure 2 the effect of the BW water temperature on TMP decrease is shown for a BW duration of 1 min, 3 min and 8 min. The TMP decrease was calculated as a percentage decrease of the TMP value using the following equation (TMP $_{before\ BW}$ − TMP $_{after\ BW}$)/ (TMP $_{before\ BW}$ − TMP $_{clean\ membrane}$). The clean membrane's TMP was 2.5 mbar at the flux of 24 L/m$^2$·h. The membranes which were treated for 1min by BW water at 8 °C resulted in a TMP decrease of about 7.1%. As the backwash water temperature was increased to 18 °C, 28 °C and 38 °C, TMP values showed a decrease of about 8.7%, 11.2% and 14.2% respectively. Consequently, the BW water temperature increase from 8 °C to 38 °C caused the doubling of the TMP reduction.

In the same figure, the impact of water temperature on the decrease of TMP is shown for BW duration of 3 min. The membranes which were fed with backwash water at 8 °C resulted in TMP decrease of 8.2%. As in the case of 1 min BW, by increasing the backwash water temperature, the TMP decrease was higher. Specifically as the BW water temperature increased at 18 °C, 28 °C and 38 °C, the TMP showed a decrease of about 12.2%, 17.1% and 20.3% respectively.

The percent TMP decrease for the four different water temperatures for BW duration of 8 min is also presented in Figure 2. The BW cleaning procedure using water at 8 °C and 18 °C have much lower improvement in terms of membrane TMP and permeability with respect to the cleaning procedure using water at 28 °C and 38 °C. Specifically the membranes which were treated by BW water at 8 °C and 18 °C, resulted in TMP decrease of about 12% and 17.5% respectively. By increasing the BW water temperature at 28 °C and 38 °C the TMP showed a decrease of about 23.7% and 30.2% respectively. Consequently, in all the experimental cycles which were performed, the increase of temperature and/or of the duration resulted in an improvement of the membrane TMP.

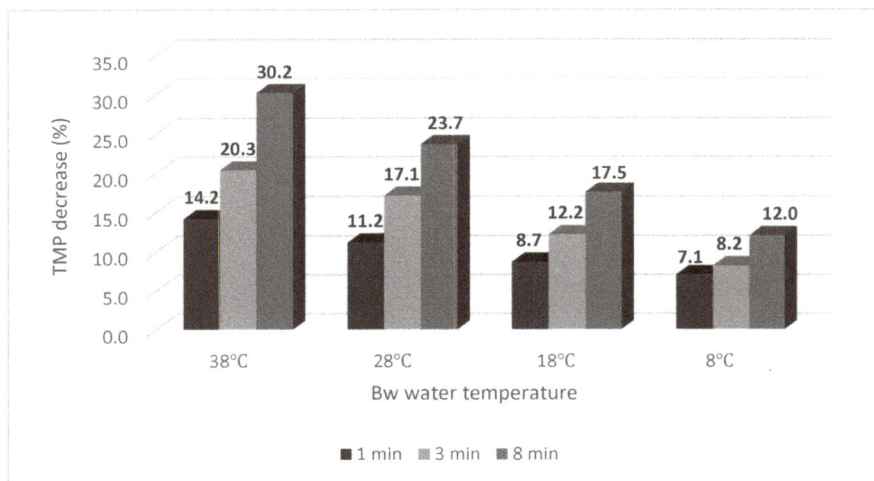

**Figure 2.** Impact of water temperature on TMP decrease for the BW durations of 1 min, 3 min and 8 min.

Figure 3a shows the monitored TMP values of the MBR before and after the backwash cleaning procedure with water at 8 °C, 18 °C, 28 °C and 38 °C for BW duration of 1 min. In this experimental scheme, only 1 h of MBR operation was required just after the BW to reach approximately the TMP value which was observed just before the BW process was applied (~210 mbar). As the BW water temperature increased to 18 °C, 28 °C and 38 °C the membrane modules operated more hours before the max TMP was reached. Specifically 3h, 9h and 18 h were required respectively.

Figure 3b presents the monitored TMP values before and after the BW cleaning procedure with BW water temperature at 8 °C, 18 °C, 28 °C and 38 °C using BW duration of 3 min. In this case, the time needed to reach the same TMP as the one recorded before the BW was 4 h (~210 mbar). The increase of backwash water temperature to 18 °C and 28 °C prolonged the time taken to reach 210 mbar to 10 h and 24 h acordingly. Finally, the membrane module which was cleaned with water at th temperature of 38 °C needed 56 h to reach a TMP above 200 mbar.

In Figure 3c the recorded TMP values before and after the BW cleaning procedure with water at 8 °C, 18 °C, 28 °C and 38 °C by applying a BW duration of 8 min. It is found that the membrane module which was cleaned with water at 8 °C needed 10 h to restore the TMP values. Apparently the BW water temperature increase to 18 °C, 28 °C and 38 °C maintained significantly increased the elapsed time which was required to reach the max TMP. Therefore, the temperature increase from 8 °C to 38 °C resulted in sixteen times more time to reach the max TMP. The utility of the above method was perceived both in short as in longer backwash period. Consequently, the increase of water temperature results in much fewer BW cycles during the operation of the MBR process and has a positive impact on membrane fouling.

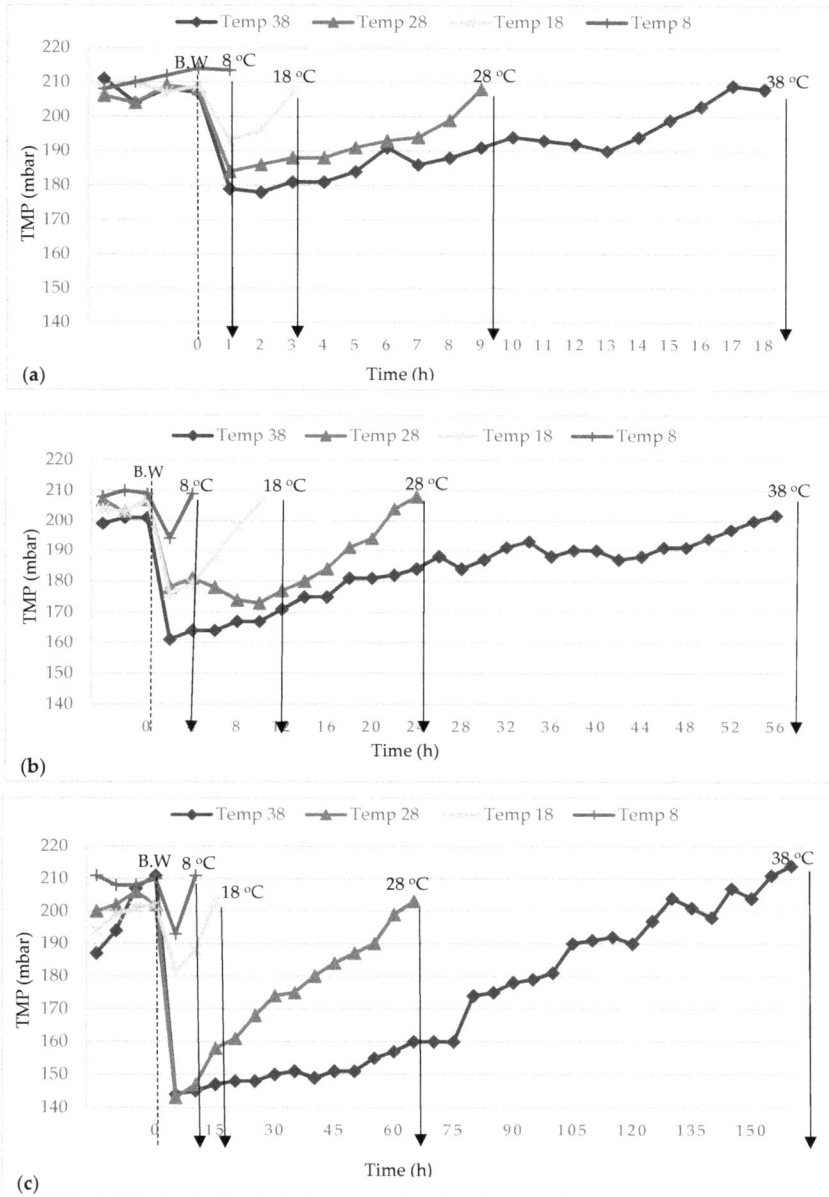

**Figure 3.** Mean TMP monitored values for the continuous operation of the MBR following BW at all examined water temperatures for BW duration of (**a**) 1 min, (**b**) 3 min and (**c**) 8 min.

Figure 4 presents the BW cleaning steps that are required in a period of a week (168 h) with 1 min, 3 min and 8 min BW period of each step. It is noticed that the increase of BW water temperature leads to decrement of BW steps. Specifically, the backwash of 1 min duration with water of 8 °C required 168 cleaning steps in order to keep TMP within the required limits. The increase of temperature at

18 °C, 28 °C and 38 °C leads to decrement of BW steps to 56, 18 and 11 accordingly. That gives a high energy and flow rate profit.

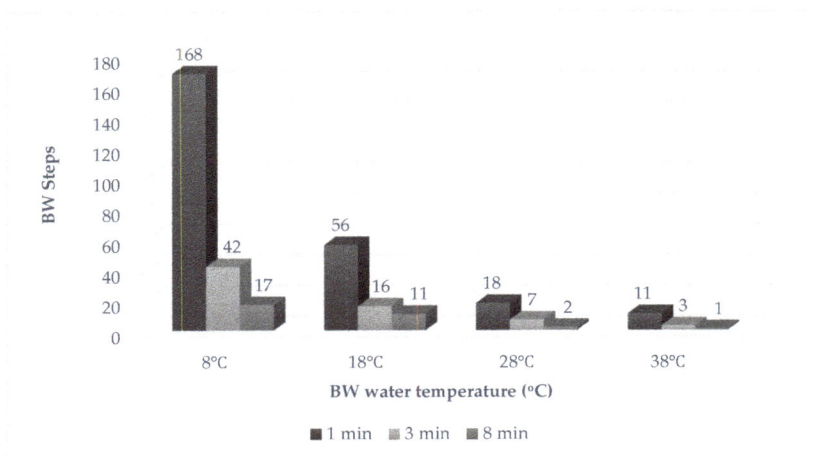

**Figure 4.** Number of BW steps required for the different water temperature in a period of a week (168 h), for BW duration of 1 min, 3 min and 8 min.

Similar results are observed in both cases of 3 min and 8 min of BW period. Specifically in the case of 8 °C the required BW steps were 42 for 3 min BW period and 17 for 8 min BW period. For 18 °C, 28 °C the respective steps were, 16 and 11 for 3 min BW period and 7 and 2 for 8 min BW period accordingly. At last, in the case of 38 °C the required steps were just 3 in the case of 3 min BW period and only one in all over the week in the case of 8 min BW duration. These results confirm the importance of the method as the decrease of required number of BW steps offer a great profit in energy consumption and a higher effluent productivity.

Table 3 shows the net flux of all the tested BW schemes for a weekly operation. It is observed that increasing the BW temperature leads to higher net flux in all the examined cases. However, in the case of lower temperatures (8 °C–18 °C) the highest flux is obtained for the 3 min BW duration. The positive impact in fouling resistance can be also observed by examining the rate of permeability decrease (Figure 5a–c). The permeability of the membrane module which was cleaned with the higher water temperature presented lower rate of permeability decrease with time and therefore required less frequent BW and chemical cleaning.

**Table 3.** Net flux (L/m²·h) of all the different BW scheme for a weekly operation.

| BW Scheme | Net Flux (L/m²·h) |
|---|---|
| 8 °C for 1 min | 18.28 |
| 8 °C for 3 min | 18.51 |
| 8 °C for 8 min | 18.46 |
| 18 °C for 1 min | 18.89 |
| 18 °C for 3 min | 18.93 |
| 18 °C for 8 min | 18.83 |
| 28 °C for 1 min | 19.10 |
| 28 °C for 3 min | 19.09 |
| 28 °C for 8 min | 19.11 |
| 38 °C for 1 min | 19.13 |
| 38 °C for 3 min | 19.15 |
| 38 °C for 8 min | 19.16 |

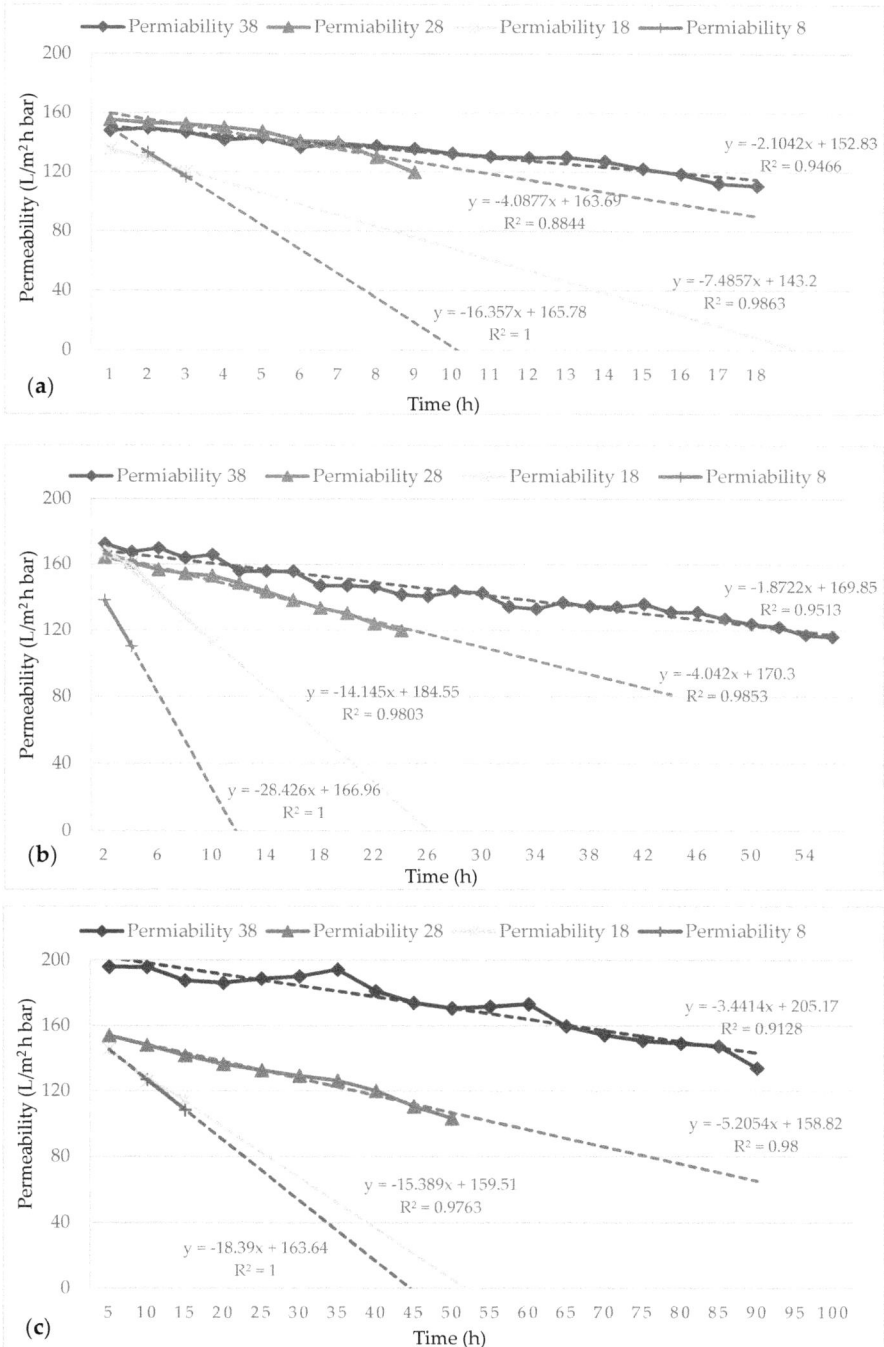

**Figure 5.** Mean membrane permeability with time for BW duration of (**a**) 1 min, (**b**) 3 min and (°) 8 min.

In Figure 6, the average values of selected physicochemical parameters (conductivity - turbidity) of the treated effluent are presented just before and after the cleaning backwash procedure for all the experimental cycles are presented.

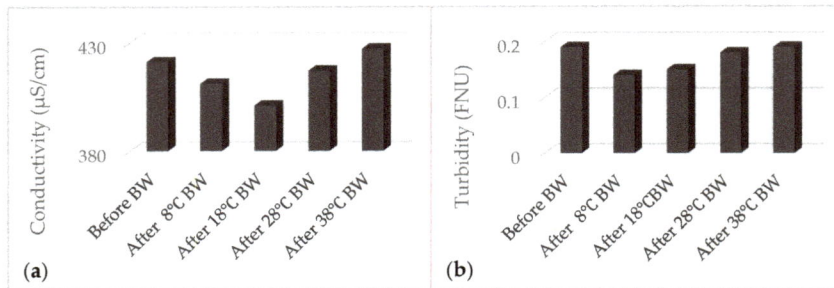

**Figure 6.** Treated effluent physicochemical parameters of (**a**) conductivity and (**b**) before and after water backwash at temperatures of 8 °C, 18 °C, 28 °C and 38 °C. Each mean value is derived from 6–10 repetitions.

From the results of Figure 6 it is observed that the BW water temperature does not influence significantly the quality of the treated effluent.

## 4. Conclusions

It was found that membranes presented better operating performance as the BW temperature water was increased. Furthermore, the increasing duration of the BW leads to better membrane performance. The increase of BW water temperature also entails an increase of the operating period before the next BW cycle is required. Increasing the BW temperature resulted in higher net flux in all the examined cases. However, in the case of lower temperatures (8 °C–18 °C) the highest flux was observed for the 3 min BW duration. In addition, the rate of membrane permeability decrease with time was much lower when the membrane was cleaned with high BW water temperature. In addition, it was found in all examined cases that the treated effluent quality is not influenced using the above cleaning procedure, by measuring certain indicative physicochemical parameters. These results are considered encouraging to the direction of the use of environmental friendly cleaning procedures in MBR units.

**Author Contributions:** L.L, K.C. and N.T. conceived and designed the experiments; L.L. performed the experiments; L.L, S.M. and K.C. analyzed the data; K.C. and S.M. contributed reagents/materials/analysis tools; L.L., S.M, K.C. and N.T wrote the paper.

**Conflicts of Interest:** The authors declare no conflict of interest.

## References

1. Meng, F.; Chae, S.-R.; Shin, H.-S.; Yang, F.; Zhou, Z. Recent advances in membrane bioreactors: Configuration development, pollutant elimination, and sludge reduction, Environ. *Eng. Sci.* **2011**, *29*, 139–160. [CrossRef]
2. Krzeminski, P.; Leverette, L.; Malamis, S.; Katsou, E. Membrane bioreactors—A review on recent developments in energy reduction, fouling control, novel configurations, LCA and market prospects (Review). *J. Membr. Sci.* **2017**, *527*, 207–227. [CrossRef]
3. Global Membrane Bioreactor (MBR) Market. Available online: https://store.frost.com/global-membrane-bioreactor-mbr-market-9809.html (accessed on 24 February 2018).
4. Naessens, W.; Maere, T.; Ratkovich, N.; Vedantam, S.; Nopens, I. Critical review of membrane bioreactor models—Part 2: Hydrodynamic and integrated models. *Bioresour. Technol.* **2012**, *122*, 107–118. [CrossRef] [PubMed]

5.   Le-Clech, P.; Chen, V.; Fane, T.A.G. Fouling in membrane bioreactors used in wastewater treatment. *J. Membr. Sci.* **2006**, *284*, 17–53. [CrossRef]
6.   Malamis, S.; Andreadakis, A. Fractionation of proteins and carbohydrates of extracellular polymeric substances in a membrane bioreactor system. *Bioresour. Technol.* **2009**, *100*, 3350–3357. [CrossRef] [PubMed]
7.   Chellam, S.; Jacangelo, J.G.; Bonacquisti, T.P. Modeling and experimental verification of pilot-scale hollow fiber, direct flow microfiltration with periodic backwashing. *Environ. Sci. Technol.* **1998**, *32*, 75–81. [CrossRef]
8.   Malamis, S.; Andreadakis, A.; Mamais, D.; Noutsopoulos, C. Comparison of alternative additives employed for membrane fouling mitigation in membrane bioreactors. *Des. Water Treat.* **2014**, *52*, 5740–5747. [CrossRef]
9.   Malamis, S. Wastewater Treatment with the Use of Membranes. Ph.D. Thesis, School of Civil Engineering, National Technical University of Athens, Zografou, Greece, 2017.
10.   Bouhabila, E.H.; Ben Aim, R.; Buisson, H. Fouling characterization in membrane bioreactors. *Sep. Purif. Technol.* **2001**, *22*, 123–132. [CrossRef]
11.   Arnal, J.M.; Garcia, B.; Sancho, M.; Instituto de Seguridad Industrial. *Radiofísica y Medioambiente (ISIRYM)*; Universitat Politècnica de València Spain: València, Spain, 2011.
12.   Water Research Australia. *Project Concept—Membrane Backwash Recycling Management*; Water Research Australia: Adelaide, Australia, 2016.
13.   Raffin, M.; Germain, E.; Judd, S.J. Influence of backwashing, flux and temperature on microfiltration for wastewater reuse. *Sep. Purif. Technol.* **2012**, *96*, 147–153. [CrossRef]
14.   Psoch, C.; Schiewer, S. Critical flux aspect of air sparging and backflushing on membrane bioreactors. *Desalin. Water Treat.* **2005**, *175*, 61–71. [CrossRef]
15.   Hai, F.I.; Yamamoto, K. Treatise on water science. In *Membrane Biological Reactors*; Wilderer, P., Ed.; Elsevier: Oxford, UK, 2011; pp. 571–613.
16.   EUROMBRA. *Membrane Bioreactor Technology (MBR) with an EU Perspective for Advanced Municipal Wastewater Treatment Strategies for the 21st Century, Sixth Framework Programme, Contract No. 018480, Deliverable Report—D5*; EUROMBRA: Dübendorf, Switzerland, 2006.
17.   Yigit, N.O.; Civelekoglu, G.; Harman, I.; Koseoglu, H.; Kitis, M. Effects of various backwash scenarios on membrane fouling in a membrane bioreactor. *Desalination* **2009**, *237*, 346–356. [CrossRef]
18.   Ye, Y.; Chen, V.; Le-Clech, P. Evolution of fouling deposition and removal on hollow fibre membrane during filtration with periodical backwash. *Desalination* **2011**, *283*, 198–205. [CrossRef]
19.   Tabraiz, S.; Sallis, P.; Nasreen, S.; Mahmood, Q.; Awais, M.; Acharya, K. Effect of cycle run time of backwash and relaxation on membrane fouling removal in submerged membrane bioreactor treating sewage at higher flux. *Water Sci. Technol.* **2017**, *76*, 963–975. [CrossRef] [PubMed]
20.   Chatzikonstantinou, K. Using High Frequency Vibration for Operational Improvement of Membrane Filtration Process. WO Patent Application No. 2015198080 A1/GR1008540, 30 December 2015.
21.   Chua, K.T.; Hawlader, M.N.A.; Malek, A. Pretreatment of seawater: Results of pilot trials in Singapore. *Desalination* **2003**, *159*, 225–243. [CrossRef]
22.   Akhondi, E.; Wicaksana, F.; Fane, G.A. Evaluation of fouling deposition, fouling reversibility and energy consumption of submerged hollow fiber membrane systems with periodic backwash. *J. Membr. Sci.* **2014**, *452*, 319–331. [CrossRef]
23.   Wu, J.; Le-Clech, P.; Stuetz, R.M.; Fane, A.G.; Chen, V. Effects of relaxation and backwashing conditions on fouling in membrane bioreactor. *J. Membr. Sci.* **2008**, *324*, 26–32. [CrossRef]
24.   Yigit, N.; Civelekoglu, G.; Harman, I.; Koseoglu, H.; Kitis, M. Effects of various backwash scenarios on membrane fouling in a membrane bioreactor. In *Survival Sustainability*; Springer: Berlin/Heidelberg, Germany, 2010; pp. 917–929.
25.   Chatzikonstantinou, K.; Tzamtzis, N.; Pappa, A.; Liodakis, S. Membrane fouling control using high-frequency power vibration, in an SMBR pilot system—Preliminary studies. *Desalin. Water Treat.* **2016**, *57*, 11550–11560. [CrossRef]

![water logo] *water*

MDPI

*Article*

# An Experimental and Theoretical Study on Separations by Vacuum Membrane Distillation Employing Hollow-Fiber Modules

**Anthoula Karanasiou [1], Margaritis Kostoglou [2],*  and Anastasios Karabelas [1]**

[1]   Chemical Process Engineering Research Institute, Centre for Research and Technology—Hellas,
     P.O. Box 60361, GR57001 Thermi-Thessaloniki, Greece; akaranasiou@cperi.certh.gr (A.K.);
     karabaj@cperi.certh.gr (A.K.)
[2]   Division of Chemical Technology, Department of Chemistry, Aristotle University of Thessaloniki,
     Univ. Box 116, GR54124 Thessaloniki, Greece
*    Correspondence: kostoglu@chem.auth.gr; Tel.: +30-231-099-7767

Received: 24 May 2018; Accepted: 13 July 2018; Published: 16 July 2018

**Abstract:** Vacuum membrane distillation (VMD) is an attractive variant of the novel membrane distillation process, which is promising for various separations, including water desalination and bioethanol recovery through fermentation of agro-industrial by-products. This publication is part of an effort to develop a capillary membrane module for various applications, as well as a model that would facilitate VMD process design. Experiments were conducted in a laboratory pilot VMD unit, comprising polypropylene capillary-membrane modules. Performance data, collected at modest temperatures (37 °C to 65 °C) with deionized and brackish water, confirmed the improved system productivity with increasing feed-water temperature; excellent salt rejection was obtained. The recovery of ethanol from ethanol-water mixtures and from fermented winery by-products was also studied, in continuous, semi-continuous, and batch operating modes. At low-feed-solution temperature (27–47 °C), ethanol-solution was concentrated 4 to 6.5 times in continuous operation and 2 to 3 times in the semi-continuous mode. Taking advantage of the small property variation in the module axial-flow direction, a simple VMD process model was developed, satisfactorily describing the experimental data. This VMD model appears to be promising for practical applications, and warrants further R&D work.

**Keywords:** vacuum membrane distillation; desalination; bioethanol recovery; modeling

---

## 1. Introduction

In membrane distillation (MD) the temperature difference between the warm feed and the cooler permeate side (Figure 1) drives the separation process. Vacuum membrane distillation (VMD) is a significant variant of the MD process, where the permeation of volatile compound through the membrane is enhanced by applying vacuum at the permeate side. Vacuum membrane distillation is applicable to single component separation (e.g., water treatment, desalination), binary mixture separations (e.g., concentration of dilute ethanol solutions), and separation of volatile compounds from multi-component mixtures (e.g., recovery of flavors from aqueous solutions). Due to the relatively low-suction pressure applied at the permeate side, and the membrane hydrophobicity, molecules of the volatile compound evaporate at the warmer feed-side of the membrane, move as vapor though the membrane pores, and are condensed in an external condenser [1–3]. The VMD mode of operation is considered [4–6] to have two advantages, in comparison to other MD variants: (a) relatively low-conductive heat loss and (b) reduced resistance to mass transfer.

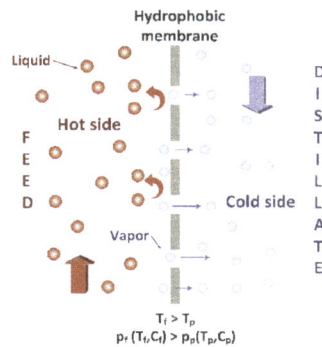

**Figure 1.** Membrane distillation (MD) principle.

Despite the above potential benefits, the development of VMD for large-scale applications appears to be lagging behind other MD versions, notably direct contact membrane distillation (DCMD) [4,7]. As in all types of MD, key issues for VMD implementation and development are the membrane type and the module design. For VMD, the application of vacuum clearly favors a *shell-and-tube* type of module, comprised of a multitude of *hollow fiber, capillary* or *tubular* membranes. A criterion for classifying these membrane types is the lumen diameter; characteristic sizes are <0.5 mm for hollow fiber (HF), 0.5 to ~2 mm for capillary, and > ~2 mm for tubular [4,7]. With HF and tubular membranes one can achieve, respectively, the greatest and smallest membrane surface area per unit volume. For this reason, HF membranes tend to be favored for module development [8,9]. However, the small hollow-fiber diameter and the densely packed bundle of HF lead to operating problems; (i.e., flow maldistribution at shell-side, increased pressure drop and related energy consumption, increased fouling, and difficulty of membrane cleaning) [4,10]. Therefore, overall, *capillary* membranes seem to hold advantages over other types for developing *VMD modules* for practical applications of a broad spectrum, including processing and separations of various aqueous product and effluent streams [4,7]. In general, very recent studies (e.g., [8–10]) suggest that further R&D work is needed to develop HF modules for practical applications.

A significant number of experimental studies have reported on various applications of VMD. These studies include the treatment of high-salinity solutions [11], desalination brines [12,13], ethanol-water separation [14,15], recovery of volatile organic compounds (e.g., ethanol, butanol, tert-methyl-butyl-ester, ethyl acetate, chloroform) from water [16–18], recovery of aromatic compounds from fruit [19,20], the concentration of juices [20] and of ginseng extracts [21,22], and the removal of radioactive compounds (e.g., cesium, strontium) [23,24], fluoride [25], arsenic [26,27], and pesticides [28] from water. Vacuum membrane distillation has also been tested for the treatment of heavily polluted wastewater such as olive mill waste-water [29], mining water [30,31], and dye solutions [32]. Although useful insights have been gained from the above studies, the aforementioned issues of selecting appropriate membrane type and module design for VMD have not been resolved.

The above considerations have motivated the authors' work toward development of a VMD module employing *capillary* membranes, for various applications. Experiments reported herein, to explore such VMD applications, include water desalination, ethanol-water separation, and recovery of a concentrated natural bio-ethanol solution from fermented winery by-products. A simple model of the VMD process was also developed focusing on utilization of the experimental data for assessment of the membrane properties. Temperature polarization was taken into account, whereas a linearization of the temperature profile along the flow allowed the use of the average temperature in the model, leading to a dimensionality reduction. The model was employed to interpret the experimental data and extract appropriate membrane permeability values [4,7].

## 2. Materials and Methods

### 2.1. Membrane Module

Commercial microporous hydrophobic polypropylene capillary membranes (Accurel® PP S6/2, Membrana Gmbh, Wuppertal, Germany) were employed in this study. The particular membranes, characterized by satisfactory durability and overall good performance, have been widely used in various literature studies [33–39]. The characteristics of the membranes as provided by the manufacturer and related literature [40] are summarized in Table A1 of the Appendix A. The MD experiments were performed with custom-made membrane contactors. The type of membrane module used is considered to be of practical interest as it can be readily up-scaled to a larger size and exhibits a rather small module-volume to membrane-surface ratio [3]. Two types of membrane modules were constructed with Plexiglas (poly-methyl methacrylate (PMMA)) housing. In the first module (M1), the vacuum was applied at the side of the shell, whereas in the second (M2), a perforated tube was placed in the core of the module for the suction of vapors. In some cases, a certain number of fibers of the membrane module M1 had to be inactivated to reduce the active membrane surface (M1A and M1B) for parametric studies. The feed solution (hot stream) was fed in the lumen of the membrane fibers. The technical data of the membrane modules are listed in Table A1 (Appendix A). A picture and a sketch of the membrane modules are included in Figure 2.

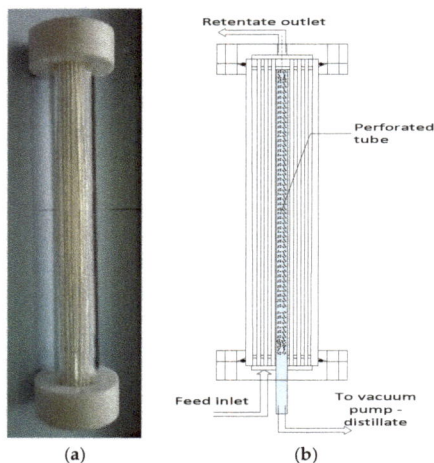

(a)  (b)

**Figure 2.** (a) A view of M1 custom-made vacuum membrane distillation (VMD) module; (b) a cross-sectional view of M2 VMD module.

### 2.2. Experimental Set-Up and Experimental Protocol

All VMD experiments were conducted in a laboratory-pilot experimental set-up, which is shown in Figure 3. This experimental set-up consisted of the "warm loop" of the feed solution and the "vacuum loop" for the recovery of the volatile component vapors from the membrane module. The warm loop was comprised of a thermostatic bath heated with a heating coil (D-79219, IKA® WERKE GMBH KG, Staufen, Germany), a primary and a secondary feed vessel, a magnetic centrifugal pump (MD-15R-230QS, IWAKI Co. Ltd., Tokyo, Japan), and a piston pump (QD, Fluid Metering Inc., Syosset, Long Island, NY, USA). The distillate side consisted of a diaphragm vacuum pump (MVP 070-3, Pfeifer Vacuum GmbH, Aßlar, Germany), a condenser where a refrigerant (ethylene glycol solution) was circulated through a chiller (LS52M21A110E, PolyScience, Niles, IL, USA), and a liquid nitrogen cold trap. Parameters such as inlet and outlet temperature, pressure, and flow rates were monitored and recorded employing a data acquisition program (GeniDAQ, Advantech Co. Ltd., Taipei, Taiwan). The membrane module was also connected to a nitrogen gas supply for drying.

The feed solution to be separated was placed in the primary feed tank, which was kept at a constant temperature using a water bath. This container could also be used as a bioreactor for continuous fermentation and in situ separation of the produced bioethanol from the fermentation broth. The feed solution was pumped via a centrifugal pump to the membrane module, passing through the fiber lumen and returned to the container. The volatile components of the feed solution tended to evaporate at the membrane surface, and the vapors were transferred (through the membrane pores) to the shell side of the membrane module by the action of vacuum. The vapor was condensed by means of a condenser and a liquid nitrogen cold-trap. The distillate solution was collected in a closed container. Due to the removal of distillate, the level in the main feed tank decreased; therefore, a level sensor, fitted to the main tank, activated the piston pump to draw solution from the auxiliary container and maintain a constant level. The fluid mass of the auxiliary container, continuously recorded with an electronic balance, provided the data required to determine the system productivity. Specifically, the permeate flux, providing a measure of system productivity, was computed as follows:

$$J = -\frac{m_{i+1} - m_i}{A\,(t_{i+1} - t_i)} \tag{1}$$

where $m_i$ is the mass in the secondary vessel at time $t_i$, $m_{i+1}$ is the mass in the auxiliary vessel at time $t_{i+1}$, and $A$ the active membrane surface area of the module.

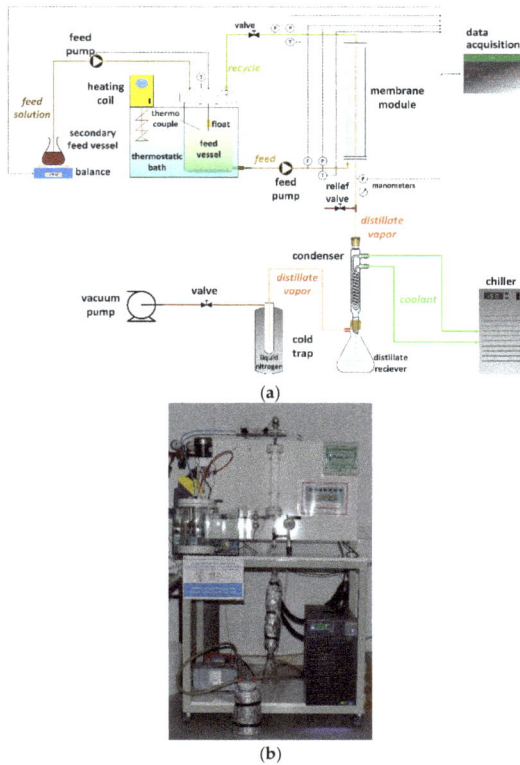

(a)

(b)

**Figure 3.** Experimental set-up: (**a**) flow diagram and (**b**) frontal view.

Usually, in the auxiliary container, a solution with the characteristics of the distillate was placed so that the concentration in the main container remained nearly constant, thus facilitating continuous operation. In some cases, the same feed solution was also placed in the auxiliary vessel (semi-continuous operation) or there was no supply to the primary vessel (batch operation). In the batch experiment, the permeate flux was determined by the overall reduction of the feed solution.

### 2.3. Types of Feed-Solutions and Experimental Conditions

To assess the VMD process, various types of feed solutions were used, such as deionized water, synthetic brackish water, ethanol binary solution, and products of fermentation of winemaking by-products for the recovery of bioethanol, as listed in Table 1. The synthetic brackish water solutions were prepared by dissolving ACS grade NaCl (Sigma–Aldrich Co., St. Louis, MO, USA) in deionized (Milli-Q, Merck Millipore, Burlington, MA, USA) water. The binary ethanol-water mixtures were prepared by dissolving denatured ethanol (Sigma–Aldrich Co., St. Louis, MO, USA) in deionized (Milli-Q) water.

**Table 1.** Types of feed solutions and experimental conditions.

| Feed Solution | Average Concentration | Type of Module | Cross Flow Velocity | Temperature | Vacuum, Mbar | Mode |
|---|---|---|---|---|---|---|
| Deionized water | - | M2 | 0.2 m/s | 34.4–58.1 °C | 28–50 | Continuous |
| Synthetic brackish water | 3000 mg/L NaCl eC = 5733 ± 155 µS/cm | M1A M1B | 0.2 m/s | 42.3–65.7 °C | 60–130 | Continuous |
| Ethanol-water mixtures | 5.2 ± 0.5% $v/v$ 12.3 ± 1.6% $v/v$ | M1B M1A | 0.2 m/s | 26.7–47.0 °C 30.4–40.7 °C | 62–114 | Continuous |
| Fermented broth + UF [1] | 5.4 ± 0.5% $v/v$ | M1B | 0.2 m/s | 29.3–38.2 °C | - | Semi-continuous |
| Fermented broth + UF | 6.5 ± 0.5% $v/v$ | M1B | 0.2 m/s | 33.0 °C | - | Batch |
| Fermented broth + UF + NF [2] | 5.1 ± 0.5% $v/v$ | M1B | 0.2 m/s | 36.1 °C | - | Batch |
| Fermented broth + UF | 12.5 ± 1% $v/v$ | M1A | 0.2 m/s | 34.0 °C | 65–122 | Semi-continuous |
| Fermented broth + UF + NF | 10.5 ± 1% $v/v$ | M1A | 0.2 m/s | 31.0 °C | - | Semi-continuous |
| Distillate | 14.0 ± 1% $v/v$ | M1A | 0.1 m/s | 32.0 °C | - | Semi-continuous |

Note: [1] UF: Ultrafiltration, [2] NF: Nanofiltration.

### 2.4. Cleaning Procedure

After the MD tests, the experimental set-up was thoroughly rinsed with deionized (DI) water for 30 min. At the end of the fermented broth experiments, the equipment was cleaned with sodium hydroxide solution (10 g/L NaOH) for 15 min and then rinsed with deionized water until the rinsing water became neutral for about half an hour. This protocol, employed in previous studies [37,41], aimed to remove organic foulants from the membranes. After the cleaning procedure, the membrane module was dried with air or dinitrogen [40].

### 2.5. Analytical Methods

During the VMD tests, samples from the feed, the distillate, and the concentrate were collected. The conductivity was measured via a laboratory multi-parameter meter (inoLab 750 pH/ION/Cond multilab, WTW, Weilheim, Germany). The concentration of ethanol (% EtOH $v/v$) was analyzed with high-pressure liquid chromatography (HPLC, Agilent Technologies, Santa Clara, CA, USA) employing a Refractive Index Detector and Hi-Plex H, 300 × 7.7 mm column (Agilent Technologies). The membrane salt rejection $R$ is given as [42]:

$$R = \frac{C_p - C_f}{C_p} = 1 - \frac{C_p}{C_f} \qquad (2)$$

where $C_p$ and $C_f$ are the permeate and the feed conductivities, respectively.

The selectivity of the volatile component *A* from a mixture is given as follows [43]:

$$S = \frac{(wt.\,\%A/wt.\,\%B)_{distillate}}{(wt.\,\%A/wt.\,\%B)_{feed}} \tag{3}$$

where *wt. %A* and *wt. %B* are the mass fraction of component *A* and *B*, respectively.

## 3. Results

### 3.1. Deionized Water

These experiments were conducted with the membrane module M2. Figure 4a depicts the distillate flux, in a steady state, as a function of the average temperature of the feed-water (i.e., the average inlet and outlet fluid temperature). As expected, the distillate flux expressed in kg/m² h, increases with the feed temperature. In Figure A1 of the Appendix B, the variation of the permeate flux with the experiment time of a test with deionized water is observed; the average value at the steady state condition was used for data interpretation. At 34.4 °C, the flux was 2.2 kg/m² h, increasing to 8.7 kg/m² h for temperature 58.1 °C. In Figure 4b, the variation is plotted of the permeate flux versus the driving force (i.e., the difference of the applied vacuum minus the feed vapor pressure (Δp)). The vapor pressure of the feed solution was determined as in Reference [44].

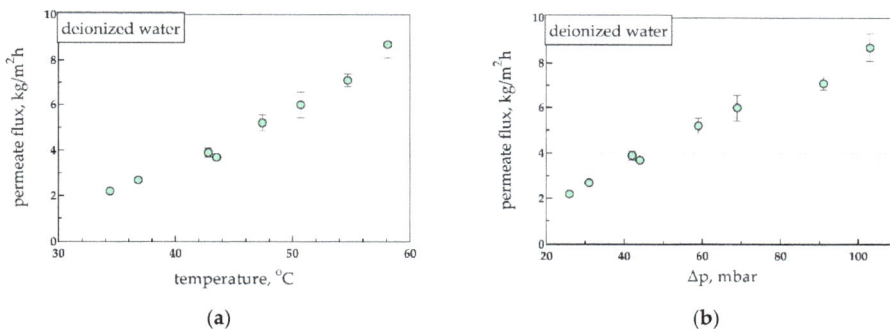

**Figure 4.** Permeate flux versus: (**a**) the feed solution temperature, and (**b**) the driving force (vapor pressure-applied vacuum). Data from deionized water distillation.

### 3.2. Desalination

Desalination experiments were performed with a synthetic brackish water solution using the membrane module M1 (types M1A and M1B). In some experiments, the element with 44 fibers was utilized (i.e., with an active surface area 0.087 m² (M1A)), while in other experiments, the element with 34 active fibers was used, having an active surface area 0.067 m² (M1B).

In all experiments conducted, it was observed that the salt rejection (as calculated from the conductivity measurements) was very high (>99.8%), and almost complete removal of NaCl was achieved. The high level of rejection indicates that the membrane performed well and was not wetted.

In Figure 5, the permeate flux is plotted for both types of experiments, with module M1A and module M1B. It is observed that the permeate flux increased with temperature and that the module with a smaller active area (M1B) performed better than the other (M1A).

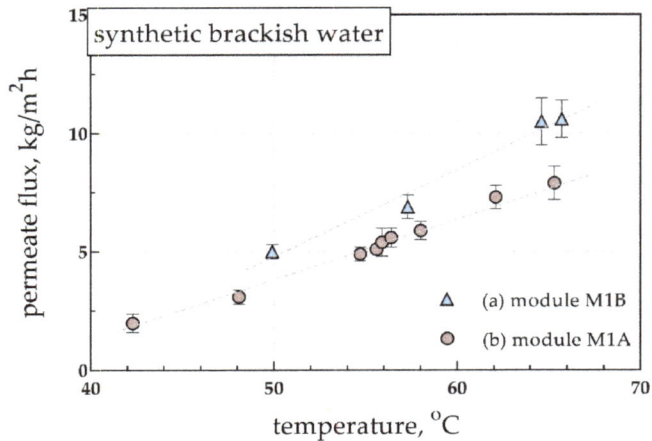

**Figure 5.** Permeate flux versus the feed solution temperature for the desalination of synthetic brackish water. Effective membrane area (**a**) module M1B, 0.067 m$^2$, and (**b**) module M1A, 0.087 m$^2$.

### 3.3. Synthetic Ethanol-Water Solutions

Binary mixtures of ethanol-deionized water were used to study the recovery of volatile components from mixtures. The experiments were conducted in a continuous mode of operation by filling the auxiliary vessel with a solution of higher concentration. Two types of feed solutions, of average ethanol concentrations 5.2% $v/v$ and 12.3% $v/v$, were used. The experiments were carried out at various temperatures, considering the average inlet and outlet temperatures of the membrane element at steady state. The performance data of these tests are included in Table 2.

**Table 2.** Conditions and performance data of the ethanol concentration tests.

| Average Feed Concentration, % $v/v$ | Average Temperature Range, °C | Average Distillate Concentration, % $v/v$ | Selectivity | Concentration Factor | Flux, kg/m$^2$ h |
|---|---|---|---|---|---|
| 5.2 | 26.7−47.0 | 29.3 ± 3.3 | 7.4 ± 1.0 | 5.6 ± 0.6 | 0.6−1.7 |
| 12.3 | 30.4−40.7 | 49.4 ± 3.8 | 6.7 ± 0.7 | 4.1 ± 0.4 | 1.0−3.2 |

For solutions with an initial concentration of 5.2% $v/v$, an average ethanol concentration factor 5.6 was estimated, corresponding to 7.4 selectivity. However, the solutions with a higher initial concentration (12.3% $v/v$) exhibited a smaller concentration factor of ethanol (i.e., 4.1), but the selectivity remained at the same level (i.e., 6.7), with the ethanol concentration reaching the value 50% $v/v$ in the distillate. As shown in Figure 6, the distillate flux increased with the feed temperature and ethanol concentration; in the case of 12.3% $v/v$ ethanol feed, the flux was somewhat higher than in the case with 5.2% $v/v$ ethanol feed, which is attributed to its higher vapor pressure. Despite the fact that the membrane area in the two types of experiments were different, the permeate flux was relatively close.

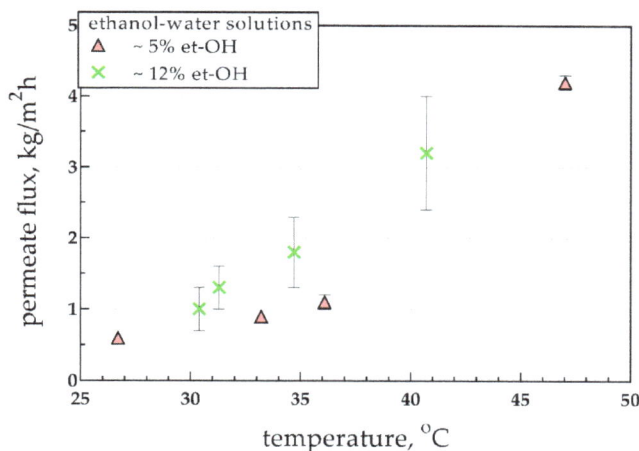

**Figure 6.** Permeate flux versus the feed solution temperature for ethanol-water mixtures distillation.

### 3.4. Fermented Broth Solutions

In the last series of experiments, bioethanol recovery from the fermentation products of winemaking by-products was studied. For the fermentation of the reducing sugars from winery by-products, a membrane bioreactor (MBR) with submerged ultrafiltration (UF) membranes was used. The final broth concentrations were in the range of 5.1 to 12.5% $v/v$. Two types of experiments were conducted: in a semi-continuous and a batch operation. In the batch experiment, the permeate flux was determined through the overall reduction of the feed solution. The performance data of the ethanol recovery from the fermented broth solutions are listed in Table 3. For all solutions used, a concentration factor in the range of 2.6 to 4.2 was observed. The degree of concentration of ethanol appeared to depend on the duration of distillation, as the concentration in the feed vessel was constantly decreasing. However, the selectivity of all the experiments was high, reaching the value of eight.

**Table 3.** Performance data for the recovery of ethanol from fermented broth solutions.

| Feed Solution | Mode | Feed Concentration, % $v/v$ | Temperature, °C | Flux, kg/m² h | Concentration Factor | Selectivity |
|---|---|---|---|---|---|---|
| Broth Fermented in MBR [1] | Semi-continuous | 5.4 | 29.3–38.2 | 1.2–2.1 | 2.9–4.2 | 5.3–7.7 |
| Broth Fermented in MBR | Batch | 6.5 | 33.0 ± 0.5 | 1.7 | 2.6 | 5.5 |
| Broth Fermented in MBR + NF | Batch | 5.1 | 36.5 ± 3.5 | 1.3 | 3.4 | 4.2 |
| Fermented Broth + UF | Semi-continuous | 12.5 | 34.0 ± 0.3 | 1.9 | 3.4 | 6.7 |
| Fermented Broth + UF + NF | Semi-continuous | 10.5 | 31 | 0.5 | 3.8 | 8 |

Note: [1] MBR: Membrane bioreactor.

For the semi-continuous experiments, as shown in Figure A2 (Appendix B), after the system attained a nearly steady-state, a rather small flux reduction occurred thereafter, due to the reduction of the ethanol concentration inside the feed vessel. The permeate flux slightly increased relative to the cases where the feed had a lower initial concentration, which is attributed to the higher content of the feed in bioethanol. Increased feed temperature led to permeate flux increase and reduction of ethanol selectivity (Figure 7a). The reduction of the selectivity is attributed to the fact that the vapor pressure of water tends to increase with increasing temperature, and therefore greater quantities of water evaporate. Furthermore, in comparison with the ethanol-water experiments (Figure 7b), there were no substantial differences between the ethanol-water solution and the fermented broth solution regarding the flux variation as a function of the temperature. Therefore, the VMD system tested did not appear to be sensitive to the type of feed solution.

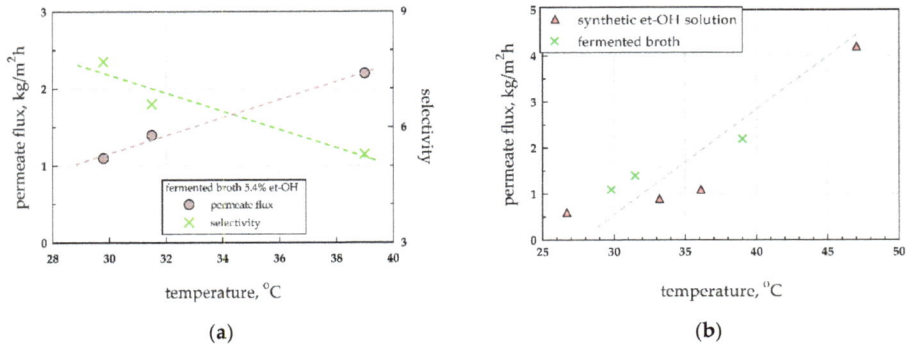

(a)                                            (b)

**Figure 7.** (a) Permeate flux and selectivity versus the feed solution temperature for fermented broth solutions; average ethanol concentration 5.4% *v/v*; and (b) comparison of synthetic ethanol solutions and fermented broth solutions.

Figure 8 depicts the variation of ethanol concentration in the main feed vessel during the batch operation. The concentration reached an asymptote, which means that the removal of ethanol becomes more difficult as the concentration of ethanol in the feed vessel decreases. However, since the concentration of the final product was quite low, for the recovery of bioethanol at higher concentration, additional distillation/MD stages should be used. Therefore, a distillation experiment of the distillate resulting from the above experiments was carried out.

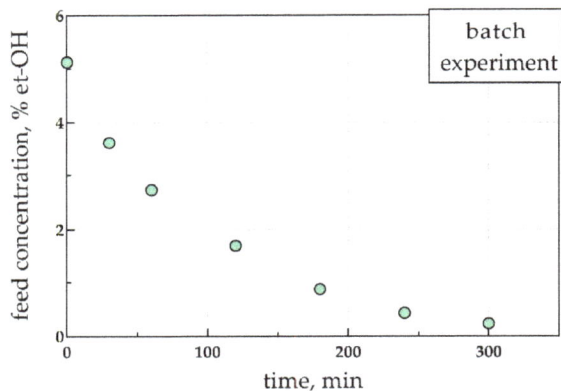

**Figure 8.** Time variation of ethanol concentration in the main feed tank for the batch operation.

*3.5. Distillate of the Fermented Broth*

In this experiment, the concentrated distillate of the fermented broth was fed to the feed vessel having a concentration of 14% *v/v* ethanol. The auxiliary container was also filled with the same solution. The performance data of this experiment are summarized in Table 4.

Table 4. Performance data of the recovery of ethanol from the distillate.

| Feed Solution | Mode | Feed Concentration, % v/v | Temperature, °C | Flux, kg/m² h | Concentration Factor | Selectivity |
|---|---|---|---|---|---|---|
| Distillate | Semi-continuous | 14 | 33 | 1.7 | 3.4 | 6 |

It is noted that for solutions with an initial concentration close to 14% $v/v$, the concentration of ethanol is 3.4 times. In this case, the experiment lasted about an hour (i.e., far less than the previous experiments). The permeate flux was higher due to the elevated initial ethanol concentration, despite the fact that the temperature of the experiment was quite low (32 °C). In experiments under similar conditions, at a temperature of 32 °C, and for an initial concentration of ethanol of 5.7% $v/v$, the permeate flux was 1.4 kg/m² h. This is because ethanol is more volatile than water, and therefore, for a feed at a higher concentration, the productivity of the distillation will increase.

## 4. Theoretical Analysis and Discussion

The experimental data for simple feed and continuous operation can be used to characterize the membrane through a simple mathematical model developed for data interpretation purposes. Mathematical modeling of the VMD process is much simpler than that of the other MD processes, since the conditions in the permeate channel are relatively easy to treat. In the small-dimension tubular module employed here, the flow field was laminar, which is also easier to model compared to flow fields with complicated geometries (e.g., spacer-filled membrane channels). The basic phenomenon in VMD is the liquid evaporation. This alters both the local concentration field on the membrane (i.e., uneven evaporation rate of the volatile species) and the local temperature field, where heat is released due to phase change. Additionally, evaporation leads to a transversely variable flow field in the tube, imposed by mass continuity. Consequently, the momentum, temperature, and species conservation equations must be solved in the two-dimensional cylindrical geometry of membrane module [7]. A first step of simplification is to reduce the dimensionality of the problem by developing cross-sectionally averaged conservation equations. In this respect, a closure is needed between the average cross-sectional properties (i.e., concentrations and temperatures) and the boundary (i.e., membranes surface) values that account for the evaporation rate. There are both concentration and temperature polarization issues in MD that adversely affect the process efficiency. The required closure for polarization is typically treated by using the film theory with the corresponding heat/mass transfer coefficients for laminar flow [15]. The resulting one-dimensional model is the one typically used in the literature to simulate the vacuum distillation process. In this particular case, further simplification is possible (as follows) by inspection of the experimental data:

1.  In all performed experiments, the reduction of the liquid mass along the flow was less than 1%, thus, it can be safely ignored for the purpose of experimental data analysis.
2.  In all experiments, the temperature difference between inlet and outlet flows never exceeded 5 °C. This means that the complete problem can be linearized around the average temperature along the flow. The linearization is certainly accurate for the required interval of only 2.5 °C (in the worst case). Then the temperature can be assumed to be uniform along the flow and equal to its average, which is calculated as the mean value of measured inlet and outlet temperatures. This approach is very accurate since the error is proportional to the second order of the Taylor expansion of the vapor pressure–temperature function which is very small for small values of $\Delta T$.
3.  Regarding temperature polarization, the temperature difference between the cup-mixing temperature and the membrane surface temperature is given by $J\Delta H/h_t$ [45], where $J$ is the evaporation flux, $\Delta H$ the evaporation enthalpy, and $h_t$ the heat transfer coefficient. It is noted that, in cases of more than one evaporating species, the numerator must be replaced by a sum over the species. A simple computation shows that this difference is of the order of 0.1 °C, thus, it can be ignored since the experimental error is certainly larger.

Regarding concentration polarization, the corresponding modulus is computed as $\exp(J\rho/h_m)$, where $h_m$ is the mass transfer coefficient and $\rho$ is the liquid density. A simple calculation shows that for the present experiments, it never exceeded 1.03, so it can be safely ignored.

The final issue to be addressed is the relation between conditions on the membrane surface and of the evaporation flux. The flow of vapor through the membrane pores can be dominated by the Knudsen mechanism (i.e., pore-vapor molecule collisions) or by the viscous (Poiseulle) flow mechanism (due to molecular collisions). In many cases, a combination of the two mechanisms (dusty gas model) is used [15]. The relative importance of the two mechanisms can be examined on the basis of the value of the ratio of molecular mean free path to pore diameter. For the membrane used here this ratio was significantly larger than unity, suggesting domination of the Knudsen mechanism. The evaporation mass flux in the case of pure water is given as [15]:

$$J = Km_w^{0.5}T^{-0.5}(P_w(T) - P_o) \tag{4}$$

where $m_w$ is the water molecular weight, $T$ is the average temperature, $P_w$ is the water vapor pressure, $P_o$ is the vacuum (permeate) pressure, and $K$ a parameter, designated as permeability, that depends on the membrane structure. Specifically, in terms of Knudsen flow theory [15], the parameter $K$ is given as $K = (2/3)(8/\pi)^{0.5}\varepsilon r/(\delta\tau R_g^{0.5})$, where $R_g$ is the ideal gas constant, $\delta$ the membrane thickness, $r$ the average pore radius, $\varepsilon$ the membrane porosity, and $\tau$ the membrane tortuosity. The value of $K$ which represents very well the present data for pure water is 0.36 kg/(m²·h·mbar·K$^{0.5}$). Comparison between predicted and experimental vapor flux data is shown in Figure 9. The fitting success is a confirmation that the Knudsen mechanism prevails, since for viscous flow the pressure dependence would be completely different. In the case of saline water, the only variation in the above equation is the multiplication of the vapor pressure by $(1 - x)$, where x is the salt molar fraction in the feed. The salt concentration for the present experiments was too small to influence the evaporation rate, thus, Equation (4) was successfully applied to the brackish water data. The deviation between experimental and computed fluxes was smaller than 3%. The fluxes appear somewhat different in Figures 4 and 5 due to differences in applied vacuum pressure.

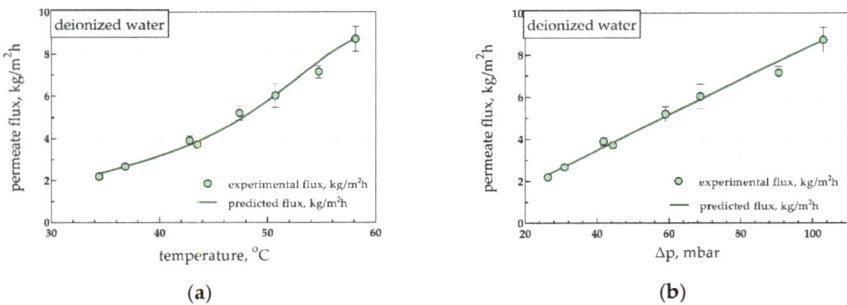

**Figure 9.** Comparison between values predicted by the model and experimental vapor flux data. Permeate flux versus: (**a**) the feed solution temperature, and (**b**) the driving force (vapor pressure-applied vacuum). Data from deionized water distillation.

To proceed from data analysis to predictive modeling, one should address the question what would be the flux for the particular membrane module for arbitrary feed mass flowrate $F$ (but of the same order of magnitude as the experimental one), feed temperature $T_{in}$, salt molar fraction x, and vacuum pressure $P_o$. Based on the preceding discussion, the model takes the form:

$$J = Km_w^{0.5}T^{-0.5}[(1 - x)P_w(T) - P_o] \tag{5}$$

$$T = (T_{in} + T_{out})/2 \tag{6}$$

$$T_{out} - T_{in} = \Delta H J A / F c_p \tag{7}$$

This model comprises a non-linear system of three algebraic differential equations. The different procedure in direct modeling, compared to the preceding data analysis, is that $T_{out}$ is an output rather than an input variable.

One can now apply the integrated method to the system ethanol-water. The generalization here is that both feed substances in solution undergo evaporation, and the permeate is a binary vapor mixture, instead of a pure vapor. The small heat of evaporation of ethanol and the low evaporation fluxes lead to a temperature decrease along the flow approximately 1 °C; in this case, the data analysis and the predictive models essentially coincide. The only difference can be the input temperature used (i.e., average for data analysis, inlet for predictions). By denoting with $x$ the ethanol molar fraction in the feed and with $y$ the same fraction in permeate, the water and ethanol evaporation fluxes (assuming Knudsen regime flow) are given as:

$$J_w = K m_w^{0.5} T^{-0.5} \cdot [P_w(T,x) - (1-y)P_o] \quad \text{(water flux)} \tag{8}$$

$$J_E = K m_e^{0.5} T^{-0.5} \cdot [P_e(T,x) - yP_o] \quad \text{(ethanol flux)} \tag{9}$$

where $P_w$ and $P_e$ are the water and ethanol vapor pressures, respectively, as functions of liquid composition and temperature. The molar weight of ethanol is denoted as $m_e$. If $y_w$ is the weight fraction of ethanol in permeate, then it can be written by employing Equations (8) and (9) as:

$$\frac{y_w}{1 - y_w} = \left(\frac{m_e}{m_w}\right)^{0.5} \frac{P_e(T,x) - yP_o}{P_w(T,x) - (1-y)P_o} \tag{10}$$

The closure is achieved by the following relation:

$$y_w = y m_e / [(1-y)m_w + y m_e] \tag{11}$$

The only unknown in the system of Equations (10) and (11) is $y$. The above equations actually represent the mathematical model for conditions similar to those of the used membrane module, and in the particular range of feed concentrations and temperatures. Employing $x$ and $y$ to compute selectivity and comparing to the respective experimental data, leads to a maximum deviation of 4%. This result is significant, meaning that, despite the smaller mean molecular free path of ethanol compared to that of water, the membrane is in the Knudsen regime for ethanol as well. It appears that for the type of membranes used in this VMD module (with very small pores), a satisfactory membrane characterization can be obtained by performing only *pure water MD* experiments. Therefore, the membrane permeability $K$ thus obtained can be employed to estimate the MD module performance for different feed types or to model more complicated operating systems of practical interest.

## 5. Conclusions

A shell-and-tube type module, comprised of capillary polypropylene membranes, was investigated for separations implementing the vacuum membrane distillation process. Employing a laboratory pilot system, based on this type of module, a variety of operating conditions and feed compositions of increasing complexity were tested (i.e., from pure water to fermented broth solutions). In all cases, the separation performance of the particular module was quite good, indicating its potential for larger-scale applications. A simplified model for the case of simple feed and continuous operation experiments was developed. Through this model it was found that the membrane can be characterized on the basis of pure water VMD experiments through a permeability-parameter $K$ value. For the particular membrane and module type employed, the data clearly show that the Knudsen mechanism of vapor permeation prevailed in the VMD tests performed. Furthermore, using the predetermined permeability $K$ value and adapting the model, a fair estimate of the membrane performance can be obtained, for any feed conditions within the range investigated. The present results suggest that the simple model and methodology developed in this work can be the basis for the design of complex separation units based on the particular VMD module-type employed here. Additional R&D work along these lines is necessary.

**Author Contributions:** A.K. (Anastasios Karabelas) and A.K. (Anthoula Karanasiou) conceived and designed the modules and the experiments; A.K. (Anthoula Karanasiou) performed the experiments; A.K. (Anthoula Karanasiou) and M.K. analyzed the data; M.K. developed the model; and all three authors wrote the paper.

**Acknowledgments:** This research has been partially funded by Project "Winery wastes exploitation for production of high added value products by environment-friendly technologies—WinWaPro", Project No. 11 ΣΥΝ_2_1992, March 2013 to October 2015, co-funded by the Hellenic Republic and the European Union—European Regional Development Fund, in the context of the O.P. Competitiveness & Entrepreneurship (EPAN II). No funds were received for publishing in open access. The authors acknowledge with thanks, the contribution of Asterios Lekkas, expert technologist in the authors NRRE Laboratory at CPERI-CERTH, in the fabrication of the VMD pilot unit.

**Conflicts of Interest:** The authors declare no conflict of interest. The founding sponsors had no role in the design of the study; in the collection, analyses, or interpretation of data; in the writing of the manuscript, and in the decision to publish the results.

## Appendix A

**Table A1.** Characteristics of the hydrophobic membranes (Accurel® PP S6/2) and the membrane module.

| Membrane Chemical Composition | PP |
|---|---|
| **membrane physical properties** | |
| thickness | 450 μm ± 50 μm |
| inside diameter, $d_{in}$ | 1800 μm ± 150 μm |
| nominal pore diameter, $d_p$ | 0.22 μm |
| maximum pore diameter | 0.6 μm |
| membrane porosity, $\varepsilon_m$ | 73% |
| **membrane performance characteristics** | |
| bubble point (isopropyl alcohol, 23 °C) | 0.95 bar |
| transmembrane flow (isopropyl alcohol, 23 °C) | $\geq$2.1 mL/(min cm$^2$ bar) |
| bacterial retension (brevundimonas diminuta) | $\geq$7 log reduction value |
| **module characteristics** | |
| **M1** | **M2** |
| mode of operation: inside out | Inside out |
| shell diameter: 40 mm | 40 mm |
| number of fibers: M1A: 44, M1B: 34 | 40 |
| fiber length: 350 mm | 350 mm |
| effective area (din): M1A: 0.087 m$^2$, M1B: 0.067 m$^2$ | 0.079 m$^2$ |

## Appendix B

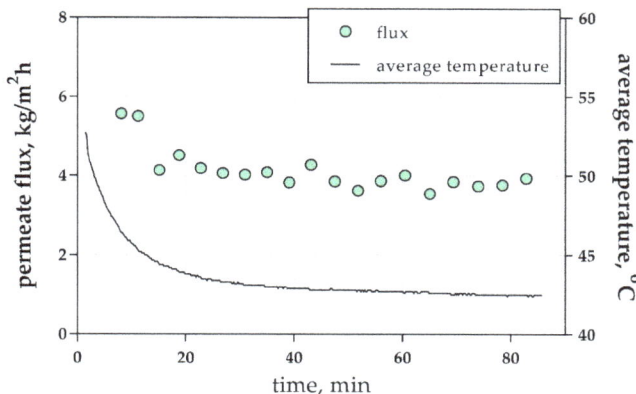

**Figure A1.** Permeate flux and average feed temperature versus the experiment time. Test with deionized water.

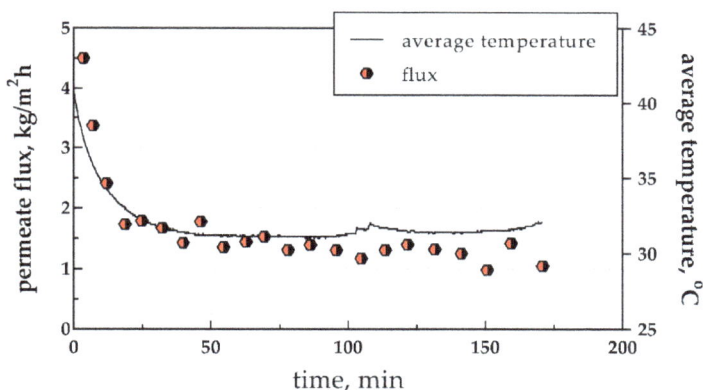

**Figure A2.** Permeate flux and average feed temperature versus the experiment time. Test with fermented broth; semi-continuous mode.

## References

1.  Drioli, E.; Ali, A.; Macedonio, F. Membrane distillation: Recent developments and perspectives. *Desalination* **2015**, *356*, 56–84. [CrossRef]
2.  Alkhudhiri, A.; Darwish, N.; Hilal, N. Membrane distillation: A comprehensive review. *Desalination* **2012**, *287*, 2–18. [CrossRef]
3.  Khayet, M.; Matsuura, T. MD membrane modules. In *Membrane Distillation*; Elsevier: Amsterdam, The Netherlands, 2011; pp. 227–247, Chapter 9.
4.  Khayet, M.; Matsuura, T. Vacuum membrane distillation. In *Membrane Distillation*; Elsevier: Amsterdam, The Netherlands, 2011; pp. 323–359, Chapter 12.
5.  Abu-Zeid, M.A.E.-R.; Zhang, Y.; Dong, H.; Zhang, L.; Chen, H.-L.; Hou, L. A comprehensive review of vacuum membrane distillation technique. *Desalination* **2015**, *356*, 1–14. [CrossRef]
6.  Susanto, H. Towards practical implementations of membrane distillation. *Chem. Eng. Process. Process Intensif.* **2011**, *50*, 139–150. [CrossRef]

7. Chiam, C.K.; Sarbatly, R. Vacuum membrane distillation processes for aqueous solution treatment—A review. *Chem. Eng. Process. Process Intensif.* **2013**, *74*, 27–54. [CrossRef]
8. Singh, D.; Li, L.; Obusckovic, G.; Chau, J.; Sirkar, K.K. Novel cylindrical cross-flow hollow fiber membrane module for direct contact membrane distillation-based desalination. *J. Membr. Sci.* **2018**, *545*, 312–322. [CrossRef]
9. Singh, D.; Sirkar, K.K. Performance of PVDF flat membranes and hollow fibers in desalination by direct contact membrane distillation at high temperatures. *Sep. Purif. Technol.* **2017**, *187*, 264–273. [CrossRef]
10. González, D.; Amigo, J.; Suárez, F. Membrane distillation: Perspectives for sustainable and improved desalination. *Renew. Sustain. Energy Rev.* **2017**, *80*, 238–259. [CrossRef]
11. Naidu, G.; Choi, Y.; Jeong, S.; Hwang, T.M.; Vigneswaran, S. Experiments and modeling of a vacuum membrane distillation for high saline water. *J. Ind. Eng. Chem.* **2014**, *20*, 2174–2183. [CrossRef]
12. Zhang, H.; Liu, M.; Sun, D.; Li, B.; Li, P. Evaluation of commercial PTFE membranes for desalination of brine water through vacuum membrane distillation. *Chem. Eng. Process. Process Intensif.* **2016**, *110*, 52–63. [CrossRef]
13. Mericq, J.P.; Laborie, S.; Cabassud, C. Vacuum membrane distillation of seawater reverse osmosis brines. *Water Res.* **2010**, *44*, 5260–5273. [CrossRef] [PubMed]
14. Rom, A.; Strommer, M.; Friedl, A. Comparison of sweepgas and vacuum membrane distillation as in-situ separation of ethanol from aqueous solutions. *Chem. Eng. Trans.* **2014**, *39*, 985–990.
15. Shi, J.Y.; Zhao, Z.P.; Zhu, C.Y. Studies on simulation and experiments of ethanol–water mixture separation by VMD using a PTFE flat membrane module. *Sep. Purif. Technol.* **2014**, *123*, 53–63. [CrossRef]
16. Sarti, G.C.; Gostoli, C.; Bandini, S. Extraction of organic components from aqueous streams by vacuum membrane distillation. *J. Membr. Sci.* **1993**, *80*, 21–33. [CrossRef]
17. Kujawski, W.; Kujawa, J.; Wierzbowska, E.; Cerneaux, S.; Bryjak, M.; Kujawski, J. Influence of hydrophobization conditions and ceramic membranes pore size on their properties in vacuum membrane distillation of water–organic solvent mixtures. *J. Membr. Sci.* **2016**, *499*, 442–451. [CrossRef]
18. Urtiaga, A.M.; Ruiz, G.; Ortiz, I. Kinetic analysis of the vacuum membrane distillation of chloroform from aqueous solutions. *J. Membr. Sci.* **2000**, *165*, 99–110. [CrossRef]
19. Bagger-Jørgensen, R.; Meyer, A.S.; Pinelo, M.; Varming, C.; Jonsson, G. Recovery of volatile fruit juice aroma compounds by membrane technology: Sweeping gas versus vacuum membrane distillation. *Innov. Food Sci. Emerg. Technol.* **2011**, *12*, 388–397. [CrossRef]
20. Hasanoğlu, A.; Rebolledo, F.; Plaza, A.; Torres, A.; Romero, J. Effect of the operating variables on the extraction and recovery of aroma compounds in an osmotic distillation process coupled to a vacuum membrane distillation system. *J. Food Eng.* **2012**, *111*, 632–641. [CrossRef]
21. Zhao, Z.P.; Ma, F.W.; Liu, W.-F.; Liu, D.Z. Concentration of ginseng extracts aqueous solution by vacuum membrane distillation. 1. Effects of operating conditions. *Desalination* **2008**, *234*, 152–157. [CrossRef]
22. Zhao, Z.P.; Zhu, C.Y.; Liu, D.Z.; Liu, W.F. Concentration of ginseng extracts aqueous solution by vacuum membrane distillation 2. Theory analysis of critical operating conditions and experimental confirmation. *Desalination* **2011**, *267*, 147–153. [CrossRef]
23. Jia, F.; Wang, J. Separation of cesium ions from aqueous solution by vacuum membrane distillation process. *Prog. Nucl. Energy* **2017**, *98*, 293–300. [CrossRef]
24. Jia, F.; Li, J.; Wang, J.; Sun, Y. Removal of strontium ions from simulated radioactive wastewater by vacuum membrane distillation. *Ann. Nucl. Energy* **2017**, *103*, 363–368. [CrossRef]
25. Plattner, J.; Naidu, G.; Wintgens, T.; Vigneswaran, S.; Kazner, C. Fluoride removal from groundwater using direct contact membrane distillation (DCMD) and vacuum enhanced DCMD (VEDCMD). *Sep. Purif. Technol.* **2017**, *180*, 125–132. [CrossRef]
26. Dao, T.D.; Laborie, S.; Cabassud, C. Direct As(iii) removal from brackish groundwater by vacuum membrane distillation: Effect of organic matter and salts on membrane fouling. *Sep. Purif. Technol.* **2016**, *157*, 35–44. [CrossRef]
27. Criscuoli, A.; Bafaro, P.; Drioli, E. Vacuum membrane distillation for purifying waters containing arsenic. *Desalination* **2013**, *323*, 17–21. [CrossRef]
28. Peydayesh, M.; Kazemi, P.; Bandegi, A.; Mohammadi, T.; Bakhtiari, O. Treatment of bentazon herbicide solutions by vacuum membrane distillation. *J. Water Process Eng.* **2015**, *8*, e17–e22. [CrossRef]

29. Carnevale, M.C.; Gnisci, E.; Hilal, J.; Criscuoli, A. Direct contact and vacuum membrane distillation application for the olive mill wastewater treatment. *Sep. Purif. Technol.* **2016**, *169*, 121–127. [CrossRef]
30. Sivakumar, M.; Ramezanianpour, M.; O'Halloran, G. Mine water treatment using a vacuum membrane distillation system. *APCBEE Procedia* **2013**, *5*, 157–162. [CrossRef]
31. Zhang, X.; Guo, Z.; Zhang, C.; Luan, J. Exploration and optimization of two-stage vacuum membrane distillation process for the treatment of saline wastewater produced by natural gas exploitation. *Desalination* **2016**, *385*, 117–125. [CrossRef]
32. Criscuoli, A.; Zhong, J.; Figoli, A.; Carnevale, M.C.; Huang, R.; Drioli, E. Treatment of dye solutions by vacuum membrane distillation. *Water Res.* **2008**, *42*, 5031–5037. [CrossRef] [PubMed]
33. Gryta, M.; Markowska-Szczupak, A.; Bastrzyk, J.; Tomczak, W. The study of membrane distillation used for separation of fermenting glycerol solutions. *J. Membr. Sci.* **2013**, *431*, 1–8. [CrossRef]
34. Tomaszewska, M.; Białończyk, L. Production of ethanol from lactose in a bioreactor integrated with membrane distillation. *Desalination* **2013**, *323*, 114–119. [CrossRef]
35. Gryta, M.; Barancewicz, M. Separation of volatile compounds from fermentation broth by membrane distillation. *Pol. J. Chem. Technol.* **2011**, *13*, 56–60. [CrossRef]
36. Tomaszewska, M.; Białończyk, L. The investigation of ethanol separation by the membrane distillation process. *Pol. J. Chem. Technol.* **2011**, *13*, 66–69. [CrossRef]
37. Lewandowicz, G.; Białas, W.; Marczewski, B.; Szymanowska, D. Application of membrane distillation for ethanol recovery during fuel ethanol production. *J. Membr. Sci.* **2011**, *375*, 212–219. [CrossRef]
38. Gryta, M. The fermentation process integrated with membrane distillation. *Sep. Purif. Technol.* **2001**, *24*, 283–296. [CrossRef]
39. Gryta, M.; Morawski, A.W.; Tomaszewska, M. Ethanol production in membrane distillation bioreactor. *Catal. Today* **2000**, *56*, 159–165. [CrossRef]
40. Gryta, M. Influence of polypropylene membrane surface porosity on the performance of membrane distillation process. *J. Membr. Sci.* **2007**, *287*, 67–78. [CrossRef]
41. Barancewicz, M.; Gryta, M. Ethanol production in a bioreactor with an integrated membrane distillation module. *Chem. Pap.* **2012**, *66*, 85–91. [CrossRef]
42. Mulder, M. *Basic Principles of Membrane Technology*; Kluwer Academic Publishers: Dordrecht, The Netherlands, 1990.
43. Smolders, K.; Franken, A.C.M. Terminology for membrane distillation. *Desalination* **1989**, *72*, 249–262. [CrossRef]
44. Poling, B.E.; Thompson, G.H.; Friend, D.G.; Rowley, R.L.; Wilding, W.V. Vapor pressures of pure substances. In *Perry's Chemical Engineers' Handbook*, 8th ed.; Green, D.W., Perry, R.H., Eds.; McGraw Hill Professional, Access Engineering: New York, NY, USA, 2008.
45. Bandini, S.; Sarti, G.C. Heat and mass transport resistances in vacuum membrane distillation per drop. *AIChE J.* **1999**, *45*, 1422–1433. [CrossRef]

*water*

MDPI

*Article*

# A Single Tube Contactor for Testing Membrane Ozonation

Garyfalia A. Zoumpouli [1,2,3], Robert Baker [1,2], Caitlin M. Taylor [1,2,3], Matthew J. Chippendale [2], Chloë Smithers [2], Sean S. X. Ho [2], Davide Mattia [2,3], Y. M. John Chew [2,3] and Jannis Wenk [2,3,*]

[1] Centre for Doctoral Training, Centre for Sustainable Chemical Technologies, University of Bath, Bath BA2 7AY, UK; g.zoumpouli@bath.ac.uk (G.A.Z.); r.baker@bath.ac.uk (R.B.); c.taylor2@bath.ac.uk (C.M.T.)
[2] Department of Chemical Engineering, University of Bath, Bath BA2 7AY, UK; mjc83@bath.ac.uk (M.J.C.); cs897@bath.ac.uk (C.S.); sxh20@bath.ac.uk (S.S.X.H.); d.mattia@bath.ac.uk (D.M.); y.m.chew@bath.ac.uk (Y.M.J.C.)
[3] Water Innovation and Research Centre (WIRC), University of Bath, Bath BA2 7AY, UK
[*] Correspondence: j.h.wenk@bath.ac.uk; Tel.: +44-1225-38-3246

Received: 13 September 2018; Accepted: 8 October 2018; Published: 10 October 2018

**Abstract:** A membrane ozonation contactor was built to investigate ozonation using tubular membranes and inform computational fluid dynamics (CFD) studies. Non-porous tubular polydimethylsiloxane (PDMS) membranes of 1.0–3.2 mm inner diameter were tested at ozone gas concentrations of 110–200 g/m$^3$ and liquid side velocities of 0.002–0.226 m/s. The dissolved ozone concentration could be adjusted to up to 14 mg $O_3$/L and increased with decreasing membrane diameter and liquid side velocity. Experimental mass transfer coefficients and molar fluxes of ozone were $2.4 \times 10^{-6}$ m/s and $1.1 \times 10^{-5}$ mol/(m$^2$ s), respectively, for the smallest membrane. CFD modelling could predict the final ozone concentrations but slightly overestimated mass transfer coefficients and molar fluxes of ozone. Model contaminant degradation experiments and UV light absorption measurements of ozonated water samples in both ozone ($O_3$) and peroxone ($H_2O_2/O_3$) reaction systems in pure water, river water, wastewater effluent, and solutions containing humic acid show that the contactor system can be used to generate information on the reactivity of ozone with different water matrices. Combining simple membrane contactors with CFD allows for prediction of ozonation performance under a variety of conditions, leading to improved bubble-less ozone systems for water treatment.

**Keywords:** ozonation; membranes; polydimethylsiloxane; mass transfer; wastewater treatment; water treatment; peroxone

---

## 1. Introduction

Ozone is a chemical oxidant used for water treatment for more than a century [1]. Ozone has a wide range of applications [2], including disinfection [3], control of disinfection byproducts [4,5], addressing taste and odour issues [6], and the removal of trace contaminants from both drinking water and during advanced wastewater treatment schemes [7–10]. Consequently, the usage of ozone has been steadily increasing for decades [11]. Ozone is an unstable gas that needs on-site production. The main costs of ozone operation relate to a combination of energy and oxygen consumption [12]. Usually, ozone is dissolved into the water phase via bubbling using various types of gas diffusers or side stream injection [13]. Knowledge and quantification of ozone mass transfer is crucial for the design of treatment operations [14]. Based on fundamental mass-transfer principles of gas–liquid systems, ozone transfer is dependent on the interfacial area, the mass transfer coefficients, the solubility of gas in liquid and the concentration gradient between the gas and the liquid phase. In practice ozone transfer is also determined by the operating conditions such as mixing, the gas diffuser type, the water matrix, and the setup of the treatment facility.

Bubble-less gas transfer by membranes permeable to ozone is an alternative to bubble-based methods [15]. Membrane contactors provide a constant interfacial surface area separating the gas phase from the liquid phase and result in readily predictable liquid flow patterns that enable straightforward control over ozone gas to liquid mass transfer. The use of membranes may result in increased process efficiencies by significantly reducing off-gas disposal volume [16,17], and diminishing practical challenges such as foaming issues [18]. The modular design of membrane contactors allows accurate responses to changing treatment needs and convenient maintenance. Particularly relevant for ozone applications is the targeted reduction of the regulated ozonation byproduct bromate that can be achieved when employing membrane contactors [19].

Despite obvious advantages and growing commercial interest for membrane ozonation, there is comparatively little available literature on ozone transfer into water through membranes, with most studies focusing on ozone as a supplementary agent within hybrid treatment processes to increase membrane performance by reducing fouling or enhancing biodegradation of contaminants in membrane bioreactors (e.g., [20–22]).

Important attributes for ozonation membranes include porosity, surface hydrophobicity, selectivity for transfer of ozone over oxygen, and stability during long-term ozone exposure. Ozone-resistant inorganic ceramic membranes are made of alumina [23–25] and porous glass [26]. To prevent wetting of the surface and flooding of membrane pores, ceramic materials usually require surface modification to increase hydrophobicity [27–29]. Most polymeric membrane materials are hydrophobic but are prone to reacting with ozone. For example, polymers such as polyethersulfone (PES) and polyetherimide (PEI) possess carbon double bonds that react with ozone, leading to decomposition [30]. Fluoropolymers such as polytetrafluoroethylene (PTFE) or polyvinylidene fluoride (PVDF) and polymers without functional groups exhibiting reactivity towards ozone such as plasticizer-free polydimethylsiloxane (PDMS) are more suitable for ozone [31]. Both PTFE and PDVF membranes have a sponge-like porous morphology and are resistant to ozone, while PDMS membranes have a dense non-porous morphology. Compared to fluoropolymers, PDMS is less resistant towards corrodants other than ozone, including UV [32–34]. Testing the long-term use of PDMS membranes in water treatment ozonation is still pending. However, given its availability in various sizes and its low purchase cost, PDMS appears as an appropriate model membrane material for smaller-scale investigations of membrane ozonation processes.

In a previous study we developed a modelling approach based on computational fluid dynamics (CFD) and convection-diffusion theory to calculate ozone and oxygen gradients and their mass transfer through non-porous PDMS membranes into the aqueous phase [35]. The known material properties of PDMS and literature values on ozone mass transfer were used as input values for the model, as our own experimental results were unavailable. The goal of this study was to create the experimental means to further verify and refine our modelling approach and enable data collection on ozonation experiments by using a simple ozone contactor system. Therefore, the objectives of this study were (1) to build a membrane contactor system as a flexible-use platform to investigate tubular membranes for ozonation; (2) to measure the mass transfer of ozone into water through different PDMS membranes; (3) to compare results with the previously developed model; and (4) to conduct tests with established probe compounds for ozonation experiments in pure water, real water samples, and under peroxone process conditions by adding hydrogen peroxide ($H_2O_2$) to the aqueous phase.

## 2. Materials and Methods

All chemicals, including solvents and analytical consumables, were purchased from commercial sources: Potassium indigo trisulfonate (CAS 67627-18-3, Sigma-Aldrich, Saint Louis, MO, USA), *para*-chlorobenzoic acid (*p*CBA, 99% purity, Acros Organics, Hampton, NH, USA), humic acid sodium salt

(CAS 68131-04-4, technical grade, Sigma Aldrich), *N,N*-diethyl-p-phenylenediamine (DPD, 99% purity, Acros Organics), and peroxidase horseradish (≥85 units/mgdw, Alfa Aesar, Haverhill, MA, USA).

Ultrapure water (resistivity > 18 MΩ/cm) was used for preparing stock solutions, analysis and experiments and was produced with Milli-Q (Merck, Kenilworth, NJ, USA) or ELGA (Veolia, Paris, France) water purification systems.

## 2.1. Membrane Contactor System

A schematic and a photograph of the lab-scale, single tube membrane contactor built for this study are shown in Figure 1. A borosilicate glass tube (length 20 cm, outer diameter (OD) 22 mm, inner diameter (ID) 18 mm) with two vertical and two horizontal screw connections (SQ24 and GL14, respectively) served as a shell for the membrane contactor. Ozone-resistant impermeable perfluoroalkoxy alkane (PFA) tubing (OD 1/8 inch, Swagelok, Solon, OH, USA) and Swagelok fittings (1/8 inch) made of 316 stainless steel were used for connecting both gas lines and liquid lines. The valves used were straight-port HAM-LET H-800 ball valves (FTI Ltd., East Sussex, UK).

An ozone generator (BMT 803N, BMT Messtechnik GmbH, Berlin, Germany) fed with high purity oxygen (industrial grade, 99.5% purity, BOC, Bristol, UK) was used to produce ozone (maximum $O_3$ concentration > 250 g/Nm$^3$, or approximately 18% $w/w$). The oxygen flow rate was controlled with a rotameter (maximum 300 mL/min, FLDO3306ST, Omega, Manchester, UK). The outlet of the ozone generator was connected to an ozone analyser (maximum 200 g/m$^3$, BMT 964, BMT, Berlin, Germany). The ozone/oxygen mixture was directed to the reactor through a plug valve. The gas mixture outlet was connected to a check valve (Swagelok) to prevent back flow, and then to a heated catalyst ozone to oxygen converter (CAT-RS, BMT). A purge line was included to flush the system off ozone when required.

The influent water was pumped using a diaphragm pump (0.25–20 mL/min, FMM 20 KPDC-P, KNF, Trenton, NJ, USA) and then directed to the reactor using a plug valve. The membrane tube ran through the centre of the reactor. It was held in place at the two ends with silicone seals that provided leak-free operation. The membrane was connected to the liquid line outside the shell with push-on fittings. The water flow was from bottom to top, in counter-flow with the gas.

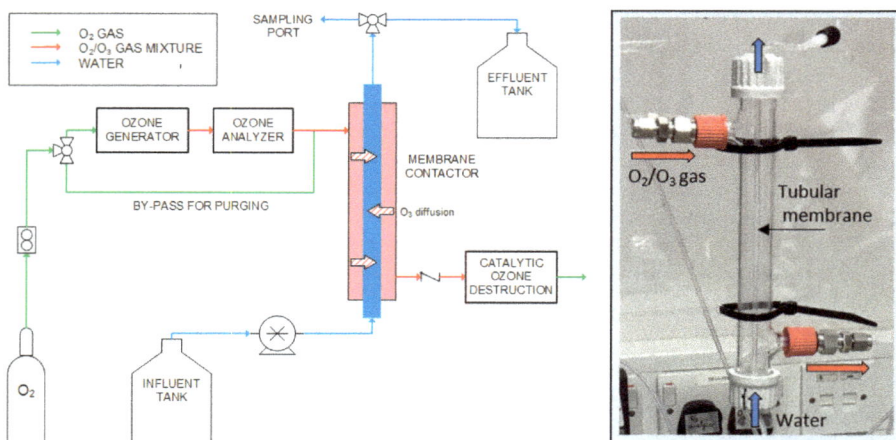

**Figure 1.** Schematic of the experimental setup and photograph of the membrane contactor.

## 2.2. Ozonation Experiments

In all the experiments, the oxygen pressure was set to approximately 0.9 bar since the maximum operational pressure of the ozone generator was 1 bar. The oxygen flow rate was set to 100 mL/min.

Experiments with different ozone concentrations in the gas phase were performed (7–14% weight $O_3$/weight gas mixture). The minimum concentration that could be achieved with the employed conditions was 110 g $O_3/m^3$ ($\pm$10%). The maximum ozone concentration was determined by the upper limit of the ozone analyser and was 200 g $O_3/m^3$. The main characteristics of the PDMS membranes used are shown in Table 1. Except for membrane longevity tests, a new membrane was used for each experiment (i.e., after a few hours of use, the membrane was replaced).

The pump flow rate was measured at the beginning of each experiment using deionized water and a balance. Based on steady-state experiments, the system was left to stabilize for at least 10 min for flow rates between 5–20 mL/min, and for at least 20 min for flow rates less than 5 mL/min, before samples were taken.

Experiments were performed with different water matrices at room temperature (21 °C $\pm$ 2 °C): (1) deionized water with or without phosphate buffer (pH = 7.1), non-spiked, or spiked with $p$CBA (concentration approximately 10 μM) or humic acid (total organic carbon, TOC of 1.3–13.7 mg/L); (2) secondary treated wastewater (wastewater effluent); or (3) river water. For peroxone experiments, the influent was supplemented with 15–100 μM $H_2O_2$.

**Table 1.** Polydimethylsiloxane (PDMS) tubing used as non-porous membranes. OD: outer diameter, ID: inner diameter.

| Material | Product Code | Supplier | OD (inch) | ID (inch) | ID (mm) | Wall Thickness (mm) |
|---|---|---|---|---|---|---|
| Silastic® (PDMS) | WZ-96115-22 | Cole Parmer | 1/4 | 1/8 | 3.2 | 1.6 |
| Silastic® (PDMS) | WZ-96155-00 | Cole Parmer | 1/8 | 1/16 | 1.6 | 0.8 |
| Silastic® (PDMS) | WZ-96115-08 | Cole Parmer | 1/12 | 1/25 | 1.0 | 0.6 |

### 2.3. Wastewater Effluent and River Water

Secondary treated wastewater was collected from a wastewater treatment plant in South-West England, UK. River water was collected from the River Avon in Bath, UK. For stabilization water samples were filtered with pre-rinsed glass microfiber filters grades GF/A or GF/C (nominal particle retention: 1.6 μm or 1.2 μm, respectively, Whatman, GE Life Sciences, Machelen, Belgium) and were stored at 4 °C until used. Water sample properties and sampling dates are shown in Table 2.

**Table 2.** Water sample characteristics. TOC: total organic carbon.

| Property | Wastewater Effluent (March 2018) | River Water I (March 2018) | River Water II (July 2017) |
|---|---|---|---|
| pH | 7.9 | 7.2 | 8.2 |
| TOC (mg/L) | 10.2 | 7.2 | 4.3 |
| UV$_{254}$ (cm$^{-1}$) | 0.14 | 0.20 | 0.10 |
| Alkalinity (mg CaCO$_3$/L) | 181 | 236 | n/a [1] |
| Nitrate (mg/L) | 31 | 22 | n/a |

[1] n/a: not analysed.

### 2.4. Analytical Methods

The concentration of dissolved ozone was measured with the indigo method [36]. The exact concentration of $H_2O_2$ in the feed water was measured with the DPD method [37]. UV-VIS measurements were performed with a UV-VIS spectrophotometer (Cary 100, Agilent Technologies, Santa Clara, CA, USA) using 1-cm quartz glass cuvettes. The pH was measured with a pH meter (FE20, Mettler Toledo). The concentration of total organic carbon (TOC), as non-purgeable organic carbon (NPOC), was measured with a TOC analyser (Shimadzu TOC 5000A, Kyoto, Japan). Alkalinity was determined by titration with 0.1 N hydrochloric acid according to ISO standard 9963-1:1994 [38]. Nitrate was measured with a HANNA® nitrate test kit (HI38050) (Woonsocket, RI, USA).

HPLC analysis of *p*CBA was performed with an Agilent HPLC System with a UV detector and an Acclaim RSLC 120 C18 column (3 μm, 120 Å, 3 × 75 mm). The mobile phase was 40% acetonitrile and 60% water with 10 mM phosphoric acid (pH = 2.5). The flow rate was 0.5 mL/min and the injection volume was 50 μL. UV absorption was measured at 240 nm.

### 2.5. Computational Modelling

Computational fluid dynamics (CFD) simulation is the generation of a numerical solution that satisfies a group of conservation equations (here, mass, momentum and species transport) over a computational domain that represents a real physical domain. The computational work was conducted using COMSOL Multiphysics V5.3 (COMSOL Inc., Shanghai, China) to determine the concentration profiles of $O_3$ (and $O_2$) in the gas, membrane, and liquid phase, so that major mass transfer resistances could be identified, i.e., using a similar approach to that reported by Berry et al. (2017) [35]. The main difference, however, was that the liquid and gas phase in this work were in the tube and shell side of the reactor respectively. The dimensions and operating conditions employed were those described in Section 2.2.

## 3. Results and Discussion

### 3.1. Ozone Concentration with Liquid Side Velocity and Membrane Size

The measured ozone concentrations at the outlet of the contactor compartment for the tested membranes at two different ozone gas concentrations and liquid side velocities ranging from 0.002 to 0.226 m/s are shown in Figure 2a. The corresponding modelled values are provided in Figure 2b, including Reynolds numbers and liquid side residence time (r.t.). The measured ozone concentration increased with smaller membrane diameter and decreasing liquid side velocity. Roughly doubling the ozone concentration from 110 g/m$^3$ to 200 g/m$^3$ in the gas phase resulted in higher dissolved ozone concentrations, especially for the thinner membranes.

Repetition experiments conducted with a different set of membranes, cut from one tube, and different flow rates indicate good reproducibility and an experimental uncertainty of ±0.2 mg/L. In addition, experiments with a preliminary batch contactor setup showed that the relative standard deviation across different experiments was smaller than 10%, and was decreasing with membrane size. Better reproducibility with smaller membranes is ascribed to their faster response upon changes of experimental conditions. At higher liquid side flow rates and larger membrane size, some initial ozone bubble formation occurred along the inside of the membrane, but bubbles dispersed with equilibration. High flow rates caused non-uniform initial internal wetting of the membrane and created hydrophobic patches across the surface with lower resistance for ozone, leading to bubble formation. Both observations underline the importance for sufficient equilibration time at system startup and when altering experimental parameters.

Modelling results based on the previously developed CFD approach [35] agree with the experimental data. The relative deviation between modelling and experimental results was normally below 15%, although higher relative differences were occasionally observed for high liquid side velocities, as the model is not applicable to transitional and turbulent flows. Similarly, for low flow rates, i.e., for Reynolds number (Re) ≤ 100, CFD slightly over-predicted the ozone concentration. A possible explanation for this disagreement is that at lower Re the dispersion of $O_3$ in the liquid phase is not uniform. Over the investigated range, this translates into an absolute deviation of less than

0.5 mg/L in the prediction of ozone concentrations, which is comparable to the experimental error of 0.2 mg/L. Generally, the modelled ozone concentration was higher than observed experimentally, indicating that the model slightly overestimates the actual ozone mass transfer.

Results confirm that liquid side velocity is the dominant parameter to determine the overall mass transfer of ozone followed by membrane thickness and ozone gas concentration. In practice, modules with bundles of thin membranes operating in parallel ensure low flow rates at large overall water fluxes to achieve required ozone concentrations [18]. Note that for data shown here both wall thickness and inner diameter simultaneously change based on the actual membrane dimensions (Table 2). Results for a hypothetical cylindrical membrane with constant inner diameter with increasing wall thickness are provided elsewhere [35].

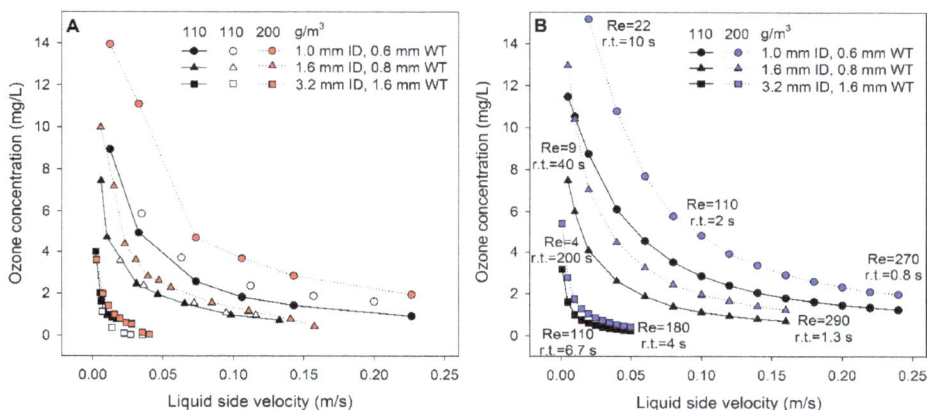

**Figure 2.** Dissolved ozone concentration in pure water in the outlet of the reactor against liquid velocity for different membranes diameters (ID: inner diameter, WT: wall thickness) and ozone gas concentrations. (**A**) Experimental results. Repeats (open symbols) were performed with a different set of membranes. (**B**) Computational fluid dynamics (CFD) modelling results, including Reynolds numbers (Re) and liquid side residence times (r.t.).

## 3.2. Overall Ozone Mass Transfer Coefficient and Molar Flux through the Membrane

To provide a better overview on mass transfer efficiency and an alternative comparison between experimental and modelling results, overall mass transfer coefficients $K_L$ and molar fluxes of ozone were calculated.

The overall mass transfer coefficient was calculated according to Equation (1):

$$K_L = \frac{u_L}{H \alpha L} \ln\left\{ \frac{\frac{C_g}{S}}{\frac{C_g}{S} - HC_{L,out}} \right\}, \tag{1}$$

where $\alpha$ is the surface area of the membrane per unit volume of liquid, L is the length of the membrane, $u_L$ is the liquid velocity, H is the solubility of ozone in water, S is the solubility of ozone in the PDMS membrane, $C_g$ is the ozone concentration in the gas phase, and $C_{L,out}$ is the measured or modelled ozone concentration at the outlet of the contactor.

The molar flux of ozone across the membrane, N, was calculated according to Equation (2):

$$N = K_L \left( \frac{C_g}{S} - HC_{L,out} \right), \tag{2}$$

The parameters used to calculate $K_L$ and N are shown in Table 3. The overall mass transfer coefficients and the molar fluxes of ozone for the three tested membranes are shown in Figure 3. Error bars for experimental data were determined using error propagation based on the uncertainties shown in Table 3. It is expected that the effect of inlet $O_3$ concentration on $K_L$ is insignificant since $K_L$ largely depends on the hydrodynamics (Re) and the transport properties of $O_3$ in water, in particular the Schmidt number (Sc) [35], which is the ratio of momentum diffusivity and mass diffusivity. In contrast, N is strongly dependent on the inlet $O_3$ concentration, which directly affects the concentration gradient. Values of N for 200 $g/m^3$ can be expected to be about twice those of 110 $g/m^3$.

Mass transfer coefficients and molar fluxes increase with increasing liquid side velocity and level off at higher velocities, while dissolved ozone concentration decreases with increasing velocity, as discussed in Section 3.1. This agrees with both experimental and computational studies employing membranes made of different materials [31,35,39,40].

**Table 3.** Parameters used for calculation of $K_L$ and N. ID: inner diameter.

| Property | Units | Value | Experimental Uncertainty ($\pm$) | Reference |
|---|---|---|---|---|
| $\alpha$ | $m^{-1}$ | 4000 (1.0 mm ID) 2500 (1.6 mm ID) 1250 (3.2 mm ID) | - | calculated |
| L | m | 0.2 | - | contactor length |
| $u_L$ | m/s | 0.002–0.224 | 0.002 | calculated |
| H | - | 0.248 | - | [41] |
| S | - | 0.881 | - | [42] |
| $C_g$ | $mol/m^3$ | 2.1–2.5 | 0.2 | measured |
| $C_{L,out}$ | $mol/m^3$ | 0.014–0.186 | 0.004 | measured (experimental) or calculated (modelled) |

In addition, the mass transfer coefficients calculated for PDMS are of the same order of magnitude as those reported for porous fluoropolymer membranes. For example, for PTFE and PVDF flat sheet membranes at Re = 60 the mass transfer coefficients of ozone ranged from $5.4 \times 10^{-6}$ m/s to $1.1 \times 10^{-5}$ m/s [31], while in a hollow fibre PVDF membrane module the mass transfer coefficients ranged from $5.30 \times 10^{-7}$ to $1.84 \times 10^{-5}$ m/s for liquid side velocities ranging from 0.01 to 0.5 m/s [39].

Mass transfer coefficient and molar flux data confirm the already discussed (Section 3.1) overestimation of ozone mass transfer by the modelling approach, which is beyond the experimental error margin. The discrepancy between modelling and experimental results appears more pronounced using the data representation of Figure 3, while the differences in actual ozone concentrations are modest from an application viewpoint. The experimental uncertainty increases for larger diameters due to the larger relative error on measured ozone concentration. In addition, at higher velocities experimental results show a decrease in both molar flux and mass transfer coefficient, which is different compared to model results and is associated to much longer equilibration times needed to achieve constant gas transfer conditions for these membrane diameters.

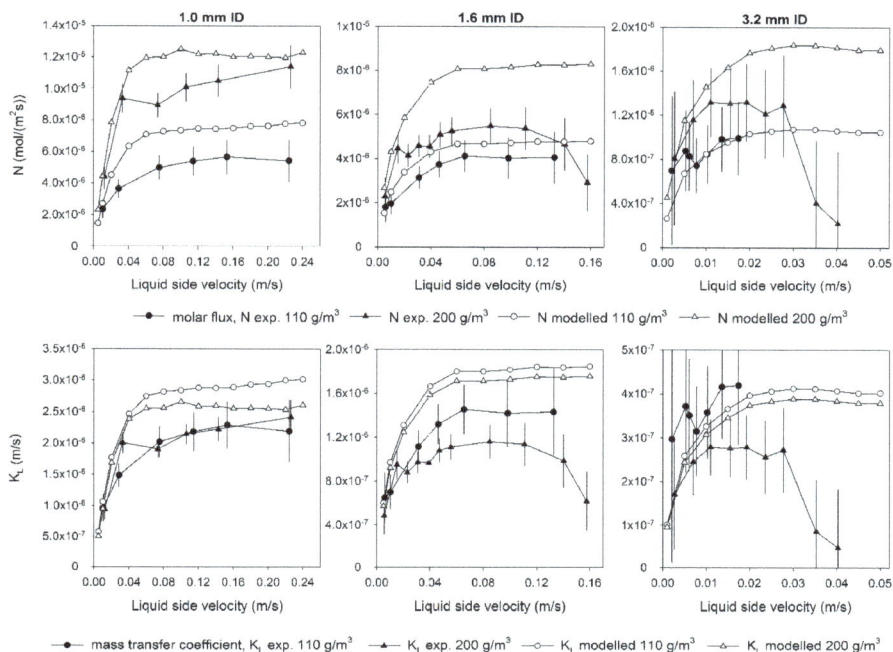

**Figure 3.** Experimental and modelled overall mass transfer coefficients $K_L$ and molar fluxes of ozone N in pure water at ozone gas concentration of 110 g/m$^3$ and 200 g/m$^3$ versus liquid side velocity.

### 3.3. Removal of pCBA by Membrane Ozonation and in Presence of Additional H$_2$O$_2$ (Peroxone Process)

The decomposition of ozone on surfaces and within the water matrix can lead to the formation of further oxidative species via radical chain reactions [43]. Ozone decomposition on the membrane surface is not considered relevant due to the low reactivity of ozone with PDMS. Ozone decomposition in water results in the formation of the OH radical, the major secondary oxidant during ozonation [44]. In contrast to ozone which is a selective oxidant, OH radicals are short lived and undergo fast reaction with most organic reactants [45]. Therefore, OH radicals play an important role in the removal of organic contaminants during ozonation because they effectively transform any organic contaminant present in water at similar reaction rates [46], while disinfection mainly occurs through ozone directly.

To assess OH radical-induced oxidation processes in the experimental setup, *p*CBA was used as an OH radical probe compound. *p*CBA is ozone-resistant [47], and an established model contaminant [48]. In addition, *p*CBA removal experiments were conducted in peroxone systems, i.e., by adding hydrogen peroxide (H$_2$O$_2$). The reaction of hydrogen peroxide with ozone creates an increased OH radical concentration [49]. The peroxone process has been widely used to improve treatment efficiency for compounds that react slowly with ozone [50].

Figure 4a shows the removal of *p*CBA in pure water for different liquid side velocities and membrane thicknesses. As expected and based on ozone concentration measurements (Section 3.1), the *p*CBA removal increased with residence time and with decreasing membrane thickness. Note that liquid side chemical reactions promote ozone transfer, by increasing the concentration gradient across the membrane.

To determine accurate mass transfer, an enhancement factor has to be considered that depends on the reaction kinetics of the target contaminants with ozone and its oxidative decomposition products [51].

The added H$_2$O$_2$ concentration to achieve an O$_3$:H$_2$O$_2$ ratio of 2:1 on a molar basis [49,52] was 15 µM (0.5 mg H$_2$O$_2$/L) and was calculated for the ozone transferred in pure water under the tested experimental conditions (1.6 mm ID membrane, v = 0.07 m/s, residence time 3 s), which was approximately 31 μM (1.5 mg $O_3$/L). A higher $H_2O_2$ concentration of 100 μM was also tested. In pure buffered water the addition of $H_2O_2$ did not lead to a significant increase in *p*CBA removal (Figure 4a). The higher $H_2O_2$ concentration tested did not have a substantial additional effect. Measurements of the residual ozone concentration confirmed that the ozone consumption was similar for both $H_2O_2$ concentrations. For lower flow rates (residence time more than 3 s) the ozone consumption was incomplete, due to the higher ozone concentration. In a PTFE hollow fibre module the increase of $H_2O_2$ concentration led to an increase in *p*CBA removal, but the experiments were performed with longer residence times and lower ozone gas concentrations [19].

The removal of *p*CBA in different types of water (Figure 4b) and a humic acid solution was lower than in pure water due to the presence of matrix compounds that act as OH radical and ozone scavengers (Table 2). With the addition of 15 μM $H_2O_2$ the *p*CBA removal showed a small increase in all four water matrices tested. Similar results have been observed in a membrane contacting system employing a ceramic membrane, where the peroxone process increased the removal of *p*CBA in river water by less than 10% [53]. A higher improvement of *p*CBA removal has been observed by $H_2O_2$ addition in batch experiments [49,54]. The difference between the conventional (batch) peroxone process and the membrane peroxone process can be attributed to the non-uniform concentration of ozone in the membrane, which influences the $O_3$:$H_2O_2$ ratio and thus the OH radical yield [53]. Adjusting flow conditions or membrane diameter to achieve a more uniform ozone concentration profile might diminish this discrepancy.

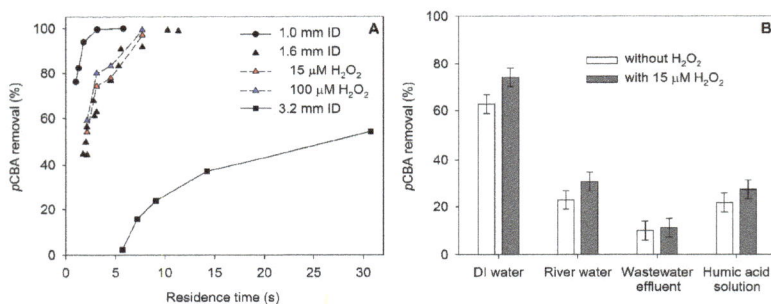

**Figure 4.** (**A**) Removal of *para*-chlorobenzoic acid (*p*CBA) in pure buffered water and with added $H_2O_2$ for different membrane sizes. Three repeats performed with a different membrane each time are shown for the 1.6 mm ID membrane without added $H_2O_2$. (**B**) Removal of *p*CBA in different water matrices, with and without addition of $H_2O_2$, for the 1.6 mm ID membrane at a residence time of three seconds. River Water I (TOC 7.2 mg/L, pH 7.2), wastewater effluent (TOC 10.2 mg/L, pH 7.9), and humic acid solution (TOC 8.3 mg/L, pH 7.1) were used. TOC: total organic carbon, ID: inner diameter.

## 3.4. Ozonation of Dissolved Organic Matter

The effect of ozone on the dissolved organic matter present in the three different water matrices (river water, wastewater effluent and humic acid solution) was determined by measuring $UV_{254}$ absorption before and after passage through the membrane module. Under low ozone doses and short residence times, ozonation increases the biodegradable fraction of dissolved organic matter, but results in little or no mineralization of organic matter (TOC removal) [55,56]. Therefore, removal of UV absorbance ($UV_{254}$) is presented instead of TOC removal, to demonstrate the transformation of organic compounds. Figure 5A provides relative changes in $UV_{254}$ and residual ozone concentrations plotted against TOC concentration. The ozone dose measured in pure water with this membrane and liquid side velocity was 2.2 mg/L (0.2 to 1.7 g $O_3$/g TOC). For humic acid, the change in absorption decreases with increasing TOC concentration, along with a decrease in residual ozone concentration.

From 3 mg/L TOC, no further reduction of the residual concentration of ozone can be observed due to the short residence time of four seconds in the reactor. At the same TOC, river water shows both a greater decrease in absorption (24%) and a higher residual ozone concentration (0.7 mg/L) than humic acid solution. This indicates that the tested river water is less reactive with ozone. The results for wastewater effluent are comparable to those for humic acid, indicating a high reactivity with ozone. In addition, the river water had a higher alkalinity than the wastewater effluent. Carbonate and bicarbonate species scavenge OH radicals in a way that inhibits the ozone decomposition cycle, leading to lower rates of ozone depletion [57], which explains the higher ozone residual in river water. The $UV_{254}$ removal of 24% at 0.3 g $O_3$/g TOC achieved for river water with PDMS membrane ozonation is comparable to a removal of 30–40% at 0.5 g $O_3$/g TOC reported for the treatment of river water with ceramic membrane ozonation [52].

Figure 5B gives the change in absorption at different liquid side velocities for river water and humic acid. The ozone dose changes with liquid side velocity as described in Section 3.1. Under the conditions employed in this experiment it ranged from 1 to 10 mg/L (measured in pure water), leading to approximate specific ozone doses of 0.2–2.8 g $O_3$/g TOC. At higher liquid side velocities, the change in absorption decreases due to the lower ozone exposure. Overall, the results for Sections 3.3 and 3.4 show that the single membrane ozonation experiments can be used to obtain information on the reactivity of the water matrix with ozone with relatively modest experimental effort as, for example, ozone exposure can be controlled via liquid side velocities.

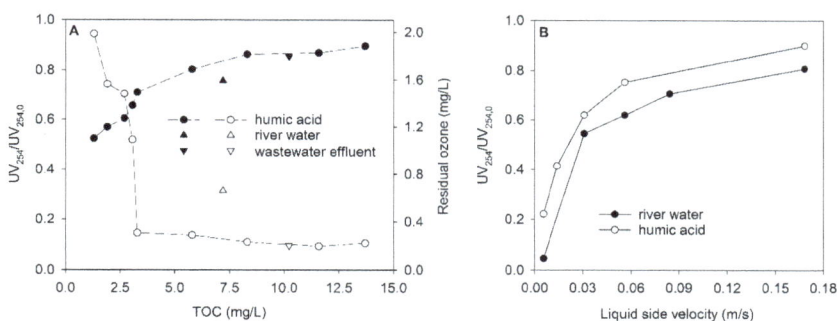

**Figure 5.** (**A**) Relative change in $UV_{254}$ absorbance (closed symbols) and residual ozone concentration (open symbols) for river water I (pH 7.2), wastewater effluent (pH 7.9) and humic acid solutions (pH 7.1) at different TOC, 1.6 mm ID membrane, 4 s residence time, 110 g/$m^3$ ozone gas concentration. (**B**) Relative change in $UV_{254}$ absorbance at different liquid side velocity for river water II (TOC 4.3 mg/L, pH 8.2) and humic acid (TOC 3.6 mg/L, pH 7) for a 1.6 mm ID membrane and 200 g/$m^3$ ozone gas concentration. TOC: total organic carbon, ID: inner diameter.

### 3.5. Membrane Longevity

During repeated use of membranes over several months in a preliminary batch setup and continuous ozonation experiments over 12 h with wastewater effluent and 24 h with pure water, no signs of reduced flexibility or visual changes of the membranes were noticed. However, a slight increase in TOC of water samples after contactor passage, which was detectable but within the analytical error, was found. Ozone exposure induces structural modifications on PDMS [30,32]. Ozone and UV combined oxidize PDMS to form $SiO_x$, by substituting methyl groups with hydroxyl groups [34,58,59], and ozone in the presence of water leads to formation of peroxides on the PDMS surface [60]. The effect of ozone on PDMS, including aging, has no implications for the data presented in this work, but should be considered when planning experimental studies.

*Water* 2018, 10, 1416

## 4. Conclusions

A single tube ozonation contactor was successfully built and tested with non-porous PDMS membranes. The overall mass transfer coefficient and the molar flux of ozone were found to increase with increasing liquid side velocity, and to level off at higher velocities. A comparison of results with a previously developed computational mass transfer model showed good agreement to predict final ozone concentrations in liquid pure water. The experimental and computational results also showed that the membrane size is an important parameter, while the concentration of ozone in the gas phase has minor importance for the ozone mass transfer. Three different types of water were tested to investigate the effect of the water matrix on the degradation of a model compound under ozone and peroxone conditions. Peroxone conditions led to little improvement of the degradation of the model compound. Overall, PDMS is a suitable material for membrane ozonation studies but its long-term stability might limit its application for water treatment. The integration of computational and experimental studies is a powerful tool to inform the design of membrane ozonation contactors with different membrane materials.

**Author Contributions:** Y.M.J.C., D.M. and J.W. conceived and designed the study; R.B. built the contactor system; C.M.T., R.B., G.A.Z., C.S. and M.J.C. conducted sampling, experiments, and analysis; S.H.S.X. and Y.M.J.C performed the computational fluid dynamics modelling; G.A.Z. and J.W. interpreted the results; G.A.Z. and J.W. conducted the literature research; G.A.Z, Y.M.J.C. and J.W. wrote the paper.

**Funding:** This research was partially funded by a Royal Society equipment grant [RG2016-150544] and C.M.T., G.A.Z., and R.B. were supported by EPSRC-funded integrated Ph.D. studentships in Sustainable Chemical Technologies [EP/L016354/1].

**Acknowledgments:** Support for undergraduate research projects of C.S. and M.J.C. by the Department of Chemical Engineering and start-up infrastructure by the Faculty of Engineering & Design for J.W. is appreciated. S.H.S.X. was supported by an Institute for Mathematical Innovation (IMI) undergraduate research internship. G.A.Z. was also supported by a University of Bath research scholarship.

**Conflicts of Interest:** The authors declare no conflict of interest.

## Abbreviations

**Nomenclature**

| | | |
|---|---|---|
| $\alpha$ | surface area of the membrane per unit volume of liquid | $m^{-1}$ |
| L | length of the membrane | m |
| $u_L$ | liquid velocity | m/s |
| H | solubility of ozone in water | - |
| S | solubility of ozone in the membrane | - |
| $C_g$ | ozone concentration in the gas phase | $g/m^3$ |
| $C_{L,out}$ | ozone concentration at the outlet of the contactor | mg/L |
| $K_L$ | Overall mass transfer coefficient | m/s |
| N | Molar flux | $mol/(m^2\ s)$ |
| Re | Reynolds number | - |
| Sc | Schmidt number | - |

**Acronyms**

| | |
|---|---|
| CFD | Computational fluid dynamics |
| PDMS | Polydimethylsiloxane |
| PES | Polyethersulfone |
| PEI | Polyetherimide |
| PFA | Perfluoroalkoxy alkane |
| PTFE | Polytetrafluoroethylene |
| PVDF | Polyvinylidene difluoride |
| DPD | *N,N*-diethyl-p-phenylenediamine |
| *p*CBA | *para*-chlorobenzoic acid |
| ID | inner diameter |
| OD | outer diameter |
| TOC | total organic carbon |

## References

1. Le Paulouë, J.; Langlais, B. State-of-the-art of ozonation in France. *Ozone Sci. Eng.* **1999**, *21*, 153–162. [CrossRef]
2. Von Gunten, U. Oxidation processes in water treatment: Are we on track? *Environ. Sci. Technol.* **2018**, *52*, 5062–5075. [CrossRef] [PubMed]
3. Gottschalk, C.; Libra, J.A.; Saupe, A. *Ozonation of Water and Waste Water: A Practical Guide to Understanding Ozone and Its Applications*; John Wiley & Sons: Hoboken, NJ, USA, 2009.
4. Chuang, Y.-H.; Mitch, W.A. Effect of ozonation and biological activated carbon treatment of wastewater effluents on formation of N-nitrosamines and halogenated disinfection byproducts. *Environ. Sci. Technol.* **2017**, *51*, 2329–2338. [CrossRef] [PubMed]
5. Hua, G.; Reckhow, D.A. Comparison of disinfection byproduct formation from chlorine and alternative disinfectants. *Water Res.* **2007**, *41*, 1667–1678. [CrossRef] [PubMed]
6. Nakamura, H.; Oya, M.; Hanamoto, T.; Nagashio, D. Reviewing the 20 years of operation of ozonation facilities in Hanshin Water Supply Authority with respect to water quality improvements. *Ozone Sci. Eng.* **2017**, *39*, 397–406. [CrossRef]
7. Camel, V.; Bermond, A. The use of ozone and associated oxidation processes in drinking water treatment. *Water Res.* **1998**, *32*, 3208–3222. [CrossRef]
8. Snyder, S.A.; Wert, E.C.; Rexing, D.J.; Zegers, R.E.; Drury, D.D. Ozone oxidation of endocrine disruptors and pharmaceuticals in surface water and wastewater. *Ozone Sci. Eng.* **2006**, *28*, 445–460. [CrossRef]
9. Loeb, B.L.; Thompson, C.M.; Drago, J.; Takahara, H.; Baig, S. Worldwide ozone capacity for treatment of drinking water and wastewater: A review. *Ozone Sci. Eng.* **2012**, *34*, 64–77. [CrossRef]
10. Bourgin, M.; Beck, B.; Boehler, M.; Borowska, E.; Fleiner, J.; Salhi, E.; Teichler, R.; von Gunten, U.; Siegrist, H.; McArdell, C.S. Evaluation of a full-scale wastewater treatment plant upgraded with ozonation and biological post-treatments: Abatement of micropollutants, formation of transformation products and oxidation by-products. *Water Res.* **2018**, *129*, 486–498. [CrossRef] [PubMed]
11. Loeb, B.L. Forty years of advances in ozone technology. A review of ozone: Science & engineering. *Ozone Sci. Eng.* **2018**, *40*, 3–20. [CrossRef]
12. Mundy, B.; Kuhnel, B.; Hunter, G.; Jarnis, R.; Funk, D.; Walker, S.; Burns, N.; Drago, J.; Nezgod, W.; Huang, J.; et al. A review of ozone systems costs for municipal applications. Report by the Municipal Committee—IOA Pan American Group. *Ozone Sci. Eng.* **2018**, *40*, 266–274. [CrossRef]
13. Rakness, K.L.; Hunter, G.; Lew, J.; Mundy, B.; Wert, E.C. Design considerations for cost-effective ozone mass transfer in sidestream systems. *Ozone Sci. Eng.* **2018**, *40*, 159–172. [CrossRef]
14. Zhou, H.; Smith, D.W. Ozone mass transfer in water and wastewater treatment: Experimental observations using a 2D laser particle dynamics analyzer. *Water Res.* **2000**, *34*, 909–921. [CrossRef]
15. Basile, A.; Cassano, A.; Rastogi, N.K. *Advances in Membrane Technologies for Water Treatment: Materials, Processes and Applications*; Woodhead Publishing: Sawston, UK, 2015; pp. 1–342.
16. Zhou, H.; Smith, D.W. Ozonation dynamics and its implication for off-gas ozone control in treating pulp mill wastewaters. *Ozone Sci. Eng.* **2000**, *22*, 31–51. [CrossRef]
17. Oneby, M.A.; Bromley, C.O.; Borchardt, J.H.; Harrison, D.S. Ozone treatment of secondary effluent at U.S. municipal wastewater treatment plants. *Ozone Sci. Eng.* **2010**, *32*, 43–55. [CrossRef]
18. Gabelman, A.; Hwang, S.-T. Hollow fiber membrane contactors. *J. Membr. Sci.* **1999**, *159*, 61–106. [CrossRef]
19. Merle, T.; Pronk, W.; von Gunten, U. MEMBRO3X, a novel combination of a membrane contactor with advanced oxidation ($O_3/H_2O_2$) for simultaneous micropollutant abatement and bromate minimization. *Environ. Sci. Technol. Lett.* **2017**, *4*, 180–185. [CrossRef]
20. Van Geluwe, S.; Braeken, L.; Van der Bruggen, B. Ozone oxidation for the alleviation of membrane fouling by natural organic matter: A review. *Water Res.* **2011**, *45*, 3551–3570. [CrossRef] [PubMed]
21. Kim, J.; Davies, S.H.R.; Baumann, M.J.; Tarabara, V.V.; Masten, S.J. Effect of ozone dosage and hydrodynamic conditions on the permeate flux in a hybrid ozonation–ceramic ultrafiltration system treating natural waters. *J. Membr. Sci.* **2008**, *311*, 165–172. [CrossRef]
22. Laera, G.; Cassano, D.; Lopez, A.; Pinto, A.; Pollice, A.; Ricco, G.; Mascolo, G. Removal of organics and degradation products from industrial wastewater by a membrane bioreactor integrated with ozone or $UV/H_2O_2$ treatment. *Environ. Sci. Technol.* **2012**, *46*, 1010–1018. [CrossRef] [PubMed]

23. Stylianou, S.K.; Sklari, S.D.; Zamboulis, D.; Zaspalis, V.T.; Zouboulis, A.I. Development of bubble-less ozonation and membrane filtration process for the treatment of contaminated water. *J. Membr. Sci.* **2015**, *492*, 40–47. [CrossRef]

24. Wenten, I.G.; Julian, H.; Panjaitan, N.T. Ozonation through ceramic membrane contactor for iodide oxidation during iodine recovery from brine water. *Desalination* **2012**, *306*, 29–34. [CrossRef]

25. Janknecht, P.; Picard, C.; Larbot, A.; Wilderer, P.A. Membrane ozonation in wastewater treatment. *Acta Hydroch. Hydrob.* **2004**, *32*, 33–39. [CrossRef]

26. Kukuzaki, M.; Fujimoto, K.; Kai, S.; Ohe, K.; Oshima, T.; Baba, Y. Ozone mass transfer in an ozone–water contacting process with Shirasu porous glass (SPG) membranes—A comparative study of hydrophilic and hydrophobic membranes. *Sep. Purif. Technol.* **2010**, *72*, 347–356. [CrossRef]

27. Stylianou, S.K.; Szymanska, K.; Katsoyiannis, I.A.; Zouboulis, A.I. Novel water treatment processes based on hybrid membrane-ozonation systems: A novel ceramic membrane contactor for bubbleless ozonation of emerging micropollutants. *J. Chem.* **2015**, *2015*, 214927. [CrossRef]

28. Mosadegh-Sedghi, S.; Rodrigue, D.; Brisson, J.; Iliuta, M.C. Wetting phenomenon in membrane contactors—Causes and prevention. *J. Membr. Sci.* **2014**, *452*, 332–353. [CrossRef]

29. Zhang, Y.; Li, K.; Wang, J.; Hou, D.; Liu, H. Ozone mass transfer behaviors on physical and chemical absorption for hollow fiber membrane contactors. *Water Sci. Technol.* **2017**, *76*, 1360–1369. [CrossRef] [PubMed]

30. Santos, F.R.A.D.; Borges, C.P.; Fonseca, F.V.D. Polymeric materials for membrane contactor devices applied to water treatment by ozonation. *Mater. Res.* **2015**, *18*, 1015–1022. [CrossRef]

31. Pines, D.S.; Min, K.-N.; Ergas, S.J.; Reckhow, D.A. Investigation of an ozone membrane contactor system. *Ozone Sci. Eng.* **2005**, *27*, 209–217. [CrossRef]

32. Shanbhag, P.V.; Sirkar, K.K. Ozone and oxygen permeation behavior of silicone capillary membranes employed in membrane ozonators. *J. Appl. Polym. Sci.* **1998**, *69*, 1263–1273. [CrossRef]

33. Shanbhag, P.V.; Guha, A.K.; Sirkar, K.K. Membrane-based ozonation of organic compounds. *Ind. Eng. Chem. Res.* **1998**, *37*, 4388–4398. [CrossRef]

34. Ouyang, M.; Yuan, C.; Muisener, R.J.; Boulares, A.; Koberstein, J.T. Conversion of some siloxane polymers to silicon oxide by UV/ozone photochemical processes. *Chem. Mater.* **2000**, *12*, 1591–1596. [CrossRef]

35. Berry, M.; Taylor, C.; King, W.; Chew, Y.; Wenk, J. Modelling of ozone mass-transfer through non-porous membranes for water treatment. *Water* **2017**, *9*, 452. [CrossRef]

36. Bader, H.; Hoigné, J. Determination of ozone in water by the indigo method. *Water Res.* **1981**, *15*, 449–456. [CrossRef]

37. Bader, H.; Sturzenegger, V.; Hoigné, J. Photometric method for the determination of low concentrations of hydrogen peroxide by the peroxidase catalyzed oxidation of N,N-diethyl-p-phenylenediamine (DPD). *Water Res.* **1988**, *22*, 1109–1115. [CrossRef]

38. International Organization for Standardization. *ISO 9963-1:1994: Water Quality—Determination of Alkalinity—Part 1: Determination of Total and Composite Alkalinity*; ISO: Geneva, Switzerland, 1994.

39. Atchariyawut, S.; Phattaranawik, J.; Leiknes, T.; Jiraratananon, R. Application of ozonation membrane contacting system for dye wastewater treatment. *Sep. Purif. Technol.* **2009**, *66*, 153–158. [CrossRef]

40. Stylianou, S.K.; Kostoglou, M.; Zouboulis, A.I. Ozone mass transfer studies in a hydrophobized ceramic membrane contactor: Experiments and analysis. *Ind. Eng. Chem. Res.* **2016**, *55*, 7587–7597. [CrossRef]

41. Sander, R. Compilation of Henry's law constants (version 4.0) for water as solvent. *Atmos. Chem. Phys.* **2015**, *15*, 4399–4981. [CrossRef]

42. Dingemans, M.; Dewulf, J.; Van Hecke, W.; Van Langenhove, H. Determination of ozone solubility in polymeric materials. *Chem. Eng. J.* **2008**, *138*, 172–178. [CrossRef]

43. Kasprzyk-Hordern, B.; Ziółek, M.; Nawrocki, J. Catalytic ozonation and methods of enhancing molecular ozone reactions in water treatment. *Appl. Catal. B Environ.* **2003**, *46*, 639–669. [CrossRef]

44. Von Gunten, U. Ozonation of drinking water: Part I. Oxidation kinetics and product formation. *Water Res.* **2003**, *37*, 1443–1467. [CrossRef]

45. Buxton, G.V.; Greenstock, C.L.; Helman, W.P.; Ross, A.B. Critical review of rate constants for reactions of hydrated electrons, hydrogen atoms and hydroxyl radicals (OH/O$^-$) in aqueous solution. *J. Phys. Chem. Ref. Data* **1988**, *17*, 513–886. [CrossRef]

46. Lee, Y.; von Gunten, U. Oxidative transformation of micropollutants during municipal wastewater treatment: Comparison of kinetic aspects of selective (chlorine, chlorine dioxide, ferrate VI, and ozone) and non-selective oxidants (hydroxyl radical). *Water Res.* **2010**, *44*, 555–566. [CrossRef] [PubMed]

47. Elovitz, M.S.; von Gunten, U. Hydroxyl radical/ozone ratios during ozonation processes. I. The $R_{ct}$ concept. *Ozone Sci. Eng.* **1999**, *21*, 239–260. [CrossRef]

48. Wenk, J.; von Gunten, U.; Canonica, S. Effect of dissolved organic matter on the transformation of contaminants induced by excited triplet states and the hydroxyl radical. *Environ. Sci. Technol.* **2011**, *45*, 1334–1340. [CrossRef] [PubMed]

49. Katsoyiannis, I.A.; Canonica, S.; von Gunten, U. Efficiency and energy requirements for the transformation of organic micropollutants by ozone, $O_3/H_2O_2$ and $UV/H_2O_2$. *Water Res.* **2011**, *45*, 3811–3822. [CrossRef] [PubMed]

50. Ferguson, D.W.; McGuire, M.J.; Koch, B.; Wolfe, R.L.; Aieta, E.M. Comparing peroxone and ozone for controlling taste and odor compounds, disinfection by-products, and microorganisms. *J. Am. Water Works Assoc.* **1990**, *82*, 181–191. [CrossRef]

51. Phattaranawik, J.; Leiknes, T.; Pronk, W. Mass transfer studies in flat-sheet membrane contactor with ozonation. *J. Membr. Sci.* **2005**, *247*, 153–167. [CrossRef]

52. Stylianou, S.K.; Katsoyiannis, I.A.; Ernst, M.; Zouboulis, A.I. Impact of $O_3$ or $O_3/H_2O_2$ treatment via a membrane contacting system on the composition and characteristics of the natural organic matter of surface waters. *Environ. Sci. Pollut. Res. Int.* **2018**, *25*, 12246–12255. [CrossRef] [PubMed]

53. Stylianou, S.K.; Katsoyiannis, I.A.; Mitrakas, M.; Zouboulis, A.I. Application of a ceramic membrane contacting process for ozone and peroxone treatment of micropollutant contaminated surface water. *J. Hazard. Mater.* **2018**, *358*, 129–135. [CrossRef] [PubMed]

54. Acero, J.L.; Von Gunten, U. Characterization of oxidation processes: Ozonationn and the AOP $O_3/H_2O_2$. *J. Am. Water Works Assoc.* **2001**, *93*, 90–100. [CrossRef]

55. Drewes, J.E.; Jekel, M. Behavior of DOC and AOX using advanced treated wastewater for groundwater recharge. *Water Res.* **1998**, *32*, 3125–3133. [CrossRef]

56. Yavich, A.A.; Lee, K.H.; Chen, K.C.; Pape, L.; Masten, S.J. Evaluation of biodegradability of NOM after ozonation. *Water Res.* **2004**, *38*, 2839–2846. [CrossRef] [PubMed]

57. Elovitz, M.S.; von Gunten, U.; Kaiser, H.-P. Hydroxyl radical/ozone ratios during ozonation processes. II. The effect of temperature, pH, alkalinity, and DOM properties. *Ozone Sci. Eng.* **2000**, *22*, 123–150. [CrossRef]

58. Fu, Y.J.; Qui, H.Z.; Liao, K.S.; Lue, S.J.; Hu, C.C.; Lee, K.R.; Lai, J.Y. Effect of UV-ozone treatment on poly (dimethylsiloxane) membranes: Surface characterization and gas separation performance. *Langmuir* **2010**, *26*, 4392–4399. [CrossRef] [PubMed]

59. Graubner, V.-M.; Jordan, R.; Nuyken, O.; Schnyder, B.; Lippert, T.; Kötz, R.; Wokaun, A. Photochemical modification of cross-linked poly (dimethylsiloxane) by irradiation at 172 nm. *Macromolecules* **2004**, *37*, 5936–5943. [CrossRef]

60. Fujimoto, K.; Takebayashi, Y.; Inoue, H.; Ikada, Y. Ozone-induced graft polymerization onto polymer surface. *J. Polym. Sci. Part A Polym. Chem.* **1993**, *31*, 1035–1043. [CrossRef]

## water

Article

# Removal of Antimony Species, Sb(III)/Sb(V), from Water by Using Iron Coagulants

**Manassis Mitrakas [1], Zoi Mantha [2], Nikos Tzollas [2], Stelios Stylianou [2], Ioannis Katsoyiannis [2] and Anastasios Zouboulis [2,*]**

[1] Laboratory of Analytical Chemistry, Department of Chemical Engineering, Aristotle University of Thessaloniki, 54124 Thessaloniki, Greece; manasis@eng.auth.gr

[2] Laboratory of Chemical & Environmental Technology, Department of Chemistry, Aristotle University of Thessaloniki, 54124 Thessaloniki, Greece; zoi.mantha92@gmail.com (Z.M.); ntzollas@gmail.com (N.T.); s.stulianou@gmail.com (S.S.); yiank@yahoo.com (I.K.)

* Correspondence: zoubouli@chem.auth.gr

Received: 23 August 2018; Accepted: 20 September 2018; Published: 25 September 2018

**Abstract:** Antimony (Sb) is classified as a toxic pollutant of high priority, because its effects on human health (toxicity) are similar to those of arsenic. However, unlike arsenic, the removal of antimony from polluted waters is still not well understood. In the present study the removal of common antimony species in water, namely, Sb(III) and Sb(V), was investigated by the addition of iron-based coagulants. The applied coagulants were Fe(II), Fe(III), and equimolar mixed Fe(II)/Fe(III) salts and the experiments were performed with realistic antimony concentrations in the range 10–100 µg/L, by using artificially polluted tap water solutions. Sb(III) removal by Fe(III) provided better adsorption capacity at a residual concentration equal to the drinking water regulation limit of 5 µg/L, that is, $Q_5 = 4.7$ µg Sb(III)/mg Fe(III) at pH 7, which was much higher than the value achieved by the addition of Fe(II) salts, that is, $Q_5 = 0.45$ µg Sb(III)/mg Fe(II), at the same pH value. Similarly, Sb(V) was more efficiently removed by Fe(III) addition, than by the other examined coagulants. However, Fe(III) uptake capacity for Sb(V) was found to be significantly lower, that is, $Q_5 = 1.82$ µg Sb(V)/mg Fe(III), than the corresponding value for Sb(III). The obtained results can give a realistic overview of the efficiency of conventionally used iron-based coagulants and of their mixture for achieving Sb concentrations below the respective drinking water regulation limit and therefore, they can be subsequently applied for the designing of real-scale water treatment units.

**Keywords:** antimony treatment; Sb(III); Sb(V); Fe-based coagulants; polluted waters

## 1. Introduction

Groundwater pollution by toxic oxyanions, such as As, Se, Sb, and so on, is considered as a major global concern, because of their adverse effects on human health; their presence in waters is mostly due to geogenic origin. Antimony is usually present in groundwaters as Sb(III) or Sb(V) species; Sb(III) was found to be 10 times more toxic than Sb(V) [1]. Antimony toxicity can accidentally occur either due to occupational exposure, or during medicinal therapy. Occupational exposure may cause respiratory irritation, pneumoconiosis, spots on the skin, and gastrointestinal symptoms, whereas as a therapeutic agent, antimony has been mostly used for the treatment of leishmaniasis and schistosomiasis [2]. The effects of antimony exposure depend on the duration of exposure, when humans are exposed at levels above the Maximum Concentration Limit (MCL). For relatively shorter periods of time U.S. E.P.A. has found that antimony can potentially cause nausea, vomiting, and diarrhea, whereas for longer periods of time antimony is considered as a human carcinogen, when the exposure exceeds constantly concentrations above MCL. Because of its toxicity, antimony concentrations are regulated by both the European Commission Drinking Water Directive and the U.S. E.P.A., with the MCLs being 5 and 6 µg/L, respectively.

Antimony speciation and distribution in freshwaters (of surface or ground origin) have not been extensively studied. Total background-Sb dissolved concentrations in groundwater have been reported in the range 0.010–1.5 µg/L [3], while anthropogenic and geothermal sources are responsible for much higher levels, in ranges of 0.7–170 µg/L and 0.06–26 µg/L, respectively [4]. The most common source of antimony in drinking water sources is the dissolution from metal plumbing and fittings. Antimony leached from Sb-containing materials would be mostly in the form of Sb(V) oxy-anion. It is most likely for this reason, that even in anoxic groundwaters that is in a reductive environment, considerable Sb(V) concentrations can be also detected [5]. However, according to thermodynamic equilibrium predictions/diagrams, dissolved antimony in water exists mainly as Sb(V) in oxic (e.g., surface) waters and as Sb(III) in anoxic (e.g., ground) waters. In pH values relevant to most natural waters, that is, between 6–8, the Sb(III) is mainly present as $Sb(OH)_3$, whereas Sb(V) exists as a negatively charged aqueous complex, $Sb(OH)_6^-$ [6].

Several treatment processes have been applied for the removal of antimony from polluted water or wastewater streams, such as reverse osmosis [7], biosorption [8], electrocoagulation [9], adsorption [10,11] and coagulation [1,12]. Nevertheless, only a few studies have focused on removing antimony from drinking water sources, aiming to achieve the residual antimony concentrations below the respective regulation limits [11,13]. It is worth noting also that most of the proposed methods for the removal of antimony from waters, have adapted treatment approaches similar to that applied for arsenic removal [14], such as adsorption and coagulation [12].

The coagulation/precipitation process usually incorporates the use of relatively low-cost ferric or aluminum salts to successively capture/remove both Sb(III) and Sb(V) species [15]. Relevant studies indicated that the removal of Sb(III) is more favorable, than that of Sb(V), due to the higher mobility of the latter (pentavalent) species at pH values above 5 [12]. Guo and co-authors (2009) [12] reported 99% Sb(V) removal from a high initial concentration of 49.2 mg Sb(V)/L, by using a ferric chloride dosage of $6 \times 10^{-4}$ mol/L. When compared to Sb(V), effective Sb(III) removal was achieved by using a significantly lower ferric dose at pH 6. Sb(III) removal became highly efficient when $4 \times 10^{-4}$ mol/L of Fe(III) was used, producing treated water in compliance with the respective drinking-water standard. Therefore, the removal of reduced antimony species, that is, Sb(III), by coagulation-precipitation is more pronounced than that of the oxidized form Sb(V) [12,13,16], unlike what happens in the case of the As(III)/As(V) system [17]. Sb(III) removal was not found to be greatly affected by groundwater composition [12], whereas Sb(V) removal was adversely affected by the presence of other anions, for example, bicarbonates, sulfates, phosphates, or humic acids, commonly encountered in waters [16].

The main advantages of coagulation/precipitation process for the drinking water treatment are the relatively low capital costs, the effectiveness over a rather wide range of pH values, the applicability to large volume of waters, and the simplicity of operation [12]. However, the disadvantages of this process are the rather low expected removal of Sb(V) species and the demand of the addition of considerable high coagulant doses, which may result in the formation of large quantities of eventually toxic sludge, difficult to be disposed of.

The aim of this work was to examine systematically the major parameters that favor effective antimony removal (both major aqueous species) by Fe-based coagulants' addition, along with the estimation of uptake capacity at the residual concentration equal to the EU drinking water regulation limit (5 µg/L), by using appropriately polluted tap water, which may allow the direct use of obtained data for upscaling purposes. To the best of our knowledge, neither the application of $FeSO_4 \cdot 7H_2O$ (a common coagulant agent), nor the application of mixed coagulants, such as the used equimolar $Fe_2(SO_4)_3$ and $FeSO_4 \cdot 7H_2O$ reagents, has been previously reported, which comprises a new approach for antimony removal and opens the field for further improvements. Fe(II) was applied in order to examine its efficiency for Sb(V) removal, through the preliminary reduction to the less soluble (and more easily removed) species of Sb(III) and to compare the results with the case of Fe(III) use for

Sb(III) and Sb(V) removal. Furthermore, experimental trials by using a mixed equimolar coagulant of $Fe_2(SO_4)_3$ and $FeSO_4$ $7H_2O$ were also conducted, in order to investigate the possible interactions between the applied coagulants, which might lead to increased overall antimony removal efficiency. To the best of our knowledge, there is no publication, estimating the $Q_5$ adsorption capacity (i.e., the necessary for lowering the concentration of Sb down to the MCL) for iron-based coagulants and for a variety of operational and physico-chemical parameters, which can provide essential data for enabling the upscale of antimony treatment/removal in drinking water treatment units.

## 2. Materials and Methods

### 2.1. Water Characteristics

Tap water of Thessaloniki city, Greece, after passing through a fixed bed of activated carbon for chlorine removal, was used in this study (Table 1). Water samples were daily spiked either with Sb(III) or Sb(V) and were used for the experiments at least 6 h after the respective antimony species addition, to allow sufficient time for them to be fully hydrolyzed, and form the species similar to those present in natural waters polluted with antimony.

**Table 1.** Major physicochemical characteristics of Thessaloniki tap water.

| Parameter | Average Value |
|---|---|
| pH | 7.30 |
| Conductivity, $\mu S/cm$ | 590 |
| Na, mg/L | 35 |
| Ca, mg/L | 80 |
| Mg, mg/L | 24 |
| $HCO_3^-$, mg/L | 342 |
| Fe, mg/L | <0.02 |
| Mn, mg/L | <0.005 |
| $NO_3^-$, mg/L | 9 |
| $SO_4^{2-}$, mg/L | 8 |
| $Cl^-$, mg/L | 13 |
| TOC, mg/L | 0.4 |

### 2.2. Reagents and Materials

Deionized water was used to prepare stock solutions of used reagents. All glassware, polyethylene bottles, and sample vessels were immersed in 15% $HNO_3$ solution and rinsed three times with deionized water before use. The 100 mg/L stock solutions of Sb(V) and Sb(III) were prepared by the dissolution of $KSb(OH)_6$ or $Sb_2O_3$ compounds (analytical grade) in 2 M HCl, respectively [12,18], whereas the initial antimony concentrations for the experiments were in the range of 100 $\mu g/L$ or lower.

### 2.3. Experimental Procedure

$FeSO_4$ $7H_2O$ and $FeCl_3$ $6H_2O$ were used for preparing daily fresh 1000 mg Fe/L stock solutions. Treatment tests were performed on a program-controlled JJ-4A jar tester with six paddles/beakers. The water pH was adjusted with the addition of 0.1 M HCl or 0.1 M NaOH. Test water (1000 mL) was transferred into a 1000 mL beaker. Under initial rapid stirring (140 rpm), a predetermined dose ranging between 1–10 mg Fe/L was added. After 2 min of rapid mixing, the stirring speed reduced to 40 rpm (duration 45 min), followed by 15 min settling time. A 100 mL supernatant sample was collected and filtered through a 0.45 mm membrane filter for further analytical determinations [13]. To determine the surface charge of FeOOH precipitates, the Iso-Electric Point (IEP) was calculated by the zeta-potential curve at 20 ± 1 °C of solid adsorbent dispersion in 0.01 M $NaNO_3$ versus the respective pH of solution, by using a Micro-electrophoresis Apparatus (Mk II device, Rank Brothers).

## 2.4. Analytical Procedure

Initial and final (effluent) antimony concentrations were determined by Atomic Absorption-Hydride Generation unit and Flow Injection Analysis (FIAs). The method's detection limit was 0.4 µg Sb/L. The used instrument was a Perkin Elmer (HG-AAS, Perkin Elmer-A Analyst 400).

## 3. Results and Discussion

### 3.1. Antimony Speciation

According to thermodynamic equilibrium predictions (Figure 1), dissolved antimony in water matrixes exists mainly either as Sb(V) in oxic waters, or as Sb(III) in anoxic groundwaters. However, the partly presence of Sb(III) in oxic waters, as well as of Sb(V) in anoxic ones, has been often reported [19]. In pH values commonly encountered in natural waters, that is, between 6–8, the Sb(III) is mainly present as $Sb(OH)_3$, whereas Sb(V) mostly exists as a negatively charged complex, $Sb(OH)_6^-$ (Figure 1).

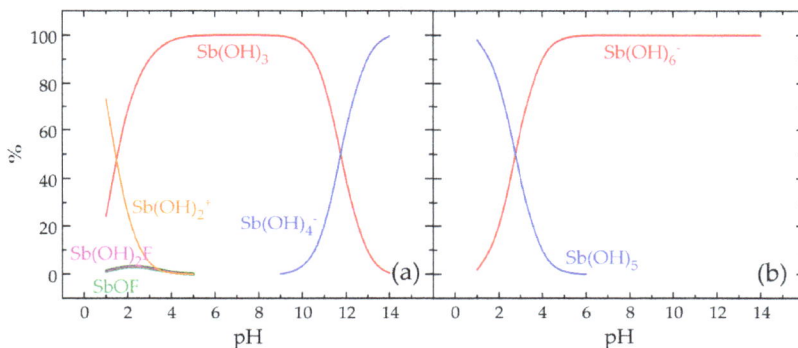

**Figure 1.** Percentage of (**a**) Sb(V) and (**b**) Sb(III) species at concentrations 100 µg/L in tap water matrix and 20 °C. Diagrams derived by Visual MINTEQ 3.0 (http://vminteq.lwr.kth.se).

### 3.2. Sb(III) Removal by Fe(III) or Fe(II) Coagulation

#### 3.2.1. Fe(III) Addition

The adsorption isotherms of Sb(III) removal by Fe(III) addition and precipitation/coagulation (Figure 2) indicated a significantly higher removal efficiency at pH 5, while the effect of pH values at the commonly encountered in natural waters range (6–8) was insignificant. Therefore, it can be concluded that pH does not play an important role on Sb(III) removal from natural waters by Fe(III) coagulation, in accordance with previous studies [18], which however referred to much higher initial antimony concentrations (at least an order of magnitude greater).

**Table 2.** Freundlich fitting parameters for the Sb(III) adsorption isotherms ($Q = K_F C^n$).

| Coagulant | pH | $K_F$ (µg/mg)/(µg/L)$^n$ | n | $R^2$ | $Q_5$ µg Sb(III)/mg Fe |
|-----------|-----|--------------------------|--------|-------|------------------------|
| Fe(III) | 5 | 2.964 | 0.7900 | 0.989 | 10.5 |
| Fe(III) | 6 | 0.985 | 1.0019 | 0.997 | 4.9 |
| Fe(III) | 7 | 0.887 | 1.0265 | 0.999 | 4.7 |
| Fe(III) | 8 | 0.995 | 0.9877 | 0.993 | 4.8 |
| Fe(II) | 7 | 0.032 | 1.6354 | 0.994 | 0.45 |

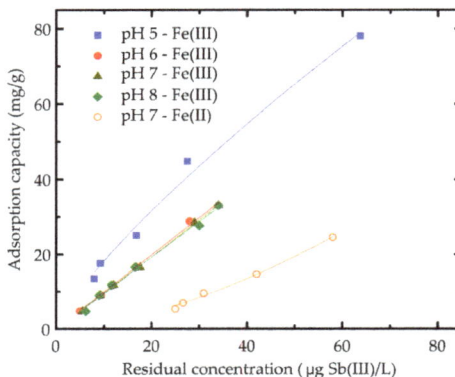

**Figure 2.** Fitting of Sb(III) adsorption (Freundlich model) onto FeOOH precipitates by using coagulation/ precipitation with Fe(III) or Fe(II) salts at various water pH values; experimental conditions: T = 22 ± 1 °C, initial antimony concentration 100 µg Sb(III)/L, iron dose range 1–10 mg/L.

The better affinity of Sb(III) with the produced Fe(III) precipitate (FeOOH) at pH 5 is also clarified by the value of the n-parameter (0.79) of the Freundlich model, while the corresponding n-values at pH range 6–8 fluctuated at 1 ± 0.02 (Table 2). The latter is partially related to the isoelectric point (IEP) of FeOOH, which was ranged at 6.9 ± 0.3; that means a low surface density of FeOOH in the pH range (6–8) close to IEP, which along with the almost neutral surface of Sb(OH$_3$) (Figure 1) results in low affinity. In contrast and at pH 5 (<IEP), where FeOOH present a higher positive surface density, the affinity ($n = 0.79$), as well the adsorption capacity, were significantly increased. Conclusively, the range (0.79–1) of n-value implies a weak chemisorption of Sb(OH$_3$) onto FeOOH.

In the majority of relevant bench-scale experiments referred in literature, the efficiency of coagulants is evaluated through the percentage of removal capacity (e.g., [18]), while the residual concentrations of Sb(III) frequently fail to meet the regulation limit (e.g., [11]). In this study, however, the added coagulants and the main parameters of the procedure, influencing the removal of Sb species, are evaluated according to their efficiency to decrease the residual (final) concentration below the drinking water regulation limit (i.e., 5 µg/L), along with their adsorption capacity, which will be abbreviated as Q$_5$ (mg/g), henceforth. The obtained adsorption data were fitted to Freundlich model (Q = K$_F$ C$^n$) and showed that Fe(III) can achieve residual Sb(III) concentrations significantly lower, than the respective regulation limit at the pH range 5–8 (Table 2). Furthermore, the adsorption capacity, Q$_5$ = 10.5 µg Sb(III)/mg Fe(III) at pH 5, was almost double in comparison to the corresponding at pH range 6–8, that is, Q$_5$ = 4.8 ± 0.1 µg Sb(III)/mg Fe(III), suggesting that the recommended dose for decreasing, for example, an initial concentration of 50 µg Sb(III)/L to the drinking water regulation limit of 5 µg/L at pH 7 is as follows:

$$\text{Fe(III) dose} = [50 - 5 \text{ µg Sb(III)/L}]/[4.7 \text{ µg Sb(III)/mg Fe(III)}] = 9.6 \text{ mg Fe(III)/L}$$

These results are in good agreement with relevant literature findings [12,16], although in these studies the results were expressed as percentage removal of antimony and, therefore, did not relate clearly with the adsorption capacity. Furthermore, the value Q$_5$ = 4.7 µg Sb(III)/mg Fe(III) at pH 7 is significantly higher in comparison with the corresponding values of other commercially available Fe-based adsorbents (mainly used for As removal), such as

- GFH (Q$_5$ = 1.4 µg Sb(III)/mg GFH, or 2.5 µg Sb(III)/mg Fe), which was supplied by SIEMENS and mainly consists of akaganeite with an iron content 55 ± 1% *w/w* at dry basis, and

- Bayoxide ($Q_5$ = 0.6 µg Sb(III)/mg Bayoxide, or 1.4 µg Sb(III)/mg Fe) [11], which was supplied by Bayer and mainly consists of goethite with an iron content 52 ± 1% *w/w* [11].

Similarly, lower adsorption capacities of solid adsorbents in comparison to freshly precipitated FeOOH (as coagulation product) were also observed for the case of As(V) removal [20]. The apparent reason for this significantly higher removal capacity for the case of in situ formed FeOOH precipitates is the formation of short-chain polymers of $Fe(OH)_y^{z+}$ with higher surface charge density, as they are gradually transformed from Fe(III) dissolved cations into FeOOH floc (solid) particles/precipitates.

### 3.2.2. Fe(II) Addition

The efficiency of Fe(II) addition on Sb(III) removal at pH 7 proved to be an order of magnitude lower, that is, $Q_5$ = 0.45 ± 0.1 µg Sb(III)/mg Fe(II), in comparison to Fe(III) (see Figure 2 and Table 2). This could be probably attributed to different hydrolysis path of iron species. The intermediate short-chain $Fe(OH)_y^{z\pm}$ polymers formed during the Fe(III) hydrolysis to FeOOH precipitates favor the Sb(III) adsorption, while the gradual Fe(II) oxidation-hydrolysis restricts the surface charge density of formed FeOOH solids [11]. Furthermore, a partial Fe(II) oxidation was observed at pH 6 and 5, which in turn resulted in overpassing the respective iron regulation limit in the treated water, along with lower Sb(III) removal (additional disadvantage). Therefore, the experimental results in this case did not fit well with the main adsorption models, since the effectiveness of Fe(II) oxidation and the removal capacity were highly influenced by the dose, that is, by the initial Fe(II) concentration.

### 3.3. Sb(V) Removal by Fe(III) or Fe(II) Coagulation

### 3.3.1. Fe(III)

The Sb(V) coordination with oxygen atoms is different from the tetrahedral formation of As(V) oxy-anions, due to its larger ionic radius and lower charge density, which in turn may favor the octahedral geometry. Therefore, $Sb(OH)_6^-$ is the dominant species in water pH values > 5 and thus, in pH range 6–8, commonly encountered in drinking water (Figure 1). Inversely to the case of Sb(III), the water pH was found to influence significantly the Sb(V) removal by the addition of Fe(III), as shown in Figure 3, which depicts that the Sb(V) removal efficiency by Fe(III) is decreasing linearly as pH rises from 5 to 7 and diminishes at pH value 8.

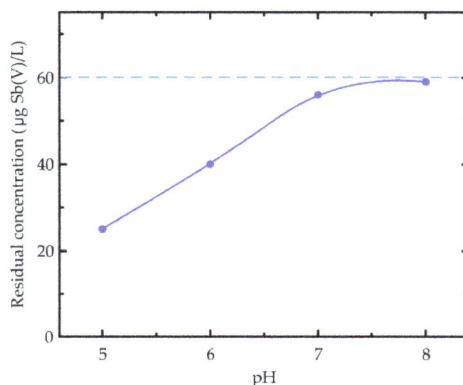

**Figure 3.** Influence of water pH on Sb(V) removal (experimental conditions: initial antimony concentration $C_o$ = 60 µg Sb(V)/L, Fe(III) dose 2.5 mg/L, T = 22 ± 1 °C).

Since the isoelectric point of FeOOH precipitates was ranged at 6.9 ± 0.3, at pH < IEP the positive charge density dominates, due to iron species $Fe(OH)^{2+}/Fe(OH)_2^+$, as illustrated in Figure S1 of supporting information, thus resulting in better uptake of negatively charged $Sb(OH)_6^-$ species. Inversely, at water pH values > IEP the dominating negatively charged $Fe(OH)_4^-$ species repulses the similar charged $Sb(OH)_6^-$ species, thus diminishing the Sb(V) uptake capacity.

The octahedral geometry of $Sb(OH)_6^-$ species results also in significantly lower effectiveness, regarding Sb(V) removal by Fe(III) coagulants, in comparison to the relevant case of As(V). The fitting attempts to the main sorption models of adsorption isotherms data at pH 7 (Figure 4) has shown that the obtained results were best described by the BET multilayer model (Table 3), whereas the attempts to fit the data according to common Freundlich or Langmuir adsorption models did not produce reasonable predictions. Noting also that the fitting according to the BET model suggests a multilayer adsorption (physisorption), where the adsorption enthalpy is the same for any layer and a new layer can start forming before the previous one is finished [21]. Nevertheless, these results are in contradiction to most published results, such as those of Ali Inam et al., 2018 [18], due to the fact that they referred to equilibrium concentrations around two orders of magnitude higher or even more than those examined in the current study. In this study the adsorption data of Figure 4 favor the accurate determination of Fe(III)-solids uptake capacity at residual concentration equal to the drinking water regulation limit (Table 3).

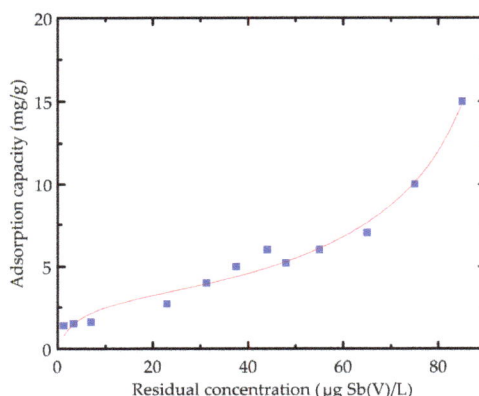

**Figure 4.** Fitting of Sb(V) adsorption onto FeOOH precipitates (BET model) by using coagulation/ precipitation with Fe(III); experimental conditions: pH 7, T = 22 ± 1 °C, initial concentrations range 15–100 µg Sb(V)/L, Fe(III) dose range 1–10 mg/L.

**Table 3.** BET fitting parameters for Sb(V) adsorption isotherms at pH 7 (according to the equation: $Q = aC/(1 + bC + dC^2)$).

| Coagulant | a | b | c | $R^2$ | $Q_5$ µg Sb(V)/mg Fe |
|---|---|---|---|---|---|
| Fe(III) (Figure 4) | 0.7732 | 0.2357 | −0.0023 | 0.981 | 1.82 |
| Fe(II) (Figure 6) | 0.2239 | 0.0102 | −0.0002 | 0.997 | 1.30 |
| [Fe(III)]/[Fe(II)] = 1 (Figure 7) | 0.4164 | 0.1251 | −0.0014 | 0.985 | 1.31 |

The uptake capacity of FeOOH-precipitates formed for the case of Sb(V) removal at the drinking water regulation limit, that is, $Q_5$ = 1.82 µg Sb(V)/mg Fe(III), proved to be equal to 39% of the corresponding value for the relevant case of Sb(III) removal, that is, $Q_5$ = 4.7 µg Sb(III)/mg Fe(III). Noting also that the uptake capacity of Fe(III) for Sb(V) is almost equal to 2.5% of the corresponding value of As(V) oxy-anions of tetrahedral geometry (i.e., in the latter case $Q_5$ = 44 µg As(V)/mg Fe(III)) [22]. Therefore, by using an initial iron coagulation dose of e.g., 10 mg/L, which is usually at the highest end for most applications in full-scale drinking water treatment plants, the maximum initial concentration of Sb(V) that can be diminished (i.e., effectively treated) down to drinking water regulation limit at pH 7 is as follows:

$$10 \text{ mg Fe(III)/L} = [C_o - 5 \text{ µg Sb(V)/L}]/[1.82 \text{ µg Sb(V)/mg Fe(III)}] \rightarrow C_o = 23 \text{ µg Sb(V)/L}$$

While the recommended dose for decreasing an initial concentration of, for example, 50 µg Sb(V)/L below the drinking water regulation limit is too high:

$$\text{Fe(III) dose} = [50 - 5 \text{ µg Sb(V)/L}]/[1.82 \text{ µg Sb(V)/mg Fe(III)}] = 24.7 \text{ mg Fe(III)/L}$$

Finally, from a techno-economical point of view, these results verify that Sb(V) removal by Fe(III) precipitation seems to be not an attractive process.

### 3.3.2. Fe(II) or Equimolar Fe(II)/Fe(III) Additions

The application of Fe(II) coagulant aimed at investigating the influence on Sb(V) uptake capacity through a preliminary chemical reduction to Sb(III) and subsequent easier removal. The redox potential of used tap water samples spiked with Sb(V) ranged at $+0.27 \pm 0.01$ V (Figure 5a), which can verify the stability of $Sb(OH)_6{}^-$ species (Figure 5b). Figure 5a indicates the immediate change in the redox potential values, when introducing in this solution the Fe(II) coagulant, whereas a smaller redox potential reduction is taking place, when using the mixed coagulants addition. More specifically, the redox potential became significantly lower than +0.1 V, where antimony is thermodynamically stable as $Sb(OH)_3$ (Figure 5b), even at the smaller applied Fe(II) dose (1 mg/L). In contrast, the Fe(III) addition does not provoke any significant changes in the redox potential values of the solution, as it is expected.

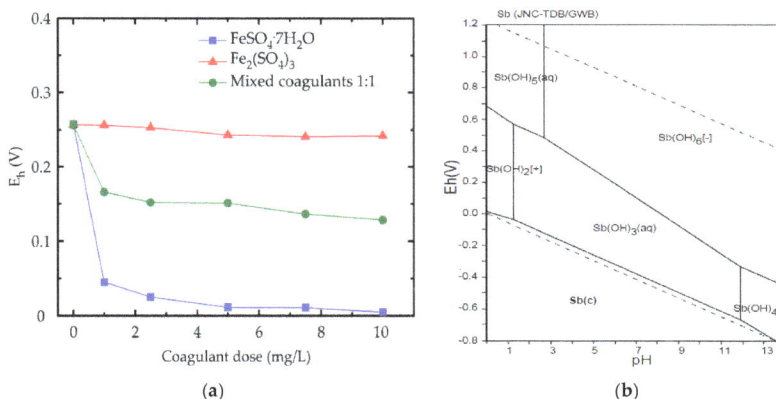

**Figure 5.** (a) Redox potential as a function of coagulant dose (at pH 7 and $C_o = 50$ µg Sb(V)/L), (b) Eh-pH diagrams of the system Sb-O-H, according to Geological Survey of Japan Open File Report No. 419 (2005).

The redox potential changes can be considered as indications that reducing conditions prevail and thus, Sb(V) reduction can eventually take place. However, the oxidation–reduction reactions are a matter of kinetics as well. The fact that the use of Fe(II) as a coagulant agent increased the uptake capacity of Sb(V) by almost three times, that is, $Q_5 = 1.30$ µg Sb(V)/mg Fe(II) (Table 3), in comparison to Sb(III) $Q_5 = 0.45$ µg Sb(III)/mg Fe(II) (Table 2), can be probably attributed to Sb(V) reduction. The mechanism of gradual electron transfer from oxidized Fe(II) to reduced Sb(V) probably can favor the affinity between them and increase the respective uptake capacity. The fitting attempts of adsorption isotherms' data of Fe(II), as well as of equimolar Fe(II)/Fe(III) addition, at pH 7 to the main sorption models showed that they were also best fitted by the BET multilayer model (Figures 6 and 7), similarly to corresponding of Fe(III) addition. The achieved uptake capacities by Fe(II) addition, that is, $Q_5$ = 1.30 µg Sb(V)/mg Fe(II), and by Fe(II)/Fe(III) addition, that is, $Q_5$ = 1.31 µg Sb(V)/mg Fe(II)/Fe(III), were equal and by 30% lower in comparison to the corresponding of Fe(III) addition.

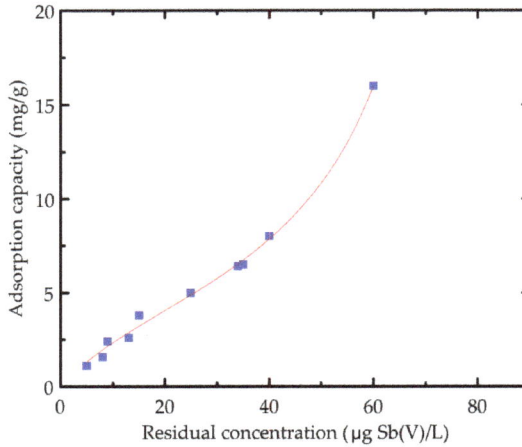

**Figure 6.** Fitting of Sb(V) adsorption onto FeOOH precipitates (BET model) by using coagulation/precipitation with Fe(II); experimental conditions: pH 7, T = 22 ± 1 °C, initial concentrations range 15–100 µg Sb(V)/L, Fe(II) dose range 1–10 mg/L.

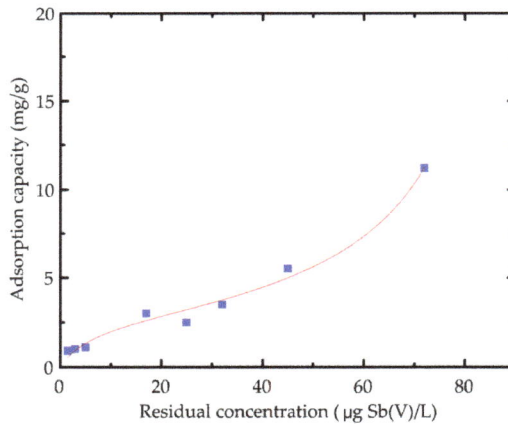

**Figure 7.** Fitting of Sb(V) adsorption onto FeOOH precipitates (BET model) by using coagulation/precipitation with equimolar Fe(III)/Fe(II); experimental conditions: pH 7, T = 22 ± 1 °C, initial concentrations range 15–100 µg Sb(V)/L, total iron dose 1–10 mg/L.

Conclusively, although the reducing conditions prevailed in this case, the kinetics of reduction-coagulation/precipitation reactions seems to be rather slow and could not complete the reduction of Sb(V) to Sb(III) within the reasonable allowed time for reaction, which is relevant to drinking water treatment (in the range of several min and not of hours). Therefore, under these conditions Fe(III) proved to be more effective also for the case of Sb(V) removal in comparison to Fe(II) and Fe(II)/Fe(III) coagulants, noting however that this uptake capacity is equal to 39% of the corresponding value for the case of Sb(III).

## 4. Conclusions

✔ Coagulation is generally an effective treatment technique for antimony removal from polluted aqueous sources, with much more efficient Sb(III) removal induced by Fe(III) coagulant, that is, $Q_5 = 4.7$ µg Sb(III)/mg Fe(III), than by Fe(II), that is, $Q_5 = 0.45$ µg Sb(III)/mg Fe(II) at pH 7. Furthermore, Fe(III)-based coagulant addition proved also more efficient than the Fe(II) or Fe(III)/Fe(II) coagulants for Sb(V) removal. However, the Fe(III) uptake capacity for Sb(V), that is, $Q_5 = 1.82$ µg Sb(V)/mg Fe(III), was found almost equal to 39% of the corresponding value for the case of Sb(III) and 2.5% of the corresponding value for the tetrahedral geometry As(V) oxy-anions (i.e., $Q_5 = 44$ µg As(V)/mg Fe(III)).

✔ Fe(II) coagulant seems to contribute to Sb(V) reduction to Sb(III), since its adsorption capacity for Sb(V), that is, $Q_5 = 1.3$ µg Sb(V)/mg Fe(II), was found to be almost three times higher than the corresponding for Sb(III), that is, $Q_5 = 0.45$ µg Sb(III)/mg Fe(II).

✔ The water pH value does not influence Sb(III) removal by the Fe(III) addition at pH range 6–8, commonly encountered in most natural waters with $Q_5 = 4.8 \pm 0.1$ µg Sb(III)/mg Fe(III), because Sb(III) is present mainly as a neutral molecule in the form of $Sb(OH)_3$. However, at pH 5 the uptake capacity proved to be significantly higher, that is, $Q_5 = 10.5$ µg Sb(III)/mg Fe(III), due to the increase of positive surface charge density of FeOOH precipitates.

✔ By lowering the water pH below the IEP value of FeOOH precipitates, the uptake of $Sb(OH)_6{}^-$ was gradually increased, due to the increase of positively charged $Fe(OH)^{2+}/Fe(OH)_2{}^+$ hydrolysis species of Fe(III), for example, for $C_o = 60$ µg Sb(V)/L and iron dose 2.5 mg Fe(III)/L the residual concentrations at water pH 8, 7, 6, 5 were found to be 59, 56, 40, 25 µg Sb(V)/L, respectively.

✔ The fitting of adsorption isotherms data to sorption models, regarding the equilibrium antimony concentrations in the range of 5–100 µg/L, showed that the Sb(III) data were better fitted to the Freundlich model, while the corresponding data for Sb(V) were better fitted to the BET model.

✔ Finally, the experimental data of this study were focused in antimony concentrations commonly found in polluted natural waters (around or lower than 100 µg/L), hence allowing the accurate determination of respective adsorption capacities by the coagulation produced precipitates-solids at the drinking water regulation limit ($Q_5$), and therefore, supplying the fundamental information for upscaling the results of this study.

**Supplementary Materials:** The following are available online at http://www.mdpi.com/2073-4441/10/10/1328/s1, Figure S1: Fe(III) speciation as a function of water pH.

**Author Contributions:** Z.M. and S.S. conducted the experimental work. Z.M. and N.T. conducted the analysis of antimony by HG-AAS in aqueous solutions. I.K., A.Z., M.M., conceived, designed, and supervised the experiments and contributed equally in writing the manuscript and providing careful explanations for the obtained results. A.Z. had the overall coordination of the project and is the corresponding author of this paper.

**Funding:** This research received no external funding

**Conflicts of Interest:** The authors declare no conflicts of interest

## References

1. Du, X.; Qu, F.; Liang, H.; Li, K.; Yu, H.; Bai, L.; Li, G. Removal of antimony(III) from polluted surface water using a hybrid coagulation–flocculation–ultrafiltration (CF–UF) process. *Chem. Eng. J.* **2014**, *254*, 293–301. [CrossRef]

2. Sundar, S.; Chakravarty, J. Antimony toxicity. *Int. J. Environ. Res. Public Health* **2010**, *7*, 4267–4277. [CrossRef] [PubMed]

3.	Filella, M.; Belzile, N.; Chen, Y.-W. Antimony in the environment: A review focused on natural waters I. Occurrence. *Earth-Sci. Rev.* **2002**, *57*, 125–176. [CrossRef]

4.	Aksoy, N.; Simsek, C.; Gunduz, O. Groundwater contamination mechanism in a geothermal field: A case study of Balcova, Turkey. *J. Contam. Hydrol.* **2009**, *103*, 13–28. [CrossRef] [PubMed]

5.	WHO. Antimony in Drinking-water Background Document for Development of WHO Guidelines for Drinking-Water Quality. WHO/SDE/WSH/03.04/74. 2003. Available online: http://www.who.int/water_sanitation_health/dwq/chemicals/antimony.pdf (accessed on 23 March 2018).

6.	Xi, J.; He, M.; Lin, C. Adsorption of antimony(III) and antimony(V) on bentonite: Kinetics, thermodynamics and anion competition. *Microchem. J.* **2011**, *97*, 85–91. [CrossRef]

7.	Kang, M.; Kawasaki, M.; Tamada, S.; Kamei, T.; Magara, Y. Effect of pH on the removal of arsenic and antimony using reverse osmosis membranes. *Desalination* **2000**, *131*, 293–298. [CrossRef]

8.	Uluozlu, O.D.; Sar, A.; Tuzen, M. Biosorption of antimony from aqueous solution by lichen (*Physcia tribacia*) biomass. *Chem. Eng. J.* **2010**, *163*, 382–388. [CrossRef]

9.	Zhu, J.; Wu, F.; Pan, X.; Guo, J.; Wen, D. Removal of antimony from antimony mine flotation wastewater by electrocoagulation with aluminum electrodes. *J. Environ. Sci.* **2011**, *23*, 1066–1071. [CrossRef]

10.	Zhao, X.Q.; Dou, X.M.; Mohan, D.; Pittman, C.U.; Ok, Y.S.; Jin, X. Antimonate and antimonite adsorption by a polyvinyl alcohol stabilized granular adsorbent containing nanoscale zero-valent iron. *Chem. Eng. J.* **2014**, *247*, 250–257. [CrossRef]

11.	Simeonidis, K.; Papadopoulou, V.; Tresintsi, S.; Kokkinos, E.; Katsoyiannis, I.A.; Zouboulis, A.I.; Mitrakas, M. Efficiency of iron-based oxy-hydroxides in removing antimony from groundwater to levels below the drinking water regulation limit. *Sustainability* **2017**, *9*, 238. [CrossRef]

12.	Guo, X.; Wu, Z.; He, M. Removal of antimony(V) and antimony(III) from drinking water by coagulation–flocculation–sedimentation (CFS). *Water Res.* **2009**, *43*, 4327–4335. [CrossRef] [PubMed]

13.	Kang, M.; Kamei, T.; Magara, Y. Comparing poly-aluminum chloride and ferric chloride for antimony removal. *Water Res.* **2003**, *37*, 4171–4179. [CrossRef]

14.	Hering, J.G.; Katsoyiannis, I.A.; Theoduloz, G.A.; Berg, M.; Hug, S.J. Arsenic removal from drinking water: Experiences with technologies and constraints in practice. *J. Environ. Eng.* **2017**, *143*, 0311700. [CrossRef]

15.	Gregor, J. Arsenic removal during conventional aluminum-based drinking water treatment. *Water Res.* **2001**, *35*, 1659–1664. [CrossRef]

16.	Wu, Z.; He, M.; Guo, X.; Zhou, R. Removal of antimony (III) and antimony (V) from drinking water by ferric chloride coagulation: Competing ion effect and the mechanism analysis. *Sep. Purif. Technol.* **2010**, *76*, 184–190. [CrossRef]

17.	Ouzounis, K.; Katsoyiannis, I.A.; Zouboulis, A.I.; Mitrakas, M. Is the coagulation-filtration process with Fe(III) efficient for As(III) removal from groundwaters? *Sep. Sci. Technol.* **2015**, *50*, 1587–1592. [CrossRef]

18.	Inam, M.A.; Rizwan, K.; Du, R.P.; Yong-Woo, L.; Ich, T.Y. Removal of Sb(III) and Sb(V) by Ferric Chloride Coagulation: Implications of Fe Solubility. *Water* **2018**, *10*, 418. [CrossRef]

19.	Niedzielski, P.; Siepak, M. The occurrence and speciation of arsenic, antimony and selenium in ground water of Poznań city (Poland). *Chem. Geol.* **2005**, *21*, 241–253. [CrossRef]

20.	Tresintsi, S.; Simeonidis, K.; Zouboulis, A.; Mitrakas, M. Comparative study of As(V) removal by ferric coagulation and oxy-hydroxides adsorption: Laboratory and full scale case studies. *Desalin. Water Treat.* **2013**, *51*, 2872–2880. [CrossRef]

21.	Ebadi, A.; Soltan Mohammadzadeh, J.S.; Khudiev, A. What is the correct form of BET isotherms for modeling liquid phase adsorption? *Adsorption* **2009**, *15*, 65–73. [CrossRef]

22.	Mitrakas, M.; Panteliadis, P.; Keramidas, V.; Tzimou-Tsitouridou, R.; Sikalidis, C. Predicting Fe$^{3+}$ dose for As(V) removal at pHs and temperatures commonly encountered in natural waters. *Chem. Eng. J.* **2009**, *155*, 716–721. [CrossRef]

*Article*

# Study of Sludge Particles Formed during Coagulation of Synthetic and Municipal Wastewater for Increasing the Sludge Dewatering Efficiency

**Lech Smoczynski [1], Slawomir Kalinowski [1], Igor Cretescu [2,\*], Michal Smoczynski [3], Harsha Ratnaweera [4], Mihaela Trifescu [1] and Marta Kosobucka [1]**

[1]    Faculty of Environmental Management and Agriculture, University of Warmia and Mazury in Olsztyn, 10719 Olsztyn, Poland; lechs@uwm.edu.pl (L.S.); kalinow@uwm.edu.pl (S.K.); trifescumihaela@gmail.com (M.T.); marta.kosobucka@uwm.edu.pl (M.K.)

[2]    Faculty of Chemical Engineering and Environmental Protection, "Gheorghe Asachi" Technical University of Iasi, 700050 Iasi, Romania

[3]    Faculty of Food Sciences, University of Warmia and Mazury in Olsztyn, 10719 Olsztyn, Poland; michal.smoczynski@uwm.edu.pl

[4]    Faculty of Sciences and Technology, Norwegian University of Life Sciences, 0454 Ås, Norway; harsha.ratnaweera@nmbu.no

\*    Correspondence: icre1@yahoo.co.uk or icre@ch.tuiasi.ro; Tel.: +40-741-914-342

Received: 30 August 2018; Accepted: 15 October 2018; Published: 9 January 2019

**Abstract:** Municipal wastewater sludge was produced by chemical coagulation of synthetic wastewater (*sww*) based on Synthene Scarlet P3GL disperse dye and real municipal wastewater (*nww*), coagulated by commercial coagulants PAX (prepolymerised aluminum coagulant) and PIX (a ferric coagulant based on $Fe_2(SO_4)_3$). An attempt was made to correlate the sludge's dewatering capacity (in terms of capillary suction time—CST) with operation parameters for wastewater treatment, size distribution and specific surface area of the sludge particles. It was found that the presence of phosphate ions in the system facilitates the removal efficiency of the above-mentioned dye (L) due to the interaction between the dye molecules and $H_2PO_4^-$ ions. Unlike *sww*, negatively charged organic substances (*sorg*) in *nww* are directly adsorbed on the surface of colloidal particles {$Fe(OH)_3$} and {$Al(OH)_3$} (*prtc*). It was also discovered that an increase in the dose of a coagulant led to an increase of CST for *sww* sludge and to a decrease of CST for *nww* sludge. It has been suggested that flocs composed of spherical {$Al(OH)_3$} units possessed more internal space for water than aggregates consisting of rod-shaped {$Fe(OH)_3$} units and, consequently, it is more difficult to remove water from Al-*sww* sludge than from Fe-*sww*. The results obtained showed that smaller particles dominate in *sww* sludge, while larger particles are prevalent in *nww* sludge. To explain this distinct difference in the size distribution of particles in sludge obtained with the use of $Al^{3+}$ and $Fe^{3+}$, simple models of aggregation and agglomeration-flocculation processes (*aaf*) of treated wastewater have been proposed. Except for PIX in *nww*, the analyzed particles of the investigated types of sludge were characterized by similar specific surface area (Sps), regardless of the kind of sludge or the applied coagulant. Slightly larger, negatively-charged *sorg* bridges, anchored directly on the surface of positive *prtc* are more effective in closing the structure of *nww* sludge than small L bridges of the dye molecules anchored on the surface of *prtc* via $H_2PO_4^-$. All the discovered aspects could lead to improved performance of wastewater treatment plants (WWTP) by increasing the efficiency of sludge dewatering.

**Keywords:** municipal wastewater sludge; dewatering; particle size distribution

## 1. Introduction

Chemical coagulation is the second step in wastewater treatment [1,2]. This process has a significant influence on the properties and structure of the resulting sludge [3–7]. The pH of sludge decreases during chemical coagulation because $Fe^{3+}$ or $Al^{3+}$ undergoes cationic hydrolysis and its first stage can be described as follows:

$$Fe^{3+} (Al^{3+}) + HOH = FeOH^{2+} (AlOH^{2+}) + H^+ \tag{1}$$

Positively-charged, colloidal particles $\{Fe(OH)_3\}$ and/or $\{Al(OH)_3\}$, described further as *prtc* (colloidal particles), are formed during the subsequent reactions which occur in wastewater [1,8]. These particles act as adsorbents [9] in the processes of aggregation and agglomeration-flocculation (*aaf*) of pollutants from wastewater. The surface charge of *prtc* is neutralized by negatively-charged components such as $H_2PO_4^-$, or by negatively-charged colloids, called *sorg* (organic substances). The removal of neutral or positively-charged pollutants takes place during the so-called flocculation [1,5,10]. This mechanism dominates also in the case of wastewater treatment by electrocoagulation [11–13], particularly during recirculating electrolysis of wastewater [14].

The quantity and quality of municipal wastewater sludge are indicators for the wastewater treatment performance and, on the other hand, determine the possibilities for further utilization of sludge [15]. For practical and technological reasons, the dewatering capability of sludge is extremely important [16]. This property of municipal wastewater sludge very much depends on the type and dose of inorganic coagulant added to the wastewater [17–19]. The structure of sludge flocs, as well as their physical and chemical characteristics, determine the efficiency of the dewatering process of municipal wastewater sludge [16,20]. Under laboratory conditions, so-called capillary suction time (CST) is a measure of the sludge's dewatering capacity [21]. Reproducibility and precision of CST measurements [22,23] are important, both for theoretical considerations [24] and in practice, e.g., for determination of an appropriate dose of an inorganic flocculant [25].

Probably the most important effect of flocs on the structure of municipal wastewater sludge is through aggregation and agglomeration-flocculation (*aaf*) processes [26–29]. Numerous papers have been published, including information about measurements, modeling and characterization of the structure of municipal wastewater sludge [30–33]. Often, the so-called fractal dimension D becomes a specific research instrument [34,35]. For self-similar objects, e.g., sludge floc-aggregates whose structure does not depend on a change in the scale, D is defined as follows:

$$M(R) \sim R^D \tag{2}$$

where M is the mass comprised in a sphere of the diameter R [34].

The value of D, either determined experimentally or calculated theoretically, has been used in many theoretical considerations, e.g., in kinetic calculations [36], and also in practical solutions, e.g., for the filtration of excessive active sludge [37]. The structure and properties of municipal wastewater sludge can also be examined directly through the determination of the distribution of the sizes of particles and their specific surface [38–40].

Since it is suspected that there is an interpretable correlation between the structure of municipal wastewater sludge and the value of CST corresponding to the sludge, this paper analyzes the observed correlations and proposes simple models of *aaf* processes for chemically-coagulated synthetic and real municipal wastewater.

The consecutive stages presented and considered in this article are based on:

(a)   results of traditional jar tests,
(b)   CST measurements,
(c)   determination of the volumetric dimension (Dv) and respectively of the specific surface area (Sps).

## 2. Materials and Methods

Synthetic dyeing wastewater (*sww*) and real municipal wastewater (*nww*) originating from a wastewater treatment plant in Reszel (North-Eastern Poland) were investigated. The municipal wastewater had the following parameters: suspended solids: SS (mg/L) = 250–800; total phosphorus: P (mg/L) = 9–13.5, chemical oxygen demand: COD (mgO$_2$/L) = 600–1800; turbidity: TU = 80–160 NTU; and pH = 6.6–7.8. In turn, each 1 L of *sww* contained 31.3 mg H$_2$PO$_4^-$ (10 mg P) and 50 mg disperse dye (L) (Synthene Scarlet P3GL) produced by the Boruta-Zachem Chemical Company, from Zgierz (Poland). Therefore, the composition of synthetic dyeing wastewater (31.3 mg H$_2$PO$_4^-$ + 50 mg disperse dye/L) is a result of the synergy effect of both components; a mixture of dye and municipal wastewaters is susceptible on coagulation and/or electrocoagulation, which has been proved and explained by previous studies [41]. It was experimentally demonstrated that in the absence of P-PO$_4$ it is impossible to remove even a small amount of dye L by chemical coagulation of its aqueous solution. However, coagulation of L solution proceeds effectively and efficiently in the presence of phosphate ions [42] and for this reason we decided to use this composition of synthetic wastewater in the present study.

The sludge samples submitted for further tests were obtained in wastewater coagulation by: (a) PIX 113, a ferric coagulant (based on Fe$_2$(SO$_4$)$_3$) widely used in Poland, and (b) PAX 18, pre-polymerised aluminum coagulant, an alternative to PIX. Both coagulants were produced by Kemipol, the Polish branch of the Kemira Chemicals in Gdansk, Poland. By adding the above-mentioned coagulants to the wastewaters, many intermediate polymeric species such as (Al(OH)$_3$)$_n$, (Fe(OH)$_3$)$_m$ or (Fe(OH)$_2$)$_p$ are produced, being responsible for colloidal sorption on their surfaces of pollutants from wastewaters.

In *sww*, the dye concentration was determined spectrophotometrically at a wavelength $\lambda$ = 460 nm, while P-PO$_4$ was assayed according to the standard method ($\lambda$ = 690 nm) using a HACH DR 3900 instrument (Hach Company, Loveland, CO, USA) with 13 mm standard cell tests.

After 30 min of sedimentation, followed by decantation, 25.0 cm$^3$ of separated sludge was collected in order to determine capillary suction time CST [21–23], which is a measurement of the dewaterability of sludge. In CST measurements, a new prototype developed in the University of Warmia and Mazury in Olsztyn, Poland (DWTEST—Dewatering tester, schematically illustrated in Figure 1), was used.

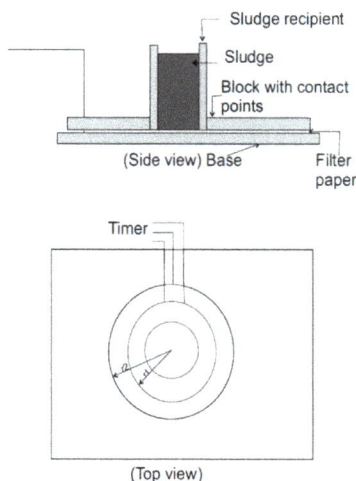

**Figure 1.** The schematic diagrame of a new DWTEST prototype for measurement of the capillary suction time (CST).

In each case, 25 cm$^3$ of separated sludge was applied onto a Whatman filter paper disc, while pressing the appropriate button on the apparatus. After several dozen seconds, the value of measured time corresponding to the flow of the liquid between electrodes in the measuring cell was displayed. The measurements were always repeated three times and the values for standard deviation (SD) (s) are indicated in the graphs. The results of CST measurements are presented in the graphical form in dependence of coagulant dose CST = f(mg Al or Fe/L).

Particle size distribution of the sludge was determined by measurement of laser light dispersion using a Mastersizer 3000 unit, Malvern Instruments, Malvern, United Kingdom [43]. Sludge samples were instilled to a measuring cell until an obscuration of 5–15% was achieved. The refractive indices for water and wastewater were 1.33 and 3.80, respectively. The particle size distribution was used to determine the mean particle size D$_{(3,2)}$ (μm) and available surface area (Sps) of particles in the sludge (m$^2$/g). Mean particle size is defined as [43]:

$$D_{(3,2)} = n_i d_i^3 / \Sigma n_i d_i^2 \tag{3}$$

where n$_i$ is the number of particles of diameter d$_i$.

Volumetric dimensions Dv10, Dv50 and Dv90 denoting the maximum particle diameter below which particles account for 10%, 50% and 90% of the volume of the analyzed sludge, respectively, were also calculated. The results represent the mean values of three replications for each type of sludge evaluated.

## 3. Results and Discussion

The results of the laboratory tests are presented in Tables 1 and 2 and some of them in Figure 2. The mean values from three repetitions were used for plotting the graphs and SD values (in % or s) were marked in each case.

Figure 2 illustrates the effects of coagulation of *sww* and *nww* by coagulants PIX and PAX.

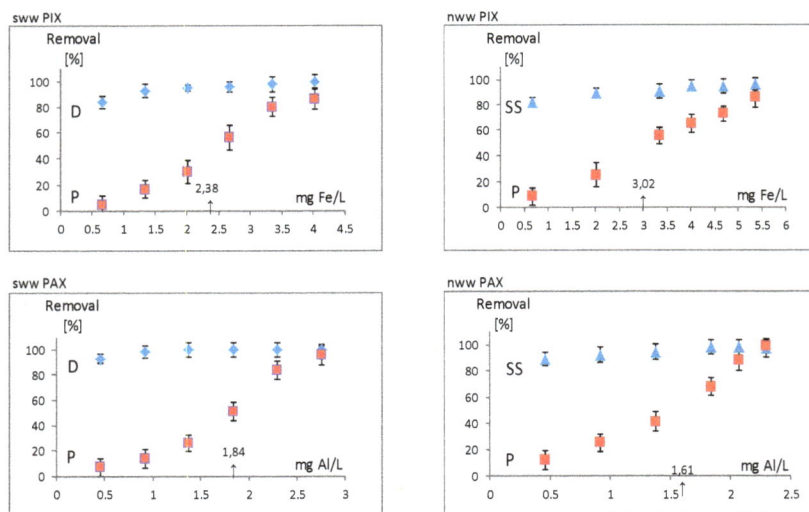

**Figure 2.** Synthetic wastewater (*sww*) and real municipal wastewater (*nww*) coagulated by PIX and PAX.

The initial ($_o$) and final ($_f$) values of the pH, concentration of total phosphorus (P), chemical oxygen demand (COD) and suspended solids (SS) are presented for both PIX and PAX coagulants used for treatment of both *sww* and *nww*:

PIX-*sww*: $pH_o = 5.2 \rightarrow pH_f = 4.75$ PAX-*sww*: $pH_o = 5.2 \rightarrow pH_f = 5.0$

PIX-*nww*: $SS_o = 750$, $P_o = 11.40$; $COD_o = 1690 \rightarrow COD_f = 360$ (mgO$_2$/L);

$$pH_o = 7.53 \rightarrow pH_f = 6.84$$

PAX-*nww*: $SS_o = 440$, $P_o = 9.75$; $COD_o = 1190 \rightarrow COD_f = 313$ (mgO$_2$/L);

$$pH_o = 7.80 \rightarrow pH_f = 7.14$$

Courses of all four dependences, where Removal = f(coagulant dose), showed that the highest dose of both aluminum ions and ferric ions ensured 100% removal of L from *sww* and SS from *nww*. Under these conditions, PIX removed 86–87%, and PAX 96–99% of phosphorus (P) from both *sww* and *nww*. Because the limit of P allowed in the effluents according to Polish legislation for wastewater treatment plants (WTTP) is 5 mg P/L, the sludge analyzed in our research was precipitated under conditions ensuring 50% removal of P from treated wastewater. It can be assumed that primary coagulation of municipal wastewater in Poland is sufficient to obtain the required P level in effluent, principally owing to the mentioned 50% coagulation of phosphorus compounds. In all of the four diagrams seen in Figure 2, an arrow on the x-axis indicates the numerical value of the dose of a coagulant (*fes, fen, als, aln*) adequate for attaining 50% of phosphorus removal from *sww* and *nww*. When these doses are added, i.e., to *sww*: *fes* = 2.38 mg Fe/L $\equiv$ 0.0425 mmol Fe/L and *als* =1.84 mg Al/L $\equiv$ 0.068 mmol Al/L; and to *nww*: *fen* = 3.02 mg Fe/L $\equiv$ 0.054 mmol Fe/L and *aln* = 1.61 mg Al/L $\equiv$ 0.060 mmol Al/L, sediments were obtained, which were subsequently tested to determine CST, Sps and Dv (Table 1) as well as percentage shares of particular particle sizes.

During the coagulation process of *sww*, the presence of phosphate ions in the system made it possible to effectively remove the dye from the liquid phase of wastewater. Phosphate anions are adsorbed on positively-charged colloidal particles {Fe(OH)$_3$} and/or {Al(OH)$_3$} (*prtc*), creating units of the type:

$$\{Fe(OH)_3\} - (H_2PO_4{}^-)_c \text{ and } \{Al(OH)_3\} - (H_2PO_4{}^-)_d \tag{4}$$

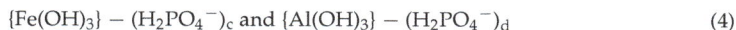

As the negative sorbate (phosphate) accumulates on *prtc*, the positive potential of the systems in Formulas (4) decreases and mutual repulsive forces weaken between particles, which determine the direction, range and intensity of Brownian motions. Thus, the stability of the colloidal system decreases, while the probability of collisions between particles rises and finally an *aaf* process may occur.

In practice, an aqueous solution of phosphate ions undergoes coagulation (both precipitation and adsorption) and almost all phosphates could be transferred into municipal wastewater sludge.

As mentioned at the beginning of the Materials and Methods Section, in the absence of P-PO$_4$ it was impossible to remove even a small amount of a dye directly, by chemical coagulation of a water solution of L, whereas the coagulation or electro-coagulation of L [42] proceeded efficiently only when "supported" by phosphate ions. To explain the reasons for this phenomenon, the following simple model of adsorption of a dye particle to systems is proposed in Formulas (5) and (6):

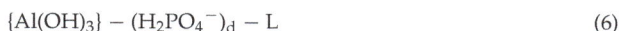

$$\{Fe(OH)_3\} - (H_2PO_4{}^-)_c - L \tag{5}$$

$$\{Al(OH)_3\} - (H_2PO_4{}^-)_d - L \tag{6}$$

This model assumes a slightly positive surface charge of a dye molecule, which repels it from a positive *prtc* while attracting it to *prtc* centers absorbing H$_2$PO$_4{}^-$. It is thus based on the assumption that dye molecules can be adsorbed to units, Formulas (4), by forming bridges with the help of previously adsorbed H$_2$PO$_4{}^-$ ions (Formulas (5) and (6)). Although the occurrence of dye molecules, bridged on the surface of *prtc* with the help of H$_2$PO$_4{}^-$ slightly decreases the process of neutralization of the system's surface, it also causes the growth of its mass and size, which is conducive to destabilization of *prtc* and leads to *aaf*. Thus, in laboratory practice, slightly more coagulant is used to achieve destabilization of a mixture of dye and H$_2$PO$_4{}^-$ than for a solution containing only H$_2$PO$_4{}^-$.

Figure 3a shows volumetric proportions for a spherical colloidal particle {Al(OH)$_3$} (sized from 86 to 206 µm, for which the diameter was assumed to be around 165 nm [44], which in fact, has a rather corrugated surface), a molecule of the dye (having an estimated 1 mole ≈ 500 g and a size of about 10 nm) and a H$_2$PO$_4$$^-$ ion (with a diameter of 1.2 nm). In the subsequent models presented in this paper (Figure 3b), for better clarity of the visualization, rod-shaped particles {Fe(OH)$_3$} [45] and spherical particles {Al(OH)$_3$} [44] are presented in appropriately diminished sizes, which is highlighted by using a broken line to draw their contours.

As expected, amounts of mg Fe/L from PIX (*fes* and *fen*) needed to remove 50% of P were higher than the respective amounts of mg Al/L from PAX (*als* and *aln*). Simultaneously, the same amounts expressed in mmol/L were higher for PAX than for PIX. In Poland PIX, rather than PAX, is a more popular coagulant in urban wastewater treatment plants (despite being less efficient) because of its much lower price per 1 kg of coagulant.

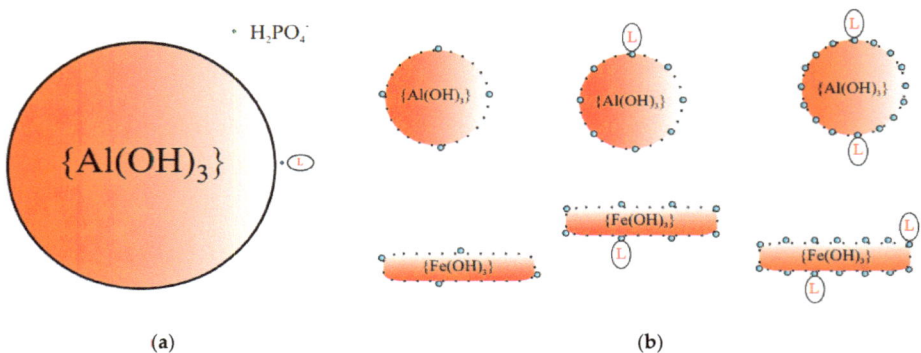

(a)                                                                 (b)

**Figure 3.** Schematic diagrams of interactions between colloidal particles and a dye molecule and H$_2$PO$_4$$^-$ ion, respectively: (**a**) Approximate proportions between sizes of: colloidal particle {Al(OH)$_3$}, dye L molecule and H$_2$PO$_4$$^-$ ion; (**b**) schemes of structures for Formulas (4) and (5) responsible for adsorption of H$_2$PO$_4$$^-$ and L from *sww* on the surface of colloidal particles (*prtc*).

Slightly more mg/L PIX was applied to remove 50% P from *nww* than from *sww*. Most probably, a part of the PIX dose in *nww* had been sacrificed to remove 1.330 mg/L *sorg* denoted as COD; COD$_o$ = 1690 → COD$_f$ = 360 mgO$_2$/L (Figure 2). Meanwhile, the same sample was characterized by a slightly higher concentration of phosphorus (P$_o$ = 11.4 → P$_f$ = 0.64 mg/L) and SS$_o$ (750 mg/L) than the sample treated by PAX (P$_o$ = 9.75 → P$_f$ = 0.22 mg/L, SS$_o$ = 440 mg/L). For *nww* a slightly lower dose of PAX in mg/L, than from *sww*, was needed to remove 50% of phosphorus, which was most probably a consequence of lower consumption of the coagulant for removing just 887 mgO$_2$/L COD from *nww* (COD$_o$ = 1190 → COD$_f$ = 313 mgO$_2$/L) at an approximately similar efficiency as PIX in the removal of L from *sww*.

Figure 4 contains a schematic representation of the removal of phosphorus as well as *sorg* from *nww* using PIX and PAX, respectively. Unlike L being bridged in *sww* by H$_2$PO$_4$$^-$, negatively charged *sorg* in *nww* are directly adsorbed on the surface of *prtc*. The schemas shown in Figure 4 also account for the hydrophilic properties of wastewater sol (*sorg*), marking in yellow the water molecules which stabilize this sol. Because of the negative charge of *sorg* here (*nww*), fewer H$_2$PO$_4$$^-$ are adsorbed on the surface of *prtc* (only 7 H$_2$PO$_4$$^-$ ions in the schema in Figure 4), compared to the respective surfaces of *prtc* formed during *sww* coagulation.

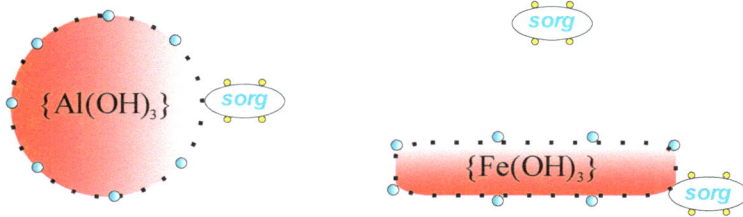

**Figure 4.** Schemas of structures responsible for adsorption of $H_2PO_4^-$ and organic substances (*sorg*) from *nww* on the surface of *prtc*.

The following graphs in Figure 5 show the process of dewatering *sww* and *nww*, presented in the form of CST = f(dose of coagulant) relationships.

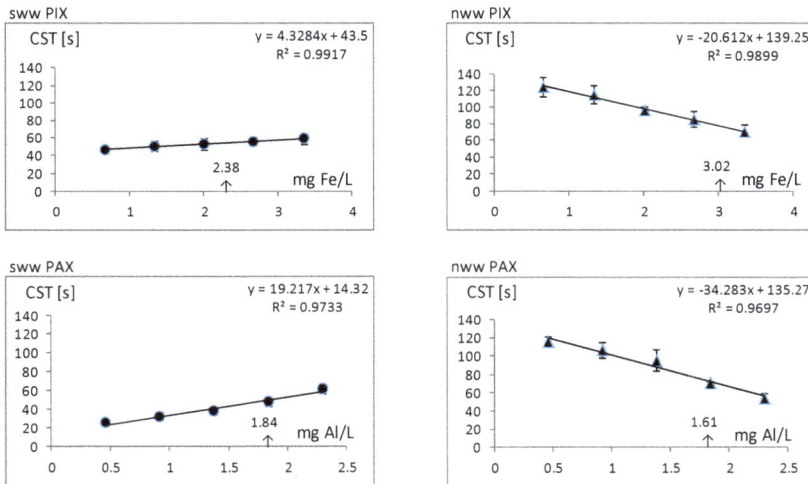

**Figure 5.** Dewatering of sludge obtained from *sww* and *nww* coagulated with PIX and PAX.

In both cases, the doses of coagulants (0.67 to 3.35 mg Fe/L and 0.46 to 2.3 mg Al/L) are within the range of doses used in the coagulation tests illustrated in Figure 2. According to Figure 5, the low values of SD (s), show a high repeatability of the measurements for capillary suction time CST, carried out by DWTEST instrument. For *sww* wastewater, the tendency of the relationship $CST_{sww}$ = f(dose of coagulant) demonstrates a linear increase in CST, when the dose of coagulant is increased. Conversely, the increase in the dose of added coagulate to *nww* causes a linear decrease in $CST_{nww}$. The high values of the regression correlation coeficient $R^2$, (which are ranged between 0.9697 and 0.9917), provide the linear character confirmation for the regression relationship CST = f(dose of coagulant). For *fes*, the $CST_{sww}$ was 53.5 s, and for *als* it was 58 s.

With increasing doses of coagulant, $CST_{sww}$ increased from 46 to 58 s for PIX and over a slightly wider range from 25 to 61 s for PAX. Undoubtedly, the structure of *sww* sludge obtained with PIX is different from the structure of sludge achieved with PAX, although the mechanism of bridging particular units is similar. As the dose of a coagulant increased, the share of these units in structures of sludge increased, which may have led to the blocking of water molecules in sludge flocs and a subsequent increase in CST. It is known that micelles $\{Fe(OH)_3\}_n$ and $\{Al(OH)_3\}_n$ differ from each other in shape and dimension [44,45]. The regression equations (Figure 5) allow the CST to be easily calculated for an identical dose of both coagulants, which is 0.05 mmol/L (within the range of *fes* = 0.045 and *als* = 0.068): $CST_{Fe}$ = 55.5 s, and $CST_{Al}$ = 40.3 s. The next calculation can be made for the same dose of both coagulants, equal to 0.1 mmol/L. Here, $CST_{Fe}$ = 67.7 s, and $CST_{Al}$ = 66.2 s, which almost the same. By extrapolating the CST values to higher coagulant doses, it can be hypothesized that *sww* sludge with PIX binds water more effectively at lower Fe doses, while *sww* sludge with PAX binds water more effectively at higher Al doses. Due to the small number of components in *sww* sludge as described in Formula (7):

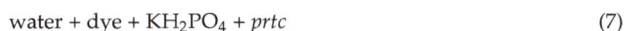

$$water + dye + KH_2PO_4 + prtc \tag{7}$$

the only explanation of the CST = f(dose of coagulant) is the structure of this sludge.

The structures illustrated in Figure 6 schematically describe the progressing *aaf* processes. It is clear that repulsion, as well as the intensity and scope of Brownian motions, decrease as the dose of a coagulant increases. The network of bridging and connections grows in the emerging aggregates-flocs of sediment *sww*. The schemas in Figure 6 may illustrate the destabilization of a single cluster (previously presented in Figure 3) due to the formation of appropriate dimers, in which a dye molecule most probably can bridge two *prtc* partly destabilized by $H_2PO_4{}^-$. At an increasing dose of a coagulant, these structures are most probably bridging, thus binding water in the internal spaces of agglomerate-flocs. It is known that rod-shaped colloidal particles, e.g., $\{Fe(OH)_3\}$, coagulate more rapidly than spherical units $\{Al(OH)_3\}$. However, the aggregation of many "rods" ultimately leads to the formation of a spherical aggregate-floc. Most probably, spherical aggregates composed of spherical units $\{Al(OH)_3\}$ leave more internal space for water than other spherical aggregates composed of rod-shaped $\{Fe(OH)_3\}$, which is why it is more difficult to remove water from Al-*sww* (longer CST) than from Fe-*sww* (shorter CST).

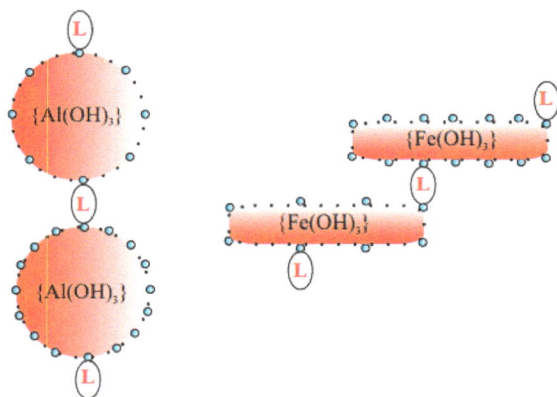

**Figure 6.** Schema of the progress of *aaf sww*.

As mentioned above, in contrast to *sww* sludge, an increase in the dose of a coagulant (both PIX and PAX) added to *nww* caused a linear decrease $CST_{nww}$. The $CST_{sww}$ for *fen* was 77 s, and for *aln* it was 80 s. As the dose of a coagulant increased, the $CST_{nww}$ decreased from 124 to 70 s for PIX, and from 115 to 54 s for PAX. In comparison with *sww*, in this case the variation of CST values for sludge obtained with PIX and PAX was distinctly smaller. From an appropriate regression equation (Figure 5), for identical doses of both coagulants (0.05 mmol/L), the following values of CST were calculated: $CST_{Fe}$ = 81.5 s and $CST_{Al}$ = 89 s. Analogously to *sww* tests, the subsequent calculations were made for the same dose of the coagulants equal to 0.1 mmol/L. The results were $CST_{Fe}$ = 23.8 s and $CST_{Al}$ = 42.7 s, which means that $CST_{Al}$ is distinctly higher than $CST_{Fe}$. When extrapolating

CST values to higher coagulant doses, it appears that at higher coagulant doses *nww* sludge with PAX may bind water more effectively than *nww* sludge with PIX. Compared to *sww*, *nww* sludge is a much more complex, multi-component system. From this point of view, the most significant constituent of *nww* sludge is *sorg*. Most probably, the participation of *sorg* in the *aaf* process, leading to the formation of *nww* sludge, is responsible for the negative regression of the course of the CST = f(dose of coagulant) function.

Figure 7 shows a suggested schema of the *aaf* processes occurring during *nww* coagulation. Similarly to *sww*, an increase in the dose of a coagulant leads to a decrease in repulsion as well as the intensity and scope of Brownian motions in the system. The network of branches and connections in the *sww* sludge is expanding. The schema in Figure 7 may illustrate the destabilization of a single unit from Figure 4 through the formation of appropriate dimers, in which *sorg* most probably bridges two *prtc* partly destabilized by $H_2PO_4^-$. At an increasing dose of the coagulant, these structures most probably begin to branch, binding water in internal spaces within agglomerates-flocs. Organic compounds, *sorg*, which in total constitute so-called the "negative wastewater colloid" of *nww*, are classified as hydrophilic colloids, stabilized by the hydration shell of water molecules. As the negative wastewater colloid is progressively destabilized, and the *aaf* processes are in progress, the concentration of *sorg* decreases and water molecules released from the hydration shell become "available", which leads to a decrease in CST. The increasing difference in the values of $CST_{Fe}$ and $CST_{Al}$ at increasing doses of a coagulant can be explained in this case as for *sww*. Gradually branching (Figure 7) structures containing spherical {Al(OH)$_3$} units absorbing components of *nww* block the "inner-network" water more effectively than rod-shaped units with {Fe(OH)$_3$}.

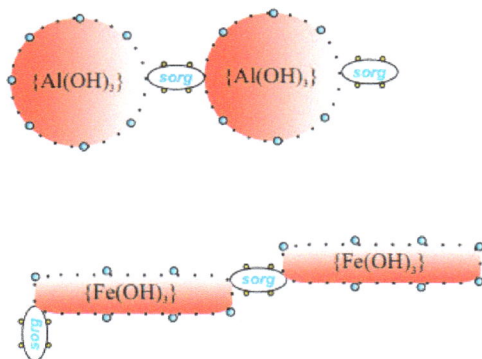

**Figure 7.** Schema of the progress of *aaf* processes in *nww*.

Figure 8 shows the percentages of specific size classes of particles in sludge obtained in the coagulation of *sww* and *nww*; using a) *fes* and *als*, and b) *fen* and *aln*. For both coagulants and in both types of wastewater, two classes of sludge particle size can be distinguished. The *sww* type of sludge is dominated by particles of a smaller size: 3–40 µm, but with a maximum of 3.72% at 12.7 µm for PIX, and 5–75 µm with a maximum of 6.38% at 24.1 µm for PAX. On the other hand, *nww* is dominated by sludge particles larger in size, in the range of 45–1300 µm for both coagulants. For the very low doses of a coagulant applied, the dominant size of *nww* sludge particles within the first range of sizes is difficult to define. For both coagulants, about 3.5% of sludge particles are sized about 130 µm, and are located within the second range.

Distributions of sludge particle size classes were very similar to the ones achieved for *nww* which were obtained in real municipal wastewater coagulated with the minimal doses of PIX (13.55 mg Fe/L) and PAX (2.45 mg Al/L), respectively [40]. It can be observed that these doses were much higher than the doses used in this study (3.02 mg Fe/L and 1.61 mg Al/L, respectively).

At the lower coagulant dose, larger size particles > 100 μm were prevalent as well. On the other hand, distributions of sludge particle classes similar to the ones identified in *sww* have also been recorded previously [39] in sludge obtained from synthetic wastewater coagulated with minimal doses of PIX (6.5% of particles about 4 μm in size) and PAX (10% of particles about 10 μm in size). Certain similarities and differences in particle size classes between *sww* and *nww* sludge discussed here are most probably a consequence of very low doses of coagulant; in this study, a coagulant dose was high enough to remove only 50% of P. Development of this issue will be completed by using the values Sps and Dv, which are collected in Table 1.

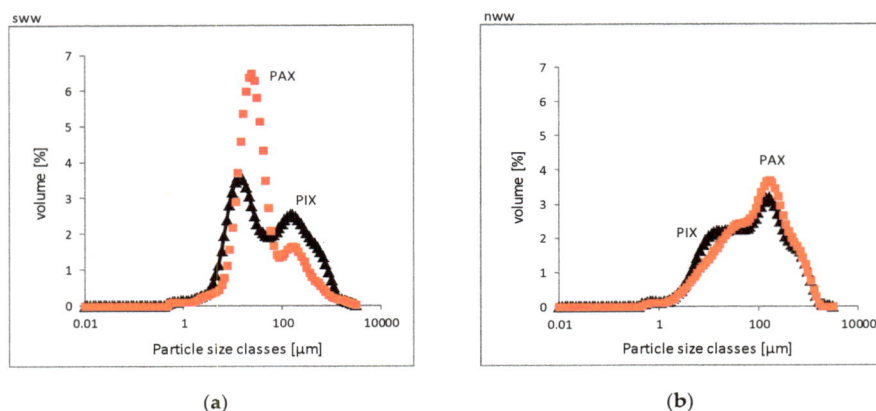

**Figure 8.** Sizes of flocs of sludge obtained from *sww* and *nww* coagulated with the help of: (**a**) *fes* and *als*, (**b**) *fen* and *aln*.

**Table 1.** The specific surface area (Sps) and volumetric dimension (Dv) values, for sludge particles illustrated in Figure 8.

| Properties | PIX | PAX | PIX | PAX |
|---|---|---|---|---|
| | *sww*/SD | *sww*/SD | *nww*/SD | *nww*/SD |
| Sps (m$^2 \cdot$g$^{-1}$) | 355/13.9 | 283/10.0 | 283/10.0 | 350/12.2 |
| Dv10 (μm) | 8.4/1.7 | 9.4/1.2 | 9.4/1.2 | 7.4/1.2 |
| Dv50 (μm) | 48.9/3.3 | 99.9/4.5 | 99.9/4.5 | 77.3/4.8 |
| Dv90 (μm) | 702/17.5 | 537/15.8 | 537/15.8 | 548/6.9 |

Table 1 presents Sps and Dv (including SD values) for particles of sludge obtained in the conditions illustrated in Figures 2 and 5 and specified in Figures 6 and 7.

The values of Sps (ranged between 283 and 355 m$^2$/g), were similar to the Sps values of 260–360 m$^2$/g, achieved for sludge obtained at coagulation of real municipal wastewater [40]. It is note worthy that such Sps values were obtained for all doses (i.e., minimal, optimal and maximal) of each coagulant PIX and PAX, respectively. Dv90 *sww* (Table 1), was slightly higher than Dv90 *nww*, which means that the applied doses of both coagulants formed slightly more uniform/homogenous particles in *nww* sludge than in *sww* sludge. It may also be suggested that the overall diversity of structures of the particles in *sww* sludge is higher than the diversity of particles in *nww* sludge.

Figure 9 contains a schematic presentation of the development of flocs in *sww* and *nww* sludge obtained with PIX. The low Dv90 value for *nww* indicated a generally higher uniformity of flocs in sludge built on *sorg* bridges than in sludge built on D bridges. The Dv90-based floc uniformity concept generally indicates the upper limit for the diameter and thus, within the specified range, limits the degree of raggedness or branching of these objects.

Since it is not an easy task to illustrate this problem, the objects presented in Figure 9 to a large extent are schematic representations. It seems likely that slightly larger, negatively charged *sorg* bridges, anchored directly on the surface of positive *prtc* more effectively close the sludge structures than the smaller L bridges anchored on the *prtc* surface by the $HPO_4^-$, because L is rather unattractive for positive *prtc*.

Thus, the left part of Figure 9 shows the closed, and the right part shows the partly open, structure of two aggregates of *nww* sludge, similar in size, and an analogous, closed and completely open structure of two aggregates of *sww* sludge. For such structures, Dv90 of *nww* would be lower than Dv90 of *sww*, represented by the "open aggregate" in the lower, right-hand corner of Figure 9. At the same time, the higher Dv90 value of *sww* sludge may generally indicate greater structural differentiation in *sww* sludge than in *nww* sludge.

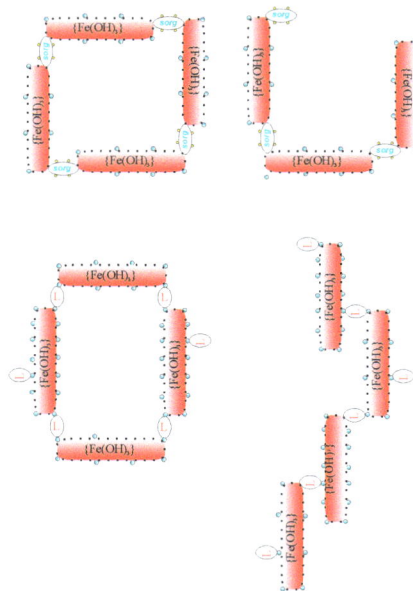

**Figure 9.** Schemas of more homogenous structures of *nww* sludge (Dv90 = 537 μm) and less homogenous *sww* sludge (Dv90 = 702 μm) obtained after coagulation with PIX.

Unquestionably, the progress of *aaf* processes is a direct consequence of the primary process forming the units illustrated in Figures 3 and 4. Due to a large deficit of the coagulant (an amount needed to remove just 50% of P), the conditions applied in this study can be referred to as "sub-stoichiometric". In line with the Langmuir's theory of adsorption, under such conditions we should expect complete saturation/use of the surface of *prtc* by adsorbed substances; i.e., L and P-PO$_4$ in *sww* flocs, and SS, *sorg* and P in *nww* flocs. In this case, individual units of flocs being formed should be characterized by higher porosity than units formed under "stoichiometric" conditions, especially in excessive amounts of a coagulant [37,39]. Porosity of floc-cluster units may influence the growth of Sps. At the same time, the larger filling of the centers on the adsorbent surface (*prtc*) should favor the *aaf* processes, as the forces of mutual repulsion decrease between the "saturated" surfaces of the units. As a result, the final porosity of these units only exerts a slight effect on the size, and on the final Sps value, for closed floc structures formed under sub-stoichiometric conditions.

It has been explained [43] that flocs of the analyzed sludge are formed from the units presented in Figure 9. Filling the space with the mass of a sludge floc can be defined with the fractal dimension, $D_f$, previously defined in Equation (2). There are more data available [19,29,46–48] on the fractal dimension values of various types of flocs, including the values of $D_f$ in sludge of municipal wastewater. Depending on the measurement method applied, the type of a coagulant and other parameters of the process, the values of $D_f$ can be equal with the following values: 1.72 [19]; 1.68–1.74 [3]; 1.50–1.87 [4]; 1.69–1.96 [20]; 1.7–1.8 [46], 1.67–1.90 [47] and 1.8 [48]. Both fundamentally and practically, it may be interesting to compare the number of units, schematically illustrated in Figures 3 and 4, in a single floc of *sww* and *nww* sludge. Based on the aforementioned $D_f$ values, the following assumptions were made for simple calculations of the number of units in such "statistical" floc: $D_f = 1.75$ and $R = Dv50$. For comparative purposes, it seems sufficient to choose $R = Dv50$, as it is implied by the definition of Dv, according to which this value can be the closest to the R diameter of a "statistical" floc. The values of the number of $n$ units in a statistical municipal wastewater sludge floc, as comprised in Table 2, were calculated from the following equation:

$$n = [R/r]^{Df} \tag{8}$$

where: $R = Dv50$ in µm, while $r = 0.17$ µm and is an approximate sum of diameters of individual components in a *sww* unit:

$$r(H_2PO_4{}^-) + r\{Al(OH)_3\} + r(H_2PO_4{}^-) + r(D)$$
$$= 0.0012 + 0.165 + 0.0012 + 0.010 = 0.1685 \ (\mu m) \tag{9}$$

Earlier, an assumption was made that $r(sorg)$ is slightly larger than $r(D)$. Simultaneously, considering the fact that the $r$ of a unit of *nww* sludge:

$$r\{Al(OH)_3\} + r(sorg) \tag{10}$$

does not contain $r(H_2PO_4{}^-)$ and in view of the lack of other data for calculations, the value of $r$ for *nww* was also adopted as being equal to 0.17 µm, because the dominant component of $r$ of a *nww* floc always has to be $r\{Al(OH)_3\}$.

**Table 2.** Number of units in a statistical floc of the analyzed sludge.

| Coagulant/Wastewater Type | Dv50 (µm) | Number of Units in a Floc |
|---|---|---|
| PIX/*sww* | 48.9 | 20,091 |
| PAX/*sww* | 32.4 | 9776 |
| PIX/*nww* | 99.9 | 70,138 |
| PAX/*nww* | 77.3 | 44,774 |

The data contained in Table 2 are not absolute values and can be used only for comparative purposes. They indicate a much denser filling of the space by units-clusters in a bigger floc (Dv50) of *nnw* sludge than that of a smaller floc (Dv50) in *sww* sludge. Within a certain range, these data confirm the general hypothesis of higher homogeneity of *nww* sludge, than of *sww* sludge (Figure 9).

The data allowed a comparison of the structures of two types of municipal wastewater sludge. In all presented case studies, the obtained sludge after chemical treatment of wastewater contained 3–5% dry solids, of which an average of 45% of the dry solids content was organic.

The results of such studies, when advanced and developed, can aid the improvement of practical wastewater treatment technologies. Simple procedures of simultaneous online measurements of Dv10, Dv50, Dv90 and Sps could be easily implemented at wastewater treatment plants. If a wastewater treatment plant is equipped with such an online measuring system, it will allow:

(1)   additional monitoring of the coagulation-flocculation process,

(2)   broader assessment of the "health"/quality of activated sludge,
(3)   better control and regulation of the process of municipal wastewater sludge dewatering.

## 4. Conclusions

Molecules of dye are adsorbed to *prtc* by bridging, with the help of $H_2PO_4^-$ ions previously adsorbed to these surfaces. Unlike in *sww*, negatively charged *sorg* in *nww* are directly adsorbed on the surface of *prtc*.

As the dose of a coagulant increased, the $CST_{sww}$ increased, while the $CST_{nww}$ decreased. The *sww* sludge with PAX binds water more effectively at higher Al doses. Aggregates composed of spherical {Al(OH)$_3$} units leave more internal space for water than aggregates built from rod-shaped {Fe(OH)$_3$} units and, in consequence, a higher dose of PAX (Al) means that it is more difficult to remove water from Al-*sww* sludge (longer CST) than from Fe-*sww* (shorter CST).

Particles that are smaller in size dominate in *sww* sludge, while larger particles prevail in *nww* sludge. Except for PIX in *nww*, the analyzed particles of the types of sludge tested were characterized by similar Sps, irrespective of the type of wastewater or the applied coagulant. Values of Dv90 for *sww* are slightly higher than Dv90 *nww*, which means that the differentiation of structures of *sww* sludge particles is greater than in *nww* sludge. Slightly larger, negatively charged *sorg* bridges, directly anchored on the surface of positive *prtc*, more effectively close structures of *nww* sludge than smaller L bridges, anchored on the surface of *prtc* via HPO$_4^-$ ions. The porosity of units has only a slight influence on the size and final value of Sps of closed structures of flocs formed under sub-stoichiometric conditions. The larger statistical floc of *nww* sludge is more densely packed with units-clusters than the space of a smaller *sww* floc.

**Author Contributions:** This article was written by L.S. and H.R. based on the investigations carried out in the frame of Polish-Norwegian Research Program operated by the National Centre for Research and Development under the Norwegian Financial Mechanism 2009–2014. These authors designed the research project and conducted the laboratory tests and analyses developed by Ph.D students M.K. and M.T. S.K. designed the equipment and provided technical knowledge to support the experimental work. M.S. carried out optical investigations and visualization and I.C. participated in reviewing and final writing of this manuscript.

**Funding:** This research was mainly funded by Polish-Norwegian Research Program, grant number, POL-NOR/196364/7/2013.; 561755-1-2015-1-NO-EPPKA2-CBHE-JP, (2015-3386/001-001); and statutory grant 20610.001-300.

**Acknowledgments:** This work was supported by a Polish-Norwegian Research Program operated by the National Centre for Research and Development under the Norwegian Financial Mechanism 2009–2014 in the framework of Project Contract No. POL-NOR/196364/7/2013; "Harmonizing water related graduate education" under grant 561755-1-2015-1-NO-EPPKA2-CBHE-JP (2015-3386/001-001); and under statutory grant 20610.001-300.

**Conflicts of Interest:** The authors declare no conflicts of interest.

## Nomenclature

| | |
|---|---|
| *sww* | synthetic wastewater |
| *nww* | real municipal wastewater (natural origin) |
| CST | capillary suction time (s) |
| *aaf* | aggregation-agglomeration-flocculation |
| SD | standard deviation of the sample (% or mg/L) |
| *fes* | number of mg $Fe^{3+}$ (PIX) which remove 50% P from 1 L of *sww* |
| *als* | umber of mg $Al^{3+}$ (PAX) which remove 50% P from 1 L of *sww* |
| *fen* | number of mg $Fe^{3+}$ (PIX) which remove 50% P from 1 L of *nww* |
| *aln* | number of mg $Fe^{3+}$ (PIX) which remove 50% P from 1 L of *nww* |
| Dv | volumetric dimension (μm) |
| Sps | specific surface area (m$^2$/g) |
| L | disperse dye (Synthene Scarlet P3GL) |
| SS | Suspended Solids (mg/L) |
| *prtc* | colloidal particle {Fe(OH)$_3$} or {Al(OH)$_3$} |
| *sorg* | organic substances (responsible for COD in *nww*) |

## References

1.  Duan, J.; Gregory, J. Coagulation by hydrolyzing metal salts. *Adv Colloid Interface Sci.* **2003**, *100–102*, 475–502. [CrossRef]
2.  The, C.Y.; Budiman, P.M.; Shak, K.P.; Wu, T.Y. Recent advancement of coagulation-flocculation and its application in wastewater treatment. *Ind. Eng. Chem. Res.* **2016**, *55*, 4363–4389.
3.  Zheng, H.; Shu, G.; Jiang, S.; Tshukudu, T.; Xiang, X.; Zhang, P.; He, Q. Investigations of coagulation-flocculation process by performance optimization, model prediction and fractal structure of flocs. *Desalination* **2011**, *269*, 148–156. [CrossRef]
4.  Chu, C.P.; Lee, D.J. Effect of pre-hydrolysis on floc structure. *J. Environ. Manag.* **2004**, *71*, 285–292. [CrossRef]
5.  Zhao, Y.X.; Shon, H.K.; Wang, Y.; Kim, J.H.; Yue, Q.Y. The effect of second coagulant dose on the regrowth of flocs formed by charge neutralization and sweep coagulation using titanium tetrachloride (TiCl$_4$). *J. Hazard Mater.* **2011**, *198*, 70–77. [CrossRef]
6.  Fytili, D.; Zabaniotou, A. Utilization of sewage sludge in EU application of old and new methods—A review. *Renew. Sustain. Energy Rev.* **2008**, *12*, 116–140. [CrossRef]
7.  Verna, S.; Prased, B.; Mishra, I.M. Pretreatment of petrochemical wastewater by coagulation and the sludge characteristics. *J. Hazard Mater.* **2010**, *178*, 1055–1064. [CrossRef]
8.  Rohrsetzer, S.; Paszli, I.; Csempesz, F.; Ban, S. Colloidal stability of electrostatically stabilized sol particles. Part I: The role of hydration in coagulation and repeptization of ferric hydroxide sol. *Colloid Polym. Sci.* **1992**, *270*, 1243–1251. [CrossRef]
9.  Smoczyński, L.; Ratnaweera, H.; Kosobucka, M.; Kvaal, K.; Smoczyński, M. Image Analysis of Sludge Aggregates Obtained at Preliminary Treatment of Sewage. *Water Sci. Technol.* **2014**, *70*, 1048–1055. [CrossRef]
10. Amirtharajah, A.; Mills, M.K. Rapid-mix design for mechanism of alum coagulation. *JAWWA* **1982**, *74*, 210–216. [CrossRef]
11. Butler, E.; Hung, Y.T.; Yu, L.R.; Al Ahmad, M. Electrocoagulation in wastewater treatment. *Water* **2011**, *3*, 395–525. [CrossRef]
12. Zaleschi, L.; Teodosiu, C.; Cretescu, I.; Rodrigo, M.A. A Comparative Study of Electrocoagulation and Chemical Coagulation Processes Applied for Wastewater Treatment. *Environ. Eng. Manag. J.* **2012**, *11*, 1517–1525.
13. Smoczynski, L.; Munska, K.; Pierozynski, B. Electrocoagulation of synthetic dairy wastewater. *Water Sci. Technol.* **2013**, *67*, 404–409. [CrossRef]
14. Groterud, O.; Smoczyński, L. Removal of phosphorus and residual aluminium by a recirculating electrolysis of wastewater. *Vatten* **1986**, *42*, 293–296.
15. Mikkelsen, L.H.; Keiding, K. Physico-chemical characteristics of full scale sewage sludge with implications to dewatering. *Water Res.* **2002**, *36*, 2451–2462. [CrossRef]
16. Jin, B.; Wilen, B.M.; Lant, P. Impact of morphological, physical and chemical properties of sludge flocs on dewaterability of activated sludge. *Chem. Eng. J.* **2004**, *98*, 115–126. [CrossRef]
17. Turchiuli, C.; Fargues, C. Influence of structural properties of alum and ferric flocs on sludge dewaterability. *Chem. Eng. J.* **2004**, *103*, 123–131. [CrossRef]
18. Niu, M.; Zhang, W.; Wang, D.; Chen, Y.; Chen, R. Correlation of physiochemical properties and sludge dewaterability under chemical conditioning using inorganic coagulants. *Bioresour. Technol.* **2013**, *144*, 337–343. [CrossRef]
19. Zhao, Y.Q. Correlations between floc physical properties and optimum polymer dosage in alum sludge conditioning and dewatering. *Chem. Eng. J.* **2003**, *92*, 227–235. [CrossRef]
20. Zhao, P.; Ge, S.; Chen, Z.; Li, X. Study on pore characteristics of flocs and sludge dewaterability based on fractal methods (pore characteristics of flocs and sludge dewatering). *Appl. Therm. Eng.* **2013**, *58*, 217–223. [CrossRef]
21. Sawalha, O.; Scholz, M. Assessment of capillary suction time (CST) test methodology. *Environ. Technol.* **2007**, *28*, 1377–1386. [CrossRef]
22. Sawalha, O.; Scholz, M. Innovative enhancement of the design and precision of the capillary suction time testing device. *Water Environ. Res.* **2009**, *81*, 2344–2352. [CrossRef]
23. Scholtz, M.; Tapp, J. Development of the revised Capillary Suction Time (CST) test. *Water Cond. Purif.* **2006**, *48*, 46–52.

24. Tuan, P.-A.; Sillanpaa, M. Migration of ions and organic matter during electro-dewatering of anaerobic sludge. *J. Hazard Mater.* **2010**, *173*, 54–61. [CrossRef]

25. Lee, C.H.; Liu, J.C. Sludge dewaterability and floc structure in dual polymer conditioning. *Adv. Environ. Res.* **2001**, *5*, 129–136. [CrossRef]

26. Meakin, P. The effects of reorganization processes on two dimensional cluster—Cluster aggregation. *J. Colloid Interface Sci.* **1986**, *112*, 187–194. [CrossRef]

27. Yu, W.; Liu, T.; Gregory, J.; Li, G.; Liu, H.; Qua, J. Aggregation of nano-sized alum–humic primary particles. *Sep. Purif. Technol.* **2012**, *99*, 44–49. [CrossRef]

28. Pastor-Satorras, R.; Rubí, J.M. Particle-cluster aggregation with dipolar interactions. *Phys. Rev. E* **1995**, *51*, 5994–6003. [CrossRef]

29. Smoczyński, L.; Bukowski, Z.; Wardzyńska, R.; Załęska-Chróst, K. Dłużyńska, B. Simulation of coagulation, flocculation and sedimentation. *Water Environ. Res.* **2009**, *81*, 348–356. [CrossRef]

30. Jullien, R.; Meakin, P. Simple models for the restructuring of 3-dimensional ballistic aggregates. *J. Colloid Interface Sci.* **1989**, *127*, 265–272. [CrossRef]

31. Jarvis, P.; Jefferson, B.; Parsons, S.A. Measuring floc structural characteristics. *Rev. Environ. Sci. Bio-Technol.* **2005**, *4*, 1–18. [CrossRef]

32. Chu, C.P.; Lee, D.J. Structural analysis of sludge flocs. *Adv. Powder Technol.* **2004**, *15*, 515–532. [CrossRef]

33. Zhao, Y.Q. Settling behaviour of polymer flocculated water treatment sludge II: Effects of floc structure and floc packing. *Sep. Purif. Technol.* **2004**, *35*, 175–183. [CrossRef]

34. Mandelbrot, B.B. *The Fractal Geometry of Nature*; Freeman: San Francisco, CA, USA, 1982.

35. Leman, J.; Smoczynski, M.; Dolgan, T.; Dziuba, Z. Fractal analysis of structure of cow and goat β-lactoglobulin preparation. *J. Food Technol.* **2005**, *42*, 428–430.

36. Yang, Z.; Yang, H.; Jiang, Z.; Huang, X.; Li, H.; Li, A.; Cheng, R. A new method for calculation of flocculation kinetics combining Smoluchowski model with fractal theory. *Colloids Surf. A* **2013**, *423*, 11–19. [CrossRef]

37. Meng, F.G.; Zhang, H.M.; Li, Y.S.; Zhang, X.W.; Yang, F.L.; Xiao, J.N. Cake layer morphology in microfiltration of activated sludge wastewater based on fractal analysis. *Sep. Purif. Technol.* **2005**, *44*, 250–257. [CrossRef]

38. Smoczynski, L.; Ratnaweera, H.; Kosobucka, M.; Smoczynski, M.; Pieczulis-Smoczynska, K.; Cretescu, I. The size of aggregates formed during coagulation and electrocoagulation of synthetic wastewater. *J. Environ. Prot. Ecol.* **2016**, *17*, 1160–1170.

39. Wang, Y.; Gao, B.Y.; Xu, X.M.; Xu, G.Y. Characterization of floc size, strength and structure in various aluminum coagulants treatment. *J. Colloid Interface Sci.* **2009**, *332*, 354–359. [CrossRef]

40. Smoczynski, L.; Kosobucka, M.; Smoczynski, M.; Ratnaweera, H.; Pieczulis-Smoczynska, K. Sizes of particles formed during municipal wastewater treatment. *Water Sci. Technol.* **2017**, *75*, 971–977.

41. Smoczyński, L.; Ratnaweera, H.; Kosobucka, M.; Smoczyński, M.; Kalinowski, S.; Kvaal, K. Modelling the structure of sludge aggregates. *Environ. Technol.* **2016**, *37*, 1122–1132. [CrossRef]

42. Smoczyński, L.; Ratnaweera, H.; Kosobucka, M.; Smoczyński, M. Image analysis of sludge aggregates. *Sep. Purif. Technol.* **2014**, *122*, 412–420. [CrossRef]

43. Malvern Instruments Limited. *A Basic Guide to Particle Characterization*; Malvern Instruments Limited: Worcestershire, UK, 2012; pp. 1–26.

44. Macedo, M.L.F.; Osawa, C.C.; Bertran, C.A. Sol-gel synthesis of transparent alumina gel and pure gamma alumina by urea hydrolysis of aluminum nitrate. *J. Sol.-Gel. Sci. Technol.* **2004**, *30*, 135–140. [CrossRef]

45. Haas, W.; Zrinyi, M.; Kilian, H.G.; Heise, B. Structural analysis of anisometric colloidal iron(III)-hydroxide particles and particle-aggregates incorporated in poly(vinyl-acetate) networks. *Colloid Polym. Sci.* **1993**, *271*, 1024–1034. [CrossRef]

46. Waite, T.D. Measurement and implications of flock structure in water and wastewater treatment. *Colloids Surf. A* **1999**, *151*, 27–41. [CrossRef]

47. Smoczynski, L.; Wardzynska, R. Study on macroscopic aggregation of silica suspensions and sewage. *J. Colloid Interface Sci.* **1996**, *183*, 309–314. [CrossRef]

48. Gregory, J. The role of floc density in solid-liquid separation. *Filtr. Sep.* **1998**, *35*, 367–371. [CrossRef]

*water*

MDPI

*Article*

# Performance Evaluation of Small Sized Powdered Ferric Hydroxide as Arsenic Adsorbent

**Muhammad Usman [1,\*], Ioannis Katsoyiannis [2], Manassis Mitrakas [3], Anastasios Zouboulis [2] and Mathias Ernst [1,\*]**

[1] Institute for Water Supply and Water Resources, Hamburg University of Technology, Am Schwarzenberg Campus 3, 21073 Hamburg, Germany

[2] Laboratory of Chemical and Environmental Technology, Department of Chemistry, Aristotle University of Thessaloniki, 54124 Thessaloniki, Greece; katsogia@chem.auth.gr (I.K.); zoubouli@chem.auth.gr (A.Z.)

[3] Analytical Chemistry Laboratory, Department of Chemical Engineering, Aristotle University of Thessaloniki, 54124 Thessaloniki, Greece; manasis@eng.auth.gr

\* Correspondence: muhammad.usman@tuhh.de (M.U.); mathias.ernst@tuhh.de (M.E.); Tel.: +49-40-42878-3177 (M.U.); +49-40-42878-3453 (M.E.)

Received: 3 July 2018; Accepted: 18 July 2018; Published: 20 July 2018

**Abstract:** The small sized powdered ferric oxy-hydroxide, termed Dust Ferric Hydroxide (DFH), was applied in batch adsorption experiments to remove arsenic species from water. The DFH was characterized in terms of zero point charge, zeta potential, surface charge density, particle size and moisture content. Batch adsorption isotherm experiments indicated that the Freundlich model described the isothermal adsorption behavior of arsenic species notably well. The results indicated that the adsorption capacity of DFH in deionized ultrapure water, applying a residual equilibrium concentration of 10 µg/L at the equilibrium pH value of 7.9 ± 0.1, with a contact time of 24 h (i.e., $Q_{10}$), was 6.9 and 3.5 µg/mg for As(V) and As(III), respectively, whereas the measured adsorption capacity of the conventionally used Granular Ferric Hydroxide (GFH), under similar conditions, was found to be 2.1 and 1.4 µg/mg for As(V) and As(III), respectively. Furthermore, the adsorption of arsenic species onto DFH in a Hamburg tap water matrix, as well as in an NSF challenge water matrix, was found to be significantly lower. The lowest recorded adsorption capacity at the same equilibrium concentration was 3.2 µg As(V)/mg and 1.1 µg As(III)/mg for the NSF water. Batch adsorption kinetics experiments were also conducted to study the impact of a water matrix on the behavior of removal kinetics for As(V) and As(III) species by DFH, and the respective data were best fitted to the second order kinetic model. The outcomes of this study confirm that the small sized iron oxide-based material, being a by-product of the production process of GFH adsorbent, has significant potential to be used for the adsorptive removal of arsenic species from water, especially when this material can be combined with the subsequent application of low-pressure membrane filtration/separation in a hybrid water treatment process.

**Keywords:** arsenic adsorption; small sized powdered ferric hydroxide; granular ferric hydroxide; water matrix; adsorption kinetics; drinking water

---

## 1. Introduction

Arsenic is globally considered as one of the major pollutants in drinking water sources and a worldwide concern because of its toxicity and carcinogenicity [1]. The presence of arsenic at elevated concentrations in natural environments can be attributed to both natural and anthropogenic inputs [2]. Arsenic pollution is primarily caused by natural processes, such as the weathering of rocks and minerals, followed by leaching and industrial activities that lead to the pollution of soil and groundwater [3]. The discharge of arsenic polluted waters from mining or mining-related activities, the pharmaceutical industry and agricultural activities plays an important role in anthropogenic arsenic pollution in Asia [4]. However, the introduction of arsenic into groundwaters is expected to occur mainly as a result of its natural geological presence in rocks [5].

Arsenite As(III) and arsenate As(V) are considered as the main oxidation states of inorganic arsenic found in natural waters. As(V) anions are predominant and stable in oxygen-rich environments, whereas the As(III) anions prevail in moderately reduced environments (i.e., anaerobic or anoxic groundwaters). Therefore, arsenic speciation mostly depends on pH and redox potential (Eh) conditions. Under oxidizing conditions and at pH values relevant to drinking water treatment, $H_3AsO_4$ is present as an oxyanion in the forms of $H_2AsO_4^-$ and/or $HAsO_4^{2-}$, whereas at low Eh values, arsenic becomes dominant as $H_3AsO_3$. Up to pH 9, $H_3AsO_3$ does not dissociate and, therefore, is present in most natural waters as the uncharged arsenious acid [6]. Therefore, As(III) species are considered as much more mobile in aquifers and cannot be easily adsorbed (and removed) onto the usually co-existing mineral surfaces, such as those of iron oxides. Moreover, As(III) is more toxic for the biological systems, as compared to As(V) [3,7].

The pollution of drinkable water sources by arsenic has been reported in more than 70 countries, where more than 150 million inhabitants are under high health risk [8]. Due to its high toxicity to humans, the World Health Organization [9] lowered the guideline value for arsenic in drinking water from 50 to 10 µg/L in 2004, aiming to minimize the health-related problems associated with arsenic pollution. The same standards also apply in the European Commission, as well as the US Environmental Protection Agency. Among other countries, the arsenic pollution of groundwater is considered as a particularly serious health-related problem in Pakistan, as a recent survey reveals [10]. Approximately 50 to 60 million people relying on groundwater as a source of drinking water in the Indus Valley are at high health risk [11]. In Punjab, more than 3% of the inhabitants are exposed to arsenic concentrations higher than 50 µg/L in drinking water, and 20% of the population is exposed to concentrations higher than 10 µg/L, while 16% and 36% of inhabitants in Sindh areas are exposed to arsenic pollution of concentrations higher than 50 µg/L and 10 µg/L, respectively [10].

Several treatment technologies to remove arsenic from drinking water have been applied worldwide [12], including adsorption using activated alumina [13] or iron oxide-based adsorbents, such as tetravalent manganese feroxyhyte [14], bayoxide [15–17], granular ferric hydroxide (GFH) [18], etc. Other treatment methods include the application of oxidation and arsenic removal using zero-valent iron (especially in Bangladesh) [19], coprecipitation of arsenic with iron or aluminum salts [2], preliminary arsenic oxidation by ozonation or biological oxidation [19], ion exchange [20], high pressure membrane separation [21,22] and electrocoagulation [23]. According to Tresintsi [14], chemical precipitation by ferric coagulation has significantly higher arsenic removal efficiencies in comparison to adsorption by iron oxy-hydroxides, and a drinking water regulation limit can be achieved at an affordable price, with operational costs estimated between 0.09 and 0.16 €/m$^3$ for initial arsenic concentrations, ranging between 19 and 208 µg/L. However, the major part (>90%) of treatment costs was attributed to the management of produced sludge, since the coagulant costs are estimated to be ≤0.01 € [15]. Previous studies have identified high pressure membrane processes as an emerging technology, due to their high removal efficiencies and easy operation features [21,22], but these high pressure membrane processes are rather energy (and cost) intensive, and subjected to the fouling of membranes. Moreover, the disposal of produced brine (high salt concentrations) is a considerable challenge.

On the other hand, for the treatment of waters with moderate arsenic concentrations, i.e., slightly higher than the regulation limits, adsorption onto iron oxide-based adsorbents has been proved to be the most economically efficient procedure [15]. The two mostly commercially applied adsorbents are the Granular Ferric oxy-Hydroxides (GFH) and the Bayoxide E33 (GFO), which are favorable in terms of cost, removal efficiency, simplicity of design, operation, maintenance and minimizing the (secondary) waste production [24]. The GFH has been tested to remove arsenic from drinking water sources under both laboratory-scale and full-scale water treatment plants [15,18,25,26].

Arsenic adsorption onto GFH is usually performed in a column filtration mode, which is a rather simple process and can be continuously operated, but the production of this material is relatively cost intensive. The cost (on dry basis) for GFH and Bayoxide materials was estimated to be 9 €/kg and 12.5 €/kg, respectively [15]. Currently, the small sized fraction (dust ferric hydroxide, DFH) generated during the industrial production of GFH cannot be employed in the common column filtration mode, since the small sized adsorbents can rapidly clog the fixed beds in filter columns, causing an increased pressure head, thereby increasing energy costs and maintenance and, hence, reducing the system performance.

Adsorption combined with the application of low-pressure membrane filtration is considered as a newly developed hybrid water treatment process. Low-pressure membrane processes, such as microfiltration or even ultrafiltration, have a reasonable energy demand and produce superior quality treated water with a rather controllable fouling of membranes and incurring quite low capital and operational costs [27]. The low-pressure membrane processes are not able to remove mono- and polyvalent ions, i.e., arsenic species, from water sources, although they can efficiently remove suspended solids, colloids, bacteria, viruses and micro-particles [28]. If the cost-effective small sized GFH adsorbent (having a substantially lower commercial price of only 1.6 €/kg on a dry basis) has the potential to remove arsenic species from drinking water sources, it might then be employed in the adsorption-microfiltration (MF) hybrid treatment scheme to economically and efficiently remove arsenic. The idea of a submerged membrane filtration adsorption hybrid system could be exploited in this regard, which allows the pollutant to be in contact with adsorbents for longer time.

The objectives of the study were: (i) To assess the adsorption potential/performance of the smaller fraction of GFH material with a particle size of <0.250 mm, which is abbreviated as DFH henceforth, for removing As(V) and As(III) species from different water matrixes; (ii) to determine the kinetics of arsenic adsorption on the studied material; (iii) to examine the effect of a water matrix on arsenic removal; and (iv) to compare the efficiencies of both major inorganic arsenic species, As(V) and As(III), with the established, conventionally applied adsorbents, such as GFH. Badruzzaman [29] studied the use of small sized GFH in packed bed columns, but investigated the adsorption potential of this material only in the case of As(V) and ultra-pure water and has found promising results. However, to the best of our knowledge, no comprehensive study concerning the application of DFH material, systematically studying the arsenic adsorption efficiency of both arsenic species and different water matrices, such as the tap water of Hamburg (HH tap water) and the NSF (National Sanitation Foundation) challenge water, used to simulate typical arsenic-containing groundwater, has yet been performed.

## 2. Materials and Experimental Methods

### 2.1. Reagents

For the preparation of As(III) or As(V) 100 mg/L stock solutions, the standard solution of As(III), as $As_2O_3$ in 2% $HNO_3$, and As(V), as $H_3AsO_4$ in $HNO_3$, with a concentration of 1 g/L, were used, obtained from Carl Roth GmbH + Co. KG (Karlsruhe, Germany) and Merck Chemicals GmbH (Darmstadt, Germany), respectively. The pH buffer solution, *N,N*-Bis-(2-hydroxyethyl)-2-aminoethane sulphonic acid (BES), used in the experiments focusing on arsenic removal from deionized (DI) water, was obtained from Carl Roth GmbH + Co. KG.

### 2.2. Material Characterization

The DFH material, with a particle size of <0.250 mm, was supplied by GEH–Wasserchemie GmbH & Co. KG, Osnabrück, Germany. The material is predominantly akaganeite, a specific form of an iron oxy-hydroxide [16]. DFH is mainly characterized by a relatively large specific surface area (252 $m^2$/g) [29] and surface charge density (Table 1).

**Table 1.** Main properties of used DFH material.

| Properties | Value | Literature Value |
|---|---|---|
| Chemical composition | β-FeOOH and Fe(OH) | |
| Dry solids content (%) | ~50 [a] | ~50 [b] |
| Moisture content (%) | ~50 [a] | ~50 [b] |
| Particle size (μm), $d_p$ | 7.4–250 [a] | 1.8–250 [b] |
| Mean particle size (μm) | 78.40 | — |
| Point of zero charge (PZC) | 5.3 ± 0.2 | ~5.5 [c] |
| Isoelectric point (IEP) | 7.8 ± 0.2 | ~7.8 [c] |
| Surface charge density | 0.9 mmol [OH$^-$]/g | — |

Note: [a] Average values from triplicate analysis, [b] Data by Ref. [29], [c] Data by Ref. [30].

Particle size distribution was determined by EyeTech[TM] instrument (combi, AmbiValue, Nijerdal, The Netherlands), ranging between 7.4 and 250 μm. The liquid flow cell of EyeTech was filled with 1 L of deionized water, and approximately 100 mg of material was added. Mechanical shaking was provided in the liquid flow cell, which keeps the material particles in suspension. Then, suspension was supplied to the optical cell and circulated through it for 5 min at a pump speed of 0.674 L/min. Three cycles of the suspension were performed to determine the particle size distributions.

To determine the surface charge of DFH in the suspension, the Isoelectric Point (IEP) and the Point-of-Zero Charge (PZC) were quantified. IEP was determined by a zeta-potential curve at $20 \pm 1$ °C of adsorbent dispersion in 0.01 M NaNO$_3$, with the respective pH of solution, using a Micro-electrophoresis Apparatus (Mk II device, Rank Brothers Ltd, Cambridge, England), while PZC and the surface charge density were defined by the application of acid/base potentiometric mass titration in suspensions of the adsorbent and for various ionic strengths [31].

### 2.3. Water Matrix

The test solution was initially prepared using deionized water (DI), spiked with either As(III) or As(V) species, at an initial concentration of 190 μg/L. 2 mM of *N,N*-Bis(2-hydroxyethyl)-2-aminoethanesulfonic acid (BES) was added to the test solution, made of DI water, for pH control at pH 7.9. In addition to DI water, As(V) and As(III) test solutions were prepared in HH tap water and NSF water with the same initial arsenic concentration, as used in the case of DI water, in order to study the effect of different water matrixes on the arsenic adsorption capacity. The major physicochemical parameters of the HH tap water and of NSF challenge water are listed in Table 2.

**Table 2.** Water quality parameters of Hamburg tap water (* data obtained from Hamburgwasser) and NSF challenge water.

| Parameter. (mg/L) | Water Matrixs. | |
|---|---|---|
| | HH Tap Water * | NSF Challenge Water |
| Na$^+$ | 14 | 73.7 |
| Ca$^{2+}$ | 42 | 40.1 |
| Mg$^{2+}$ | 4 | 12.6 |
| HCO$_3^-$ | 150–300 | 183.0 |
| Cl$^-$ | 19 | 71.0 |
| SO$_4^{2-}$ | 23 | 50.0 |
| NO$_3^-$ | 0.62 | 2.0 |
| F$^-$ | 0.13 | 1.0 |
| PO$_4^{3-}$ | 0.05–0.15 | 0.123 |
| SiO$_2$ | 16.6–18.5 | 20 |
| DOC | 0.8 ± 0.2 | — |

The NSF challenge water was prepared according to the National Sanitation Foundation (NSF) international and contains the following: 252 mg $NaHCO_3$, 12.14 mg $NaNO_3$, 0.178 mg $NaH_2PO_4 \cdot H_2O$, 2.21 mg NaF, 70.6 mg $NaSiO_3 \cdot 5H_2O$, 147 mg $CaCl_2 \cdot 2H_2O$ and 128.3 mg $MgSO_4 \cdot 7H_2O$ in 1 L of DI water. Prior to adsorption experiments, the pH was adjusted to 7.9 by adding either NaOH or HCl standard solutions (0.1 N) [32].

### 2.4. Batch Adsorption Procedure

Batch equilibrium and kinetic adsorption tests were performed to study the adsorption potential of DFH for removing arsenic species from the different examined test solutions/water matrixes. To derive the adsorption isotherms, the method of adding various quantities of adsorbent to a constant solution volume (500 mL), having the same initial concentration of As(V) or As(III) species, was adopted. Additionally, As(III) test solutions were preliminary bubbled for 30 min with pure $N_2$ gas at 0.1 bar (flowrate 11.25 mL/min) to minimize the influence of dissolved oxygen on As(III) potential oxidation and adsorption, and the flasks were immediately closed and placed on the platform shaker in darkness in the thermostate cabinet (20 ± 0.5 °C) to insure the stability of As(III) species during and after adsorption onto the examined iron oxide-based adsorbents.

The evaluation of the examined adsorbent material focused on its ability to decrease the residual arsenic concentration below the drinking water regulation limit of 10 μg/L (termed $Q_{10}$ hereafter), rather than studying the (more convenient) maximum capacity ($Q_{max}$) at higher residual arsenic concentrations. If efficiency of the adsorbent was evaluated through $Q_{max}$, which usually points to high residual concentrations and, indeed, brings high adsorption capacities, but provides marginal information on its ability to reach low concentrations, such as the regulation limits [33].

Different adsorbent dosages were placed in flasks for the three different water matrixes, while only adsorbent dosages, ranging between 5–40 mg/L, 6–50 mg/L and 10–80 mg/L, provided equilibrium As(V) concentrations between 1 and 120 μg/L in DI water, HH Tap water and NSF water, respectively. Adsorbent dosages of 10–60 mg/L, 15–100 mg/L and 40–200 mg/L were found to provide the same range of equilibrium concentrations in the experiments focusing on the removal of As(III) in DI water, HH Tap water and NSF water, respectively. For comparison, batch adsorption isotherm studies were also conducted with the GFH material, using DI water to compare the adsorption characteristics of GFH with those obtained when using DFH, i.e., to examine the efficiency of both particle size fractions of this adsorbent. GFH dosages, ranging between 10–80 and 20–120 mg/L, were carefully placed in flasks for the removal of As(V) or As(III) species, respectively. For each experimental test focusing on adsorption isotherm, a reference blank sample (i.e., without the presence of an adsorbent) was filled. The flasks were stirred using a platform shaker for 24 h at 20 ± 0.5 °C. The equilibration time was determined for the corresponding kinetic experiments. At the end of the equilibration time, the suspensions were immediately filtered through a 0.45 μm membrane syringe filter (PVDF, Carl Roth GmbH + Co. KG), and the filtrates were collected and stored for the subsequent analytical determination of residual (still dissolved and removed) arsenic.

In the kinetic studies, the initial arsenic concentration and adsorbent quantity was kept at 190 μg/L and 50 mg/L, respectively. The initial concentrations of either only As(V) or only As(III) species were the same as in the respective isotherm studies. Batch adsorption kinetics tests were conducted at the initial pH value of 7.9. Unlike the isotherm studies, a magnetic stirrer (100 rpm) was used in the kinetic studies experiments. The samples were collected at regular time intervals and the residual arsenic in the solution was analyzed. Each set of adsorption batch isotherm and kinetics experiments was replicated at least twice, and the average values are reported.

### 2.5. Chemical Analytics

Initial and residual arsenic concentrations were determined by Graphite Furnace Atomic Absorption Spectrophotometry (Perkin-Elmer 4100ZL, Baesweiler, Germany), using a Perkin-Elmer 4100ZL instrument [34]. The limit of detection was 0.5 μg/L. Prior to analysis, As(III) water samples from the isotherm experiments were acidified (2 < pH < 4) and passed through a 30 mL column (with ID = 2 cm), containing an anion exchange resin (Dowex® 1 × 8–100, mesh size 50–100, Sigma-aldrich Chemie GmbH, Taufkirchen, Germany), which retained As(V), whereas the total arsenic concentration of water samples from the adsorption kinetics experiments were analyzed. This method of arsenic speciation needs approx. 50 mL of water sample, noting that, in the kinetics experiments, only small volumes (~7 mL) of water samples were collected at regular intervals; accordingly, arsenic speciation using this method was not possible. Therefore, only the total arsenic concentration of the water samples from adsorption kinetics was analyzed, presenting the concentration of individual arsenic species in the water samples. The initial concentration of phosphate in HH tap water was measured using ICP-MS (NexION 300D, PerkinElmer, Baesweiler, Germany).

## 3. Results and Discussion

### 3.1. Particle Size Distribution

The particle size has a strong effect on the removal kinetics of arsenic. Banerjee [35] observed that the removal of As(III) by the pulverized/powdered GFH (with $d_p$ < 63 µm) was faster than that of as-received GFH (0.320 mm < $d_p$ < 2 mm) at same experimental conditions. A similar trend was also recorded by Tresintsi [36] during the adsorption of arsenic species onto an iron oxide-based adsorbent. The length-based and volume-based particle size distributions of DFH are shown in Figure 1a,b. The major fraction of this material has a length-based particle size ranging from 7 to less than 65 µm, while the volume-based particle size has two peaks ranging between (i) 65 and 100 µm, and (ii) 200 and 250 µm. As DFH has a constant density, the volume-based distribution gives an indication of mass distribution. The average length-based particle size of DFH particles, as determined by the EyeTech instrument, is 78.4 µm.

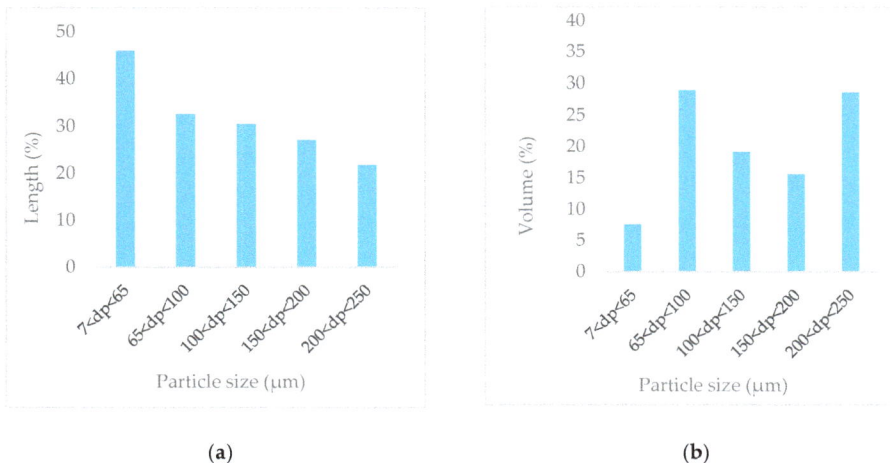

(a)                                              (b)

**Figure 1.** (a) Length-based particle size distribution, and (b) volume-based particle size distribution of DFH particles, as measured by the EyeTech instrument.

Conventional GFH, which is applied in fixed bed adsorption columns (commercially available), has particle size ranging from 0.32 to 2 mm, while the particle size of tetravalent manganese feroxyhyte media, produced by Tresintsi [14] in a kilogram-scale continuous process, is of non-uniform size. A fine fraction of adsorbent media (with a particle size of <250 µm) is also generated during its production at the kilogram-scale in a laboratory two-stage continuous flow reactor.

## 3.2. Batch Isotherm Studies

The batch equilibrium adsorption experiments are conducted to evaluate the adsorption potential of arsenic species onto small sized powdered ferric hydroxide (DFH). The amount of arsenic adsorbed at the equilibrium stage is calculated using the mass balance of an adsorption system:

$$Q_e = \frac{(C_o - C_e)\, V}{m},$$

(1)

where $Q_e$ is the amount of arsenic adsorbed at the equilibrium stage per mass of adsorbent, $C_o$ and $C_e$ are the initial and equilibrium concentrations of arsenic in the test solution, respectively; $m$ is the quantity (mass) of the adsorbent used and $V$ is the volume of test solution.

The non-linear form of the Freundlich and Langmuir isotherm model is used to describe the adsorption behavior:

$$Q_e = K_F\, C_e^{1/n},$$

(2)

$$Q_e = Q_m \frac{K_L C_e}{1 + K_L C_e},$$

(3)

where $K_F$ and $n$ are constants, explaining the adsorption capacity and the intensity, respectively; $K_L$ and $Q_m$ are the Langmuir adsorption constant and maximum adsorption capacity per unit mass of adsorbent, respectively. To identify the fitting of the isotherm model to the experimental data, a chi-squared value ($x^2$, Equation (4)) was also calculated, in addition to the calculation of correlation coefficients, for the non-linear form of the isotherm model. According to Tran [37], this indicates the bias in the experimental and model results. Its value is close to zero, if the data obtained using a model are similar to the experimental data, whereas its high value indicates the high biasness between the experimental data and the model estimations.

$$x^2 = \sum \frac{\left(Q_{e,exp} - Q_{e,cal}\right)^2}{Q_{e,cal}},$$

(4)

where $Q_{e,exp}$ is the amount of arsenic adsorbed at equilibrium, and $Q_{e,cal}$ is the amount of arsenic adsorbed as calculated from the isotherm model.

### 3.2.1. As(V) Adsorption

The major parameters of the Freundlich isotherm in the case of As(V), along with the correlation coefficients and the respective chi-squared values, are presented in Table 3, while the Langmuir isotherm parameters are shown in Table S15 (supplementary information). The correlation coefficients and the chi-squared values indicated that the Freundlich model described the isothermal adsorption behavior of arsenic species notably well. The $R^2$ of the Freundlich model was greater than 0.91 and $x^2$ was less than 1 for all of the (3) water matrixes. Badruzzaman [29] also reported the fitting of the Freundlich model for small fractions of GFH with an $R^2$ of greater than 0.92.

**Table 3.** Parameters of the Freundlich isotherm for As(V), along with the correlation coefficients and the respective chi-squared values.

| Water Matrix | Adsorbent | n (−) | $K_F$ * | $Q_{10}$ (µg/mg) | $R^2$ | $x^2$ |
|---|---|---|---|---|---|---|
| DI water | GFH | 2.03 | 0.68 | 2.1 | 0.960 | 0.136 |
| DI water | DFH | 3.10 | 3.25 | 6.9 | 0.991 | 0.117 |
| HH tap water | DFH | 3.41 | 3.20 | 6.3 | 0.941 | 0.551 |
| NSF water | DFH | 2.39 | 1.22 | 3.2 | 0.918 | 0.367 |

Note: * (µg/mg)/(µg/L)$^{1/n}$.

The As(V) adsorption isotherms for DFH and GFH in DI water at 20 °C, with an equilibrium pH value of 7.9 ± 0.1 after a contact time of 24 h, are shown in Figure 2a. The adsorption characteristics of the DFH and GFH in DI water were analyzed and evaluated using the Freundlich isotherm equation. The $K_F$ value for the case of DI water was 3.25 (µg As(V)/mg DFH)/(µg/L)$^{1/n}$ and the n value was 3.10, while the corresponding $K_F$ and n values, in the case of As(V) adsorption onto GFH in DI water, were found to be 0.68 (µg As(V)/mg GFH)/(µg/L)$^{1/n}$ and 2.03, respectively. The adsorption capacity at an equilibrium liquid phase As(V) concentration of 10 µg/L ($Q_{10}$) and an equilibrium pH value of 7.9 ± 0.1, produced by setting the isotherm parameters in the Freundlich model, was found to be 6.9 As(V)/mg DFH and 2.1 mg As(V)/mg GFH. At the equilibrium As(V) concentration of 10 µg/L, the adsorption capacity ($Q_{10}$) of DFH for As(V) was almost triple that of GFH, which is also shown in Figure 2a. However, the DFH/GFH ratio of adsorption capacity diminishes from 3.2 times to 2.4 and 2.2 times as the equilibrium As(V) concentration was increased to 50 µg/L and to 100 µg/L, respectively. Banerjee [35] also reported a higher adsorption capacity of the pulverized/powdered (particle size <63 µm) GFH during the adsorption of As(V) in comparison to the as-received GFH material (with a particle size of 0.32–2.0 mm) after a contact time of 24 h and at the equilibrium pH value of 7.0–7.5. The $Q_{10}$ value reported by Banerjee [35] for As(V) is approximately 4 times higher for pulverized GFH compared to the as-received GFH, whereas the $Q_{10}$ value of DFH is 3.2 times higher than the as-received GFH in the current study. These results can be considered to be in agreement, since the observed small differences are considered negligible and could be attributed to the respective difference in the initial material used, since the pulverized GFH used by Banerjee [35] presented a particle size smaller than the examined DFH in this study.

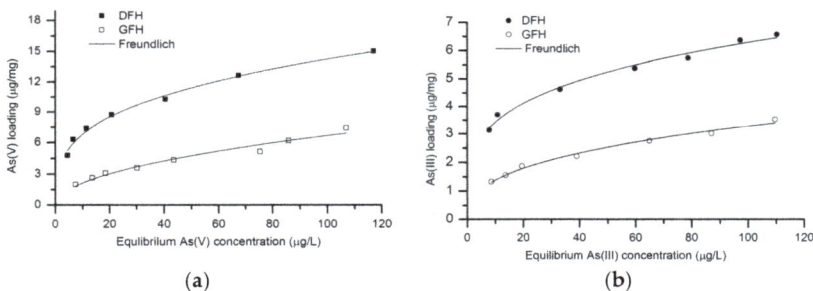

**Figure 2.** (**a**) As(V) and (**b**) As(III) adsorption isotherms for DFH and GFH materials in DI water. Solid lines represent the Freundlich model using non-linear fitting. Experimental conditions: Initial (As(V)) 190 µg/L, initial (As(III)) 190 µg/L, equilibrium pH value 7.9 ± 0.1 and temperature 20 °C.

In another study, Badruzzaman [29] obtained $K_F$ (4.45 (µg As(V)/mg DFH)/(µg/L)$^{1/n}$) and n (3.57) values of a similar magnitude using the same fraction of the same adsorbent at the equilibrium pH value of 7 and 24 ± 0.5 °C after 18 days of contact time. The calculated $Q_{10}$ value in this case was 8.5 µg As(V)/mg, which is higher than the recorded value in the current study. The divergence in $Q_{10}$ value between the current study and Badruzzaman [29] could be ascribed to the differences of experimental conditions (equilibrium pH value, temperature, water matrix and longer contact time). At pH 7, As(V) is present as an oxyanion in the form of $H_2AsO_4^-$, while it transforms into $HAsO_4^{2-}$ at pH 8. The latter requires two active adsorption sites to be adsorbed on the absorbent surface. In addition, Badruzzaman [29] used bicarbonate as a pH buffer. In the current study, BES was used as a pH buffer to facilitate the required constant pH condition, since no influence on arsenic adsorption was observed, which is in agreement with the results reported by Banerjee [35].

In the case of granular GFH with particle sizes ranging from 0.32 to 2 mm, Banerjee [38] obtained a $K_F$ value of 3.13 (µg As(V)/mg GFH)/(µg/L)1/n and an n value of 0.23 at the equilibrium pH of 6.5

and at 20 °C. An adsorption capacity of 5.3 μg As(V)/mg GFH is calculated by the Freundlich model, setting $K_F$ and n values at the equilibrium liquid phase, with an As(V) concentration of 10 μg/L.

### 3.2.2. As(III) Adsorption

The adsorption capacity of DFH in the case of As(III), which is also higher than that obtained by the commercially-used GFH material, is shown in Figure 2b. The calculated $Q_{10}$ value in DI water, under experimental conditions similar to As(V) adsorption isotherm experiments, was 3.5 μg As(III)/mg DFH (Table 4). Consequently, the $Q_{10}$ value for As(III) was almost half of the corresponding value for As(V). The difference in the adsorption efficiencies between As(V) and As(III) could be attributed to the different behavior of arsenic species at the equilibrium pH value, because at pH 8, As(V) species predominantly exist as $HAsO_4^{2-}$ in aqueous solutions, while As(III) species predominantly exists as undissociated $H_3AsO_3$ (pKa = 9.2). As(V) adsorption onto GFH takes place via electrostatic attraction and Lewis acid–base interactions (ligand exchange reactions) [38,39]. The higher adsorption capacity of As(V) is possibly due to As(V) presenting a greater electrostatic attraction to the charged DFH particles, as compared to the electrically neutral As(III) at circumneutral pH values. Therefore, As(V) adsorbs better onto DFH than As(III).

**Table 4.** Parameters of the Freundlich isotherm in the case of As(III), along with the correlation coefficients and the chi-squared values.

| Water Matrix | Adsorbent | n (−) | $K_F$ * | $Q_{10}$ (μg/mg) | $R^2$ | $\chi^2$ |
|---|---|---|---|---|---|---|
| DI water | GFH | 2.66 | 0.58 | 1.4 | 0.985 | 0.01 |
| DI water | DFH | 3.96 | 1.96 | 3.5 | 0.987 | 0.023 |
| HH tap water | DFH | 4.34 | 1.64 | 2.8 | 0.972 | 0.036 |
| NSF water | DFH | 4.39 | 0.64 | 1.1 | 0.970 | 0.005 |

Note: * $(\mu g/mg)/(\mu g/L)^{1/n.}$

As shown in Figure 2b, DFH has a higher As(III) adsorption capacity than GFH within the investigated concentration range. In particular, the obtained $Q_{10}$ value of DFH was 2.5 times higher, than the respective values obtained by GFH. This difference was, however, reduced to 2.1 and 1.9 times at the equilibrium As(III) concentrations of 50 μg/L and of 100 μg/L, respectively. The $Q_{10}$ value of pulverized GFH for As(III), reported by Banerjee [35], is approximately 1.8 times higher than that obtained by the granular GFH experiments. The divergence in $Q_{10}$ value might be attributed to difference in the initial concentration, particle size and equilibrium pH.

The results of this study can be compared with similar studies using very advanced nanomaterials to achieve efficient arsenic adsorption. The study of Bolisetty [40] shows that amyloid–carbon hybrid membranes containing 10% (by weight) amyloid fibrils indeed diminished the arsenic concentration in ultrapure water within the drinking water regulation limit, but the adsorption capacity is lower than 0.3 and 1.2 μg/mg for As(V) and As(III), respectively, and the adsorption efficiency of amyloid–carbon hybrid membranes for As(V) is almost 25 times lower than that of DFH (recorded from adsorption isotherms) and 3 times lower in case of As(III). DFH media is a by-product, otherwise useless in the water industry, and can find real scale applications in a short period of time with a rather higher adsorption capacity and lower operational costs.

### 3.3. Effect of Water Matrix on Arsenic Adsorption

DFH batch adsorption isotherms studies were also conducted with HH tap water and NSF water to assess the real and practical adsorption potential for removing As(III) and As(V) from drinking water. The adsorption isotherms for As(V) and As(III) onto DFH at 20 °C and at the equilibrium pH value of 7.9 ± 0.1 in three different water matrixes after a contact time of 24 h are shown in Figure 3a,b.

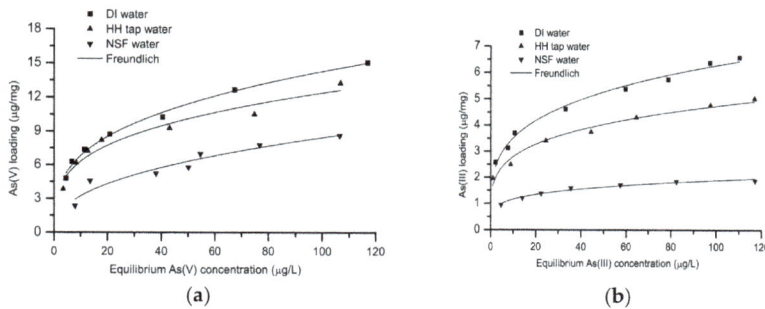

**Figure 3.** (**a**) As(V) and (**b**) As(III) adsorption isotherms for DFH using three different water matrixes. Solid lines represent the Freundlich model by using non-linear fitting. Experimental conditions: Initial (As(V)) 190 µg/L, initial (As(III)) 190 µg/L, equilibrium pH value 7.9 ± 0.1 and temperature 20 °C.

The adsorption capacity of DFH in HH tap water and in NSF water decreased, as compared to DI water, over the entire range of equilibrium As(V) concentrations is shown in Figure 3a. However, the decrease of the adsorption capacity was found to be more significant in the case of NSF water. For example, the $Q_{10}$ value for As(V) in HH tap water and in NSF at the equilibrium pH of 7.9 ± 0.1 was 6.3 µg As(V)/mg DFH and 3.2 µg As(V)/mg DFH, respectively. Due to the presence of different competing interfering ions, such as phosphate and silica, the reduction of 8.2% and 53.4% in $Q_{10}$ values for As(V) was observed regarding the cases of HH tap water and of NSF water, respectively.

In the case of As(III), the $Q_{10}$ value of 3.5 µg As(III)/mg observed in DI water at the equilibrium pH of 7.9 ± 0.1 was reduced to 2.8 and 1.1 µg As(III)/mg in HH tap water and in NSF water, respectively. The observed reduction of $Q_{10}$ values in the cases of HH tap water and of NSF water, can be attributed to As(III) speciation, because As(III) is electrically neutral at the set pH value of 7.9, resulting in nearly negligible electrostatic attraction, while the co-presence of their anions may have significant electrostatic attraction with the charged surface of DFH. Therefore, reductions of 20.1% and 69.0% were recorded in the $Q_{10}$ values for As(III) in HH tap water and in NSF water, respectively, indicating also that in NSF water, the concentration of competing and interfering ions is higher, as compared with HH tap water.

Especially, the presence of phosphates and silicates showed the most adverse effect on the arsenic adsorption capacity of iron-based adsorbents [41,42]. At pH 8.2, GFH has a strong affinity with phosphate, existing as $HPO_4{}^{2-}$ [43], and strongly competes with arsenic species for similar adsorption sites. Amy [16] reported a reduction of 3% in the As(V) adsorption capacity of GFH in the presence of only 125 µg/L phosphate at pH 8 during batch tests. However, the reduction of the $Q_{10}$ value increased to 36% when the phosphate concentration was increased to 250 µg/L. In the case of silica competition under the same experimental conditions, Amy [16] reported a reduction of 25% and 60% in the $Q_{10}$ values for As(V) when silica was present with 13.5 and 22 mg/L concentrations, respectively. During the adsorption of As(V) onto tetravalent manganese feroxyhyte (TMF) in batch adsorption tests at the equilibrium pH of 8, the measured $Q_{10}$ values of 10.3 µg As(V)/mg and 10.9 µg As(III)/mg in DI water were reduced to 5.4 µg As(V)/mg and 4.6 µg As(III)/mg in the case of NSF water, resulting in reductions of 47.6% and 57.8% for As(V) and As(III), respectively [14]. In the case of arsenic adsorption onto Bayoxide in NSF water, Amy [16] reported reductions of 54.6% and 96.9% at the equilibrium pH of 7.5 for As(V) and As(III), respectively. Silicate also presents strong competition with arsenic species for similar adsorption sites because it exists as $H_3SiO_4{}^-$ at pH 8, which requires one active site for adsorption [44].

### 3.4. Arsenic Removal Kinetics

The rate adsorptions of As(V) and As(III) onto DFH in three different water matrixes, at 20 °C and an initial pH of 7.9, respectively, are shown in Figure 4a,b. First-order and second-order kinetic models were considered to analyze the removal rates of As(V) and As(III) from these aqueous solutions. Banerjee [38] used the first-order kinetic equation to study the adsorption of arsenic species onto GFH, while Eljamal [45] and Saldaña-Robles [46] employed the second-order kinetic equation for the adsorption of As(V). The simple forms of the first- and second-order kinetic models can be expressed as [38,46]:

$$\ln\left(\frac{As_t}{As_o}\right) = -k_1 t, \tag{5}$$

$$\frac{1}{As_t} - \frac{1}{As_o} = k_2 t, \tag{6}$$

$$SE = \sqrt{\left[\frac{\sum(C_e - C_p)^2}{n-2}\right]} \tag{7}$$

where $As_o$ is the initial concentration of arsenic species (either As(V) or As(III)), $As_t$ is the liquid phase arsenic concentration remaining in the solution at time t, and $k_1$ and $k_2$ are the first- and second-order rate constants, respectively. $C_e$ and $C_p$ are the experimental and the predicted solid phase arsenic concentrations, and n is the total number of data points. The fitting of the kinetic model was determined by $R^2$ and by the standard error of estimation (SE, Equation (7)).

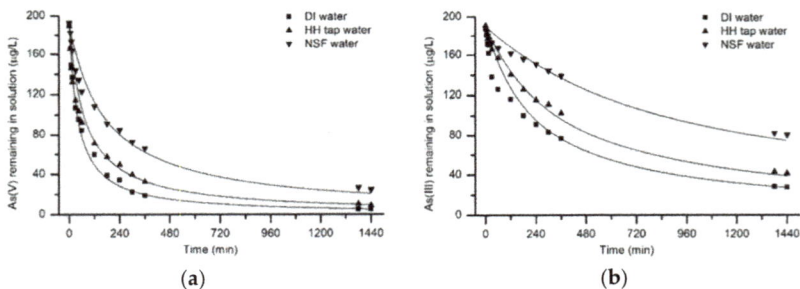

**Figure 4.** Rate of adsorption of (**a**) As(V) and (**b**) As(III) in different water matrixes at an initial pH of 7.9 and temperature of 20 °C, using an adsorbent dosage of 50 mg/L and an initial (As(V)) 190 µg/L and initial (As(III)) 190 µg/L.

$R^2$ and SE coefficients for both kinetic models, for As(V) and As(III) species, in all (3) examined water matrixes are shown in Table 5. The calculated values of SE are lower, whereas the $R^2$ values are higher for the second-order equation, providing a closer fit. Accordingly, the second-order kinetic model was used to further analyze the kinetics data of all water matrixes. The second-order rate constants ($k_2$) for As(V) and As(III) in DI water, in HH tap water and in NSF water, are presented in Table 6. The $R^2$ ranged from 0.992 to 0.998 and from 0.980 to 0.996 for As(V) and As(III), respectively, whereas the values of $k_2$ are in the range of $1.82 \times 10^{-3}$ to $7.51 \times 10^{-3}$ h$^{-1}$ and of $0.34 \times 10^{-3}$ to $1.34 \times 10^{-3}$ h$^{-1}$, as calculated in the case of As(V) and As(III), respectively. The interfering ions present in the HH tap water and in the NSF water results in significantly lower $k_2$ values, while the lowest $k_2$ value was calculated in the case of NSF water.

**Table 5.** Correlation coefficient ($R^2$) for the first- and second-order kinetic models, along with the standard error of estimation (SE).

| Water Matrix | As(V) | | | | As(III) | | | |
|---|---|---|---|---|---|---|---|---|
| | First Order | | Second Order | | First Order | | Second Order | |
| | $R^2$ | SE | $R^2$ | SE | $R^2$ | SE | $R^2$ | SE |
| DI water | 0.529 | 59.99 | 0.998 | 4.93 | 0.802 | 35.31 | 0.996 | 14.52 |
| HH tap water | 0.579 | 57.62 | 0.994 | 14.31 | 0.769 | 21.59 | 0.969 | 6.72 |
| NSF water | 0.714 | 23.76 | 0.992 | 10.68 | 0.905 | 14.45 | 0.980 | 8.12 |

**Table 6.** Second-order rate constant ($k_2$) for the three examined different water matrixes.

| Water Matrix | As(V) | As(III) |
|---|---|---|
| | $k_2$ ($h^{-1}$) | $k_2$ ($h^{-1}$) |
| DI water | $7.51 \times 10^{-3}$ | $1.34 \times 10^{-3}$ |
| HH tap water | $4.23 \times 10^{-3}$ | $0.85 \times 10^{-3}$ |
| NSF water | $1.82 \times 10^{-3}$ | $0.34 \times 10^{-3}$ |

The results from batch adsorption kinetics reveal that the rate of the adsorption of As(V) onto DFH is initially fast, followed by a slower rate of adsorption, which eventually approaches an equilibrium plateau. A similar trend was observed during the adsorption of As(III) for the same experimental conditions and initial concentration of adsorbate. However, the observed uptake rate of As(III) onto DFH is slower than that of As(V), possibly because of the insignificant presence of electrostatic attraction (Coulombic interaction) in the case of As(III) adsorption [38,39]. For example, 55% and 33% adsorption of As(V) and As(III) occurred within the first 1 h of contact time, respectively, which increased up to 90% and 59% after at the end of 6 h of contact time. The slower adsorption rate of both arsenic species after 6 h of contact time can be attributed to the majority of adsorption sites, already occupied by the adsorbate species, leaving a relatively small number of adsorption sites still available for adsorption.

The results reveal that the competitive interfering ions present in HH tap water and in NSF water (in comparison with the respective experiments of DI water) can substantially reduce the removal rate of both As(V) and As(III). More specifically, the presence of interfering ions has a significant influence on the behavior of adsorption kinetics. In the case of HH tap water and NSF water, as shown in Figure 4, the strong interference of competing ions resulted in a lower adsorption rate of As(V) onto DFH, as concluded from the significant decrease of $k_2$ values, which decreases from $7.51 \times 10^{-3}$ $h^{-1}$ (for DI water) to $4.23 \times 10^{-3}$ $h^{-1}$ and $1.82 \times 10^{-3}$ $h^{-1}$ in the cases of HH tap water and of NSF water, respectively (Table 6). A similar negative influence, regarding the presence of phosphate and silicate ions on the As(V) adsorption rate by GFH, was also observed by Xie [41] and Nguyen [42]. Furthermore, during the adsorption of lead by iron-based adsorbents, Smith [47] also reported that the behavior of adsorption kinetics is influenced by the accessibility and availability of adsorbent surface sites, as well as by the relative surface charge, adsorbed species, and complexation rate of dissolved species with the surface sites.

## 4. Conclusions

In this study, the potential of a cost-effective DFH adsorbent, considered as a by-product of GFH production, for As(V) and As(III) was investigated in a systemic and detailed batch-scale study. The results show that the adsorption isotherm data obtained for As(V) and As(III) at an initial pH of 7.9 were well described by the Freundlich isotherm equation. The calculated adsorption capacity in deionized ultrapure water at the equilibrium liquid phase concentration of 10 µg/L was 6.9 and 3.5 µg/mg for As(V) and As(III), respectively. The calculated adsorption capacity of GFH at the same pH value and equilibrium liquid phase concentration, as determined in the present study, was lower than that of DFH under the applied experimental conditions. At the equilibrium pH value of $7.9 \pm 1$, DFH has a considerably higher adsorbent capacity and, therefore, can bind more arsenic within the given contact time in a technical installation. However, the presence of different competing interfering ions reduces the adsorption capacity significantly, and the lowest adsorption capacity was measured in the case of NSF water that has rather elevated levels of silicate and phosphate anions. The adsorption kinetic data for both As(V) and As(III) fitted well to a second-order kinetic model. The different interfering ions of HH tap water and NSF artificial water matrixes strongly decrease the rate of uptake of As(V) and As(III), and the latter is even more greatly affected by the water matrix. To conclude, this study suggests that DFH might be successfully employed for arsenic removal from groundwaters, for example, in an adsorption-low pressure membrane hybrid system, which will be investigated in ongoing research.

**Supplementary Materials:** The following are available online at http://www.mdpi.com/2073-4441/10/7/957/s1, Tables S7–S14: Used adsorbent dosages for adsorption isotherm experiments. Tables S15 and S16: Langmuir isotherms parameters for arsenic adsorption.

**Author Contributions:** This article was written by M.U. within his PhD project. M.E. supervised the overall activities of the research project. These authors designed the research project and conducted the laboratory tests and analyses. I.K. participated in editing the paper. M.M. and A.Z. participated in the final write up.

**Acknowledgments:** The authors are grateful to the Higher Education Commission (HEC) of Pakistan, German Academic Exchange Service (DAAD), Greece State Scholarships Foundation and the German Society for Academic Exchanges (IKYDA), and the Technische Universität Hamburg.

**Conflicts of Interest:** The authors declare no conflict of interest.

## References

1. Smith, A.H.; Hopenhayn-Rich, C.; Bates, M.N.; Goeden, H.M.; Hertz-Picciotto, I.; Duggan, H.M.; Wood, R.; Kosnett, M.J.; Smith, M.T. Cancer risks from arsenic in drinking water. *Environ. Health Perspect.* **1992**, *97*, 259. [CrossRef] [PubMed]

2. Violante, A.; Ricciardella, M.; Del Gaudio, S.; Pigna, M. Coprecipitation of arsenate with metal oxides: Nature, mineralogy, and reactivity of aluminum precipitates. *Environ. Sci. Technol.* **2006**, *40*, 4961–4967. [CrossRef] [PubMed]

3. Tantry, B.A.; Shrivastava, D.; Taher, I.; Nabi Tantry, M. Arsenic exposure: Mechanisms of action and related health effects. *J. Environ. Anal. Toxicol.* **2015**, *5*, 1. [CrossRef]

4. Mukherjee, A.; Sengupta, M.K.; Hossain, M.A.; Ahamed, S.; Das, B.; Nayak, B.; Lodh, D.; Rahman, M.M.; Chakraborti, D. Arsenic contamination in groundwater: A global perspective with emphasis on the Asian scenario. *J. Health Popul. Nutr.* **2006**, 142–163.

5. Garelick, H.; Jones, H.; Dybowska, A.; Valsami-Jones, E. Arsenic pollution sources. In *Reviews of Environmental Contamination*; Springer: New York, NY, USA, 2009; Volume 197, pp. 17–60.

6. Ware, G.W.; Albert, L.A.; Crosby, D.G.; Voogt, d.P.; Hutzinger, O.; Knaak, J.B.; Mayer, F.L.; Morgan, D.P.; Park, D.L.; Tjeerdema, R.S.; et al. *Reviews of Environmental Contamination and Toxicology*; Springer: New York, NY, USA, 2005.

7. Mandal, S.; Sahu, M.K.; Patel, R.K. Adsorption studies of arsenic(III) removal from water by zirconium polyacrylamide hybrid material (ZrPACM-43). *Water Resour. Ind.* **2013**, *4*, 51–67. [CrossRef]

8. Abejón, R.; Garea, A. A bibliometric analysis of research on arsenic in drinking water during the 1992–2012 period: An outlook to treatment alternatives for arsenic removal. *J. Water Process Eng.* **2015**, *6*, 105–119. [CrossRef]

9. Organization, W.H. *Guidelines for Drinking-Water Quality*; World Health Organization: Geneva, Switzerland, 2004.

10. Sanjrani, M.A.; Mek, T.; Sanjrani, N.D.; Leghari, S.J.; Moryani, H.T.; Shabnam, A.B. Current situation of aqueous arsenic contamination in Pakistan, focused on Sindh and Punjab Province, Pakistan: A review. *J. Pollut. Eff. Cont.* **2017**, *5*, 207.

11. Podgorski, J.E.; Eqani, S.A.M.A.S.; Khanam, T.; Ullah, R.; Shen, H.; Berg, M. Extensive arsenic contamination in high-pH unconfined aquifers in the Indus Valley. *Sci. Adv.* **2017**, *3*, e1700935. [CrossRef] [PubMed]
12. Hering, J.G.; Katsoyiannis, I.A.; Theoduloz, G.A.; Berg, M.; Hug, S.J. Arsenic removal from drinking water: Experiences with technologies and constraints in practice. *J. Environ. Eng.* **2017**, *143*, 3117002. [CrossRef]
13. Wang, L.; Chen, A.S.C.; Sorg, T.J.; Fields, K.A. Field evaluation of as removal by IX and AA. *J. Am. Water Work. Assoc.* **2002**, *94*, 161–173. [CrossRef]
14. Tresintsi, S.; Simeonidis, K.; Estradé, S.; Martinez-Boubeta, C.; Vourlias, G.; Pinakidou, F.; Katsikini, M.; Paloura, E.C.; Stavropoulos, G.; Mitrakas, M. Tetravalent manganese feroxyhyte: A novel nanoadsorbent equally selective for As(III) and As(V) removal from drinking water. *Environ. Sci. Technol.* **2013**, *47*, 9699–9705. [CrossRef] [PubMed]
15. Tresintsi, S.; Simeonidis, K.; Zouboulis, A.; Mitrakas, M. Comparative study of As(V) removal by ferric coagulation and oxy-hydroxides adsorption: Laboratory and full-scale case studies. *Desalt. Water Treat.* **2013**, *51*, 2872–2880. [CrossRef]
16. Amy, G.L.; Chen, H.-W.; Dinzo, A.; Brandhuber, P. *Adsorbent Treatment Technologies for Arsenic Removal*; American Water Works Association: Denver, CO, USA, 2005.
17. Ćurko, J.; Matošić, M.; Crnek, V.; Stulić, V.; Mijatović, I. Adsorption characteristics of different adsorbents and iron(III) salt for removing As(V) from water. *Food Technol. Biotechnol.* **2016**, *54*, 250–255. [CrossRef] [PubMed]
18. Thirunavukkarasu, O.S.; Viraraghavan, T.; Subramanian, K.S. Arsenic removal from drinking water using granular ferric hydroxide. *Water Sa* **2003**, *29*, 161–170. [CrossRef]
19. Katsoyiannis, I.A.; Mitrakas, M.; Zouboulis, A.I. Arsenic occurrence in Europe: Emphasis in Greece and description of the applied full-scale treatment plants. *Desalt. Water Treat.* **2015**, *54*, 2100–2107. [CrossRef]
20. An, B.; Steinwinder, T.R.; Zhao, D. Selective removal of arsenate from drinking water using a polymeric ligand exchanger. *Water Res.* **2005**, *39*, 4993–5004. [CrossRef] [PubMed]
21. Abejón, A.; Garea, A.; Irabien, A. Arsenic removal from drinking water by reverse osmosis: Minimization of costs and energy consumption. *Sep. Purif. Technol.* **2015**, *144*, 46–53. [CrossRef]
22. Víctor-Ortega, M.D.; Ratnaweera, H.C. Double filtration as an effective system for removal of arsenate and arsenite from drinking water through reverse osmosis. *Process Saf. Environ. Prot.* **2017**, *111*, 399–408. [CrossRef]
23. Nidheesh, P.V.; Singh, T.S.A. Arsenic removal by electrocoagulation process: Recent trends and removal mechanism. *Chemosphere* **2017**, *181*, 418–432. [CrossRef] [PubMed]
24. Lata, S.; Samadder, S.R. Removal of arsenic from water using nano adsorbents and challenges: A review. *J. Environ. Manag.* **2016**, *166*, 387–406. [CrossRef] [PubMed]
25. Driehaus, W.; Jekel, M.; Hildebrandt, U. Granular ferric hydroxide—A new adsorbent for the removal of arsenic from natural water. *J. Water Supply Res. Technol. AQUA* **1998**, *47*, 30–35. [CrossRef]
26. Pal, B.N. Granular ferric hydroxide for elimination of arsenic from drinking water. *Technol. Arsen. Remov. Drink. Water* **2001**, 59–68.
27. Katsoyiannis, I.A.; Zouboulis, A.I.; Mitrakas, M.; Althoff, H.W.; Bartel, H. A hybrid system incorporating a pipe reactor and microfiltration for biological iron, manganese and arsenic removal from anaerobic groundwater. *Fresenius Environ. Bull.* **2013**, *22*, 3848–3853.
28. Kalaruban, M.; Loganathan, P.; Shim, W.; Kandasamy, J.; Vigneswaran, S. Mathematical modelling of nitrate removal from water using a submerged membrane adsorption hybrid system with four adsorbents. *Appl. Sci.* **2018**, *8*, 194. [CrossRef]
29. Badruzzaman, M.; Westerhoff, P.; Knappe, D.R.U. Intraparticle diffusion and adsorption of arsenate onto granular ferric hydroxide (GFH). *Water Res.* **2004**, *38*, 4002–4012. [CrossRef] [PubMed]
30. Kersten, M.; Karabacheva, S.; Vlasova, N.; Branscheid, R.; Schurk, K.; Stanjek, H. Surface complexation modeling of arsenate adsorption by akagenéite (β-FeOOH)-dominant granular ferric hydroxide. *Colloids Surf. A Physicochem. Eng. Asp.* **2014**, *448*, 73–80. [CrossRef]
31. Kosmulski, M. *Surface Charging and Points of Zero Charge*; CRC Press: Boca Raton, FL, USA, 2009.
32. Simeonidis, K.; Papadopoulou, V.; Tresintsi, S.; Kokkinos, E.; Katsoyiannis, I.; Zouboulis, A.; Mitrakas, M. Efficiency of iron-based oxy-hydroxides in removing antimony from groundwater to levels below the drinking water regulation limits. *Sustainability* **2017**, *9*, 238. [CrossRef]

33. Simeonidis, K.; Mourdikoudis, S.; Kaprara, E.; Mitrakas, M.; Polavarapu, L. Inorganic engineered nanoparticles in drinking water treatment: A critical review. *Environ. Sci. Water Res. Technol.* **2016**, *2*, 43–70. [CrossRef]

34. Skoog, D.A.; Leary, J.J. Principles of instrumental analysis. *Clin. Chem.-Ref. Ed.* **1994**, *40*, 1612.

35. Banerjee, K.; Nour, S.; Selbie, M.; Prevost, M.; Blumenschein, C.D.; Chen, H.W.; Amy, G.L. Optimization of process parameters for arsenic treatment with granular ferric hydroxide. In Proceedings of the AWWA Annual Conference, Anaheim, CA, USA, 15–19 June 2003.

36. Tresintsi, S.; Mitrakas, M.; Simeonidis, K.; Kostoglou, M. Kinetic modeling of AS(III) and AS(V) adsorption by a novel tetravalent manganese feroxyhyte. *J. Colloid Interface Sci.* **2015**, *460*, 1–7. [CrossRef] [PubMed]

37. Tran, H.N.; You, S.-J.; Hosseini-Bandegharaei, A.; Chao, H.-P. Mistakes and inconsistencies regarding adsorption of contaminants from aqueous solutions: A critical review. *Water Res.* **2017**, *120*, 88–116. [CrossRef] [PubMed]

38. Banerjee, K.; Amy, G.L.; Prevost, M.; Nour, S.; Jekel, M.; Gallagher, P.M.; Blumenschein, C.D. Kinetic and thermodynamic aspects of adsorption of arsenic onto granular ferric hydroxide (GFH). *Water Res.* **2008**, *42*, 3371–3378. [CrossRef] [PubMed]

39. Manning, B.A.; Fendorf, S.E.; Goldberg, S. Surface structures and stability of arsenic(III) on goethite: Spectroscopic evidence for inner-sphere complexes. *Environ. Sci. Technol.* **1998**, *32*, 2383–2388. [CrossRef]

40. Bolisetty, S.; Reinhold, N.; Zeder, C.; Orozco, M.N.; Mezzenga, R. Efficient purification of arsenic-contaminated water using amyloid-carbon hybrid membranes. *Chem. Commun.* **2017**, *53*, 5714–5717. [CrossRef] [PubMed]

41. Xie, B.; Fan, M.; Banerjee, K. Modeling of arsenic (V) adsorption onto granular ferric hydroxide. *J. Am. Water Works Assoc.* **2007**, *99*, 92–102. [CrossRef]

42. Nguyen, V.L.; Chen, W.-H.; Young, T.; Darby, J. Effect of interferences on the breakthrough of arsenic: Rapid small scale column tests. *Water Res.* **2011**, *45*, 4069–4080. [CrossRef] [PubMed]

43. Genz, A.; Kornmüller, A.; Jekel, M. Advanced phosphorus removal from membrane filtrates by adsorption on activated aluminium oxide and granulated ferric hydroxide. *Water Res.* **2004**, *38*, 3523–3530. [CrossRef] [PubMed]

44. Tresintsi, S.; Simeonidis, K.; Vourlias, G.; Stavropoulos, G.; Mitrakas, M. Kilogram-scale synthesis of iron oxy-hydroxides with improved arsenic removal capacity: study of Fe(II) oxidation—Precipitation parameters. *Water Res.* **2012**, *46*, 5255–5267. [CrossRef] [PubMed]

45. Eljamal, O.; Sasaki, K.; Tsuruyama, S.; Hirajima, T. Kinetic model of arsenic sorption onto zero-valent iron (ZVI). *Water Qual. Expo. Health* **2011**, *2*, 125–132. [CrossRef]

46. Saldaña-Robles, A.; Saldaña-Robles, N.; Saldaña-Robles, A.L.; Damian-Ascencio, C.; Rangel-Hernández, V.H.; Guerra-Sanchez, R. Arsenic removal from aqueous solutions and the impact of humic and fulvic acids. *J. Clean. Prod.* **2017**, *159*, 425–431. [CrossRef]

47. Smith, E.H. Surface complexation modeling of metal removal by recycled iron sorbent. *J. Environ. Eng.* **1998**, *124*, 913–920. [CrossRef]

*water*

MDPI

Article

# Simultaneous Treatment of Agro-Industrial and Industrial Wastewaters: Case Studies of Cr(VI)/Second Cheese Whey and Cr(VI)/Winery Effluents

Triantafyllos I. Tatoulis [1], Michail K. Michailides [1], Athanasia G. Tekerlekopoulou [1,*], Christos S. Akratos [1], Stavros Pavlou [2,3] and Dimitrios V. Vayenas [2,3]

[1]  Department of Environmental and Natural Resources Management, University of Patras, 2 G. Seferi Str., GR-30100 Agrinio, Greece; ttatoulis@upatras.gr (T.I.T.); mmixail@upatras.gr (M.K.M.); cakratos@upatras.gr (C.S.A.)
[2]  Institute of Chemical Engineering Sciences (FORTH/ICE-HT), Stadiou Str., Platani, P.O. Box 1414, GR-26504 Patras, Greece; sp@chemeng.upatras.gr (S.P.); dvagenas@chemeng.upatras.gr (D.V.V.)
[3]  Department of Chemical Engineering, University of Patras, GR-26504 Patras, Greece
*  Correspondence: atekerle@upatras.gr; Tel.: +30-26410-74204

Received: 14 February 2018; Accepted: 23 March 2018; Published: 25 March 2018

**Abstract:** Hexavalent chromium (Cr(VI)) was co-treated either with second cheese whey (SCW) or winery effluents (WE) using pilot-scale biological trickling filters in series under different operating conditions. Two pilot-scale filters in series using plastic support media were used in each case. The first filter (i.e., Cr-SCW-filter or Cr-WE-filter) aimed at Cr(VI) reduction and the partial removal of dissolved chemical oxygen demand (d-COD) from SCW or WE and was inoculated with indigenous microorganisms originating from industrial sludge. The second filter in series (i.e., SCW-filter or WE-filter) aimed at further d-COD removal and was inoculated with indigenous microorganisms that were isolated from SCW or WE. Various Cr(VI) (5–100 mg L$^{-1}$) and SCW or WE (d-COD, 1000–25,000 mg L$^{-1}$) feed concentrations were tested. Based on the experimental results, the sequencing batch reactor operating mode with recirculation of 0.5 L min$^{-1}$ proved very efficient since it led to complete Cr(VI) reduction in the first filter in series and achieved high Cr(VI) reduction rates (up to 36 and 43 mg L$^{-1}$ d$^{-1}$, for SCW and WW, respectively). Percentage d-COD removal for SCW and WE in the first filter was rather low, ranging from 14 to 42.5% and from 4 to 29% in the Cr-SCW-filter and Cr-WE-filter, respectively. However, the addition of the second filter in series enhanced total d-COD removal to above 97% and 90.5% for SCW and WE, respectively. The above results indicate that agro-industrial wastewater could be used as a carbon source for Cr(VI) reduction, while the use of two trickling filters in series could effectively treat both industrial and agro-industrial wastewaters with very low installation and operational costs.

**Keywords:** hexavalent chromium; second cheese whey; winery effluents; co-treatment; trickling biofilter

## 1. Introduction

Nowadays, the removal of inorganic (heavy metals, such as Cr, Pb, and Cd) and organic pollutants (proteins, carbohydrates, fats, and nucleic acids) from industrial/agro-industrial wastewaters remains a huge challenge, leading to an important problem in the field of wastewater purification [1].

Chromium has found a wide range of applications and under normal conditions exists in two stable oxidation states: hexavalent (Cr(VI)) and trivalent (Cr(III)). Cr(VI) is much more mobile in the environment than Cr(III). In contrast, Cr(III) has low solubility in water (<1 μg L$^{-1}$) and readily precipitates (formation of insoluble hydroxide, K$_{sp}$ of Cr(OH)$_3$ = 6.3 × 10$^{-31}$, in the pH range between

7 and 10) [2]. Of the two, Cr(VI) compounds have the most significant effects on health. On the contrary, Cr(III) is less toxic than Cr(VI) and is listed as an essential element and a micronutrient. It is poisonous only at high concentrations [3]. Consequently, the removal or reduction of Cr(VI) to Cr(III) from contaminated waters and wastewaters is a very important process [4–6]. Most countries, including those in the European Union, have regulated the permitted limit for the total chromium in drinking and surface waters to 0.05 mg $L^{-1}$ [7,8].

Several methods have been reported for chromium removal. The most widely used techniques for Cr(VI) removal are: adsorption, biosorption, reduction, filtration, ion-exchange, electrochemical, phyto-remediation, flotation, and solvent extraction [9]. Various chromium-contaminated waters such as groundwater, drinking water, tannery wastewater, electroplating wastewater, and synthetic industrial wastewater were used. The success of the applied technologies, in terms of Cr(VI) removal rate and efficiency, depends on the physicochemical conditions that are occurring in the aquatic environment. Treatment systems based on Cr(VI) reduction can be biological (microorganisms can catalyze redox reactions by a combination of several mechanisms, including enzymatic extra-cellular reduction, nonmetabolic reduction by bacterial surfaces, and intra-cellular reduction and precipitation) or abiotic (reduction by iron salts, sulfur compounds, and metals), or a combination of these [10]. Reductive precipitation, commonly by adding Fe(II), is a well-known treatment process for the removal of Cr(VI) from groundwaters that usually contain relatively low Cr(VI) concentrations. Firstly, Cr(VI) is reduced to Cr(III), while Fe(II) is oxidized to ferric iron (Fe(III)). This process has proven to reduce Cr(VI) concentration to sub-ppb levels in relatively short reaction time (<20 min), significantly reducing the capital and operating costs [11]. Similar to the other treatment technologies, the greatest challenge with the operation of this process is disposal of the waste backwash water, which will contain all the Cr removed from the water as well as all the iron added. According to the literature, for a RCF (Reduction-Coagulation-Filtration) system using ferrous sulfate (FeSO$_4$) to reduce Cr(VI) to Cr(III), the total water (treatment) cost is 4.5\$ kgal$^{-1}$ (0.216 € m$^{-3}$) for 250 gpm (1362.75 m$^3$ d$^{-1}$) and initial Cr(VI) concentration of 25 μg L$^{-1}$ [12]. Membrane filtration (ultrafiltration–UF, nanofiltration–NF, and reverse osmosis–RO) has received significant attention, however it cannot be easily applied for the removal of Cr(VI) from water sources at sub-ppb levels. These processes appear to offer significant efficiency, relatively simple operation, and save space [10]. However, their high capital and operating costs has become the greatest obstacle to the development of this technology in full-scale systems.

Microbiological Cr(VI) reduction is regarded a safe and sustainable process which has been focused on extensively even in wastewaters with high Cr(VI) concentrations [1,13,14]. Taking into account the nutritional requirements of the microorganisms involved, the addition of an external organic carbon source is necessary to enhance Cr(VI) reduction [15]. Various carbon sources have been tested (mainly in pure rather than mixed cultures) including lactose, maltose, glucose, fructose, humic acid, sodium acetate, sugar, molasses, etc., with significant Cr(VI) reduction rates [16–20]. However, costs for some of these feedstocks (which serve as an electron donor) might become prohibitive in actual field applications. The use of industrial-grade molasses may provide a more cost-effective alternative to any of the above-mentioned carbon and energy sources [21–25]. According to the literature, the treatment cost when using molasses (0.13 € (Kg molasses)$^{-1}$ in Greece) amounts to about 0.39 € (m$^{-3}$ of wastewater) [25].

To further reduce treatment cost, researchers have used agricultural wastewater as feed for bacterial growth [26]. Cheese whey, an organic by-product (resulting from cheese production), has been examined as a low-cost carbon source for Cr(VI) reduction. According to Panousi et al. [16], the cost of treatment using cheese whey (in a suspended growth system operating as a Sequencing Batch Reactor, SBR, using initial Cr(VI) concentration of 1.0–2.0 mg L$^{-1}$ and low feed chemical oxygen demand-COD of just 100–200 mg L$^{-1}$) varies between 0.16–0.34 € and 0.68–0.86 € m$^{-3}$ for treatment capacities of 100 m$^3$ d$^{-1}$ and 10 m$^3$ d$^{-1}$, respectively. Cheese whey is considered to be the most important pollutant of dairy wastewaters. This is not only because of its high organic load, but also because of the large volume generated. Only a few studies have been published using cheese whey for Cr(VI) reduction [16,22,27–33] and most of these were performed for in-situ groundwater and soil remediation [22,30–33], while the remainder took place in pilot-scale suspended growth systems (activated sludge technology) [16,27–29]. Another low-cost carbon source that was examined for Cr(VI) reduction was liquid pineapple waste by Zakaria et al. [20]. Zakaria et al. [26] studied the ability of a pure culture (*Acinetobacter haemolyticus*) to reduce Cr(VI) in a packed bed reactor using various initial Cr(VI) concentrations (15–240 mg L$^{-1}$). Successful application of cheese whey and liquid pineapple waste as organic substrates to promote the biotic reduction of Cr(VI) was reported in the above-mentioned studies, however limited information exists on the use of attached growth systems and the organic load (d-COD) removal of wastewater.

Second cheese whey (SCW), which is a by-product of cottage cheese production, is characterized as a high strength organic pollutant [34]. The use of SCW for Cr(VI) bioreduction has been examined only by our research group in a previous work, using pilot-scale horizontal subsurface flow constructed wetlands [35]. Constructed wetlands proved to be effective as they achieved 100% and 50–70% removal of Cr(VI) and dissolved chemical oxygen demand (d-COD), respectively. However, the initial Cr(VI) and d-COD concentrations that were used in the constructed wetlands were relatively low (0.4–5 mg L$^{-1}$ and 1300–4100 mg L$^{-1}$, respectively).

Winery effluents (WE) also contain high concentrations of readily biodegradable soluble organic matter (BOD is about 0.4–0.9 of the COD value) [36–38], however, to our knowledge, it has not yet been used as a carbon source for hexavalent chromium reduction.

The present paper focuses on the co-treatment of Cr(VI) (in high concentrations in the range of mg L$^{-1}$, present mainly in industrial wastewaters) with SCW or winery effluents (WE) in attached growth (trickling filter) aerobic biological systems. To achieve both Cr(VI) and d-COD removal, two pilot-scale trickling filters were operated in series in each case. The first filter was used for complete Cr(VI) reduction to Cr(III) and partial COD removal from SCW or WE, while the second filter in series was used for further COD removal. Indigenous microorganisms originating from industrial sludge were used for Cr(VI) reduction and indigenous microorganisms that were isolated from SCW or WE were also used for COD removal. To the best of our knowledge, the present study is the first to report on Cr(VI) reduction using SCW and WE in trickling filters, and the simultaneous removal of d-COD from the wastewater.

## 2. Materials and Methods

### 2.1. Wastewater Characterization

SCW and WE were obtained from the "Papathanasiou" cheese factory that is located in Agrinio city (western Greece) and the local "Grivas winery" in Agrinio, respectively. SCW had acidic characteristics (pH 4.5–6) with high salinity (about 5.25 ppt), conductivity of 8455 ± 1000 μS cm$^{-1}$, and high d-COD (43,000 ± 2000 mg L$^{-1}$). WE derived from water used for washing tanks after grape processing. The WE presented pH values between 6 and 7, conductivity of about 300–700 μS cm$^{-1}$, and high d-COD (60,000 ± 3000 mg L$^{-1}$).

### 2.2. Indigenous Microorganisms/Enrichment and Culture Conditions

Mixed cultures of microorganisms were used for the reduction of Cr(VI) and degradation of d-COD. Initially, industrial sludge obtained from the Hellenic Aerospace Industry S.A. was used to grow indigenous microorganisms able to reduce Cr(VI). In two different Erlenmeyer flasks (SCW-flask and WE-flask of 2 L) 10 g of sludge were diluted with SCW or WE (concentration of 1000 mg d-COD L$^{-1}$) and 1.5 mg L$^{-1}$ Cr(VI) (added as K$_2$Cr$_2$O$_7$). When Cr(VI) concentration dropped below 0.2 mg L$^{-1}$, 0.9 L of the SCW-flask's and CW-flask's liquid was replaced with fresh SCW or WE and chromium (45% volume exchange), in order to achieve the desired final d-COD and Cr(VI) concentrations of 1000 mg L$^{-1}$ and 1.5 mg L$^{-1}$, respectively, within the flasks. This process completed one operating cycle. A series of operating cycles was performed while a start-up time of approximately three weeks was necessary to reach maximum Cr(VI) reduction rates. A minimum period of about 2.0 and 1.90 h$^{-1}$ was needed, using SCW and WE, respectively, for Cr(VI) concentration to fall below the maximum permitted limit of 0.2 mg L$^{-1}$. The liquid in the flasks was kept under oxic conditions through aeration (Dissolved Oxygen D.O. > 5 mg L$^{-1}$), while agitation was at 400 rpm with a magnetic stirrer. It is known that the pH of a solution has a very significant effect on bioreduction experiments. According to the literature, many microbes are capable of successfully reducing Cr(VI) from aqueous solutions at a wide range of pH [1,10]. The pH of the media affects the solubility and ionization state of Cr(VI), and bacterial growth is also affected by change in the pH of the media [9]. The most important redox species at pH values >6.5 is chromate (CrO$_4{}^{2-}$), while Cr$_2$O$_7{}^-$ and HCrO$_4{}^-$ are mainly relevant to pH values <6.5 and its speciation depends on Cr(VI) concentration and pH value [2]. The pH range in this study was about 4.5–6.0, without adjustment (making the process easier to apply in full-scale units), while the temperature was maintained at 22 ± 2 °C. In this way, a natural selection process led to the predominance of species that are able to reduce Cr(VI) by using SCW or WE as a carbon source. These two enrichment cultures (Cr-SCW-EC and CR-WE-EC) were used to set up the pilot-scale packed bed reactors for Cr(VI) removal (Cr-SCW-filter or Cr-WE-filter).

Indigenous microorganisms from SCW or WE were isolated for the aerobic biological degradation of residual d-COD in the post-treatment step (SCW-filter or WE-filter). The procedure described by Tatoulis et al. [34] was followed, in which a natural selection process led to the predominance of indigenous species able to biodegrade SCW or WE in the SCW-filter or WE-filter, respectively. Specifically, 0.5 L of SCW or WE (concentration of 23,000 ± 1500 mg L$^{-1}$) was placed into two different 1 L Erlenmeyer flasks. The solution in the flasks was kept under aerobic conditions (D.O. > 5 mg L$^{-1}$) and was mixed constantly. As soon as d-COD degradation increased, 90% of the liquid volume was discarded and the flask content was then brought back to 0.5 L volume by adding tap water and SCW or WE in appropriate amounts that were sufficient to achieve the desired final concentration of 23,000 ± 1500 mg L$^{-1}$. A series of operating cycles were performed while a start-up time of approximately one month was necessary to reach maximum d-COD degradation rates. The minimum duration of the operating cycle was about 72 and 70 h$^{-1}$ for SCW and WE, respectively. These two enrichment cultures (SCW-EC and WE-EC) were used to set up the pilot-scale packed bed reactors for d-COD removal (SCW-filter or WE-filter).

### 2.3. Packed Bed Reactors (Pilot-Scale Filters)

The packed-bed reactors that were used in this work have been described in detail by Michailides et al. [25] and Tatoulis et al. [34]. All of the packed-bed reactors consisted of a Plexiglas tube, 1.6 m high and 9 cm internal diameter (Figure 1). Plastic tubes (1.6 cm internal diameter, 3 cm length, specific surface area (A$_s$) of 500 m$^2$ m$^{-3}$, and filter porosity ($\varepsilon$) of 0.8) were used as support material inside the reactors. The depth of the support media in the filters was 1.43 m. For each wastewater used (SCW or WE) two pilot-scale trickling filters were operated in series. The first filter was used for complete Cr(VI) reduction and partial d-COD removal of SCW or WE, while the second filter in series was used for further d-COD degradation. Initially, 0.5 L volume of each enrichment culture (Cr-SCW-EC or CR-WE-EC and SCW-EC or WE-EC, Section 2.2) was used to set up the corresponding filter (Cr-SCW-filter or Cr-WE-filter and SCW-filter or WE-filter, respectively). The pilot-scale filters were first operated as batch reactors to ensure microorganism attachment (development of a biofilm layer) onto the support material. The sequencing batch reactor (SBR) operating mode with recirculation (SBR with recirculation) was then applied. Three different recirculation rates were tested in the Cr-SCW-filter or the Cr-WE-filter (0.5, 1.0, and 2 L min$^{-1}$), while the SCW-filter or WE-filter was operated as batch reactor at a constant recirculation rate of 0.5 L min$^{-1}$. The Cr-SCW-filter or Cr-WE-filter were loaded with various concentrations of Cr(VI) (5, 10, 20, 30, 60, and 100 mg L$^{-1}$) and SCW or WE (1000, 5000, 13,000, 21,500, and 25,000 mg d-COD L$^{-1}$). According to Owlad et al., Cr(VI) concentrations in industrial wastewaters are estimated to be between 0.1 and 200 mg L$^{-1}$ [9]. It should be mentioned that the concentrations of Cr(VI) in the wastewaters of many industries can reach up to

100 mg L$^{-1}$, such as from chrome tanning plants (3.7 mg L$^{-1}$), electropolishing plants (20.7 mg L$^{-1}$, 42.8 mg L$^{-1}$), hardware factories (60 mg L$^{-1}$), and tanneries (100 mg L$^{-1}$) [9]. The feed SCW or WE concentrations of 1000 to 25,000 mg d-COD L$^{-1}$ examined in this work simulate cheese factory and winery wastewater effluents (including factory washing waters) [34]. Finally, experiments under continuous operating mode and continuous operating mode with recirculation were also performed. The total duration of these experiments was about twenty months.

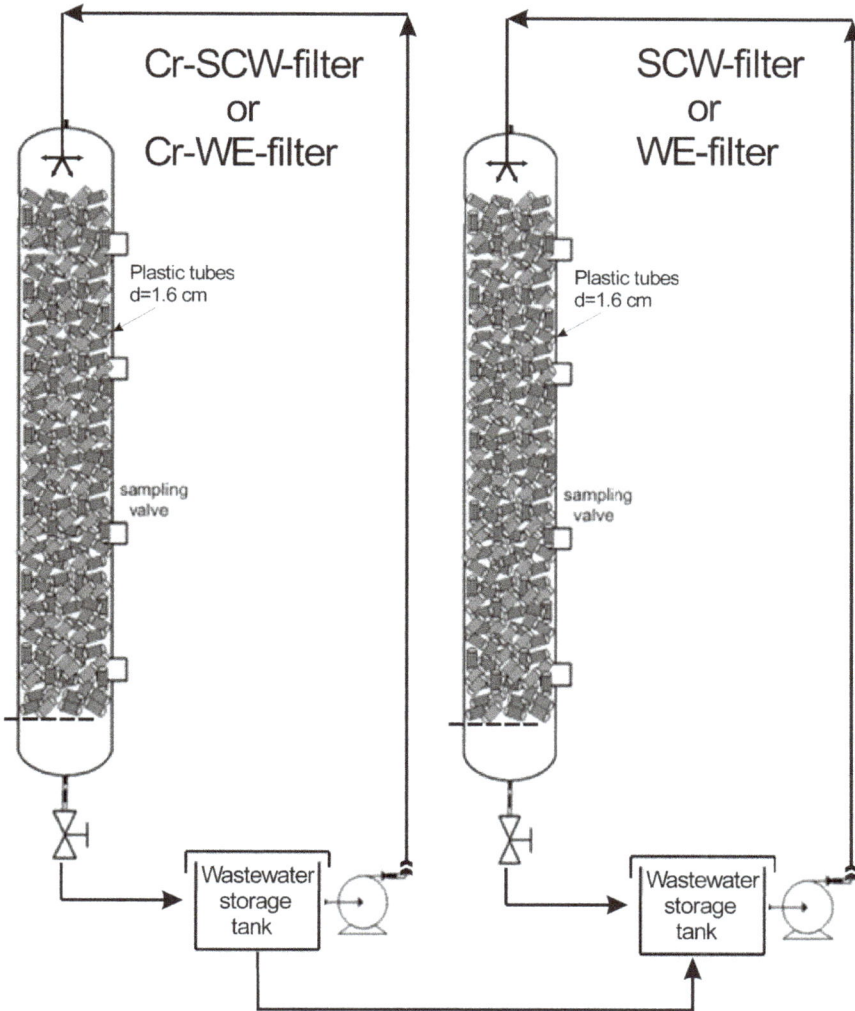

**Figure 1.** Scheme of the pilot-scale bio-filters.

The working volume (pore volume after biofilm formation) of all filters used was 6 L. Filter backwashing due to pore clogging was necessary using water (8 L min$^{-1}$ for all filters used, Cr-SCW-filter, Cr-WE-filter, SCW-filter, and WE-filter) and air (upflow velocities of 10 L min$^{-1}$). Backwash frequency for the Cr-SCW-filter or Cr-WE-filter depended on the initial hexavalent chromium concentration tested (i.e., once every 3–4 days for Cr(VI) concentrations above 30 mg L$^{-1}$ and once every eight days for lower concentrations). For the SCW-filter or WE-filter, backwash frequency depended on the SCW or WE feed concentration and was determined by the loading time (less than 1 min) of the working volume of polluted water into the filter, thus indicating pore clogging.

## 2.4. Sample Collection and Analyses

Samples were collected daily and were filtered through 0.45 μm-Millipore filters (GN-6 Metricel Grid 47 mm, Pall Corporation). Cr(VI) concentration was determined according to the "Standard Methods for the Examination of Water and Wastewater" (3500-Cr D Colorimetric method at 540 nm) [39], using a spectrophotometer (Boeco S-20). Total Cr concentrations were also measured using an atomic absorption spectrophotometer (Perkin-Elmer AAS-700), following the "Standard Methods for the Examination of Water and Wastewater" [39]. Cr(III) was estimated as the difference between the total chromium and Cr(VI). Measurements of total chromium concentrations were taken at the beginning and end of each experiment and their values were almost identical to those of the Cr(VI) (measurements of total chromium not presented here). d-COD concentration was measured by the closed reflux, colorimetric method using a Multiparameter Bench Photometer (HANNA C99). Total phenolic compounds (with respect to syringic acid) were determined spectrophotometrically using a Boeco (Germany, S-20) spectrophotometer, according to the Folin-Ciocalteu method [40]. A HANNA HI9828 multi-parameter meter was used to measure dissolved oxygen (D.O.) and pH.

## 2.5. Statistical Analysis

All of the analyses were carried out in triplicate with a relative standard deviation not exceeding 5%. Results were expressed as mean ± standard deviation. Statistically significant differences between data were evaluated using the *t*-student confidence interval, for 95% probability. In this statistical analysis, the data was considered to follow the *t*-student distribution and the confidence interval for the difference of a pair of mean values was calculated.

## 3. Results and Discussion

### 3.1. Packed Bed Reactor Experiments in Cr-SCW-Filter and Cr-WE-Filter

#### 3.1.1. Sequencing Batch Mode Operation

Initially, abiotic chromium reduction was investigated. Clean Cr-SCW-filter and Cr-WE-filter (without microorganisms) were loaded with SCW or WE (of concentration 5000 mg/L) and chromium at a final concentration of 10 mg Cr(VI) L$^{-1}$ and operated in batch mode. The system was aerated by an air pump and for an operation period of 24 h no significant change in the concentration of Cr(VI) was observed thus indicating the absence of abiotic Cr(VI) reduction. Therefore, the addition of microorganisms was necessary to achieve Cr(VI) reduction.

The pilot-scale Cr-SCW-filter and Cr-WE-filter were then inoculated with the enrichment cultures Cr-SCW-EC and Cr-WE-EC, respectively (as described in Section 2.2). The filter was first operated in batch operating mode with initial Cr(VI) and d-COD concentrations of 5 and 1000 mg L$^{-1}$, respectively. As soon as Cr(VI) reduction was completed the solution was discharged (and then used as an influent solution for the SCW-filter or WE-filter) and the new feed medium with the same Cr(VI) and d-COD concentrations was added into the filter. This process comprised one complete operating cycle. Cycles were repeated until the maximum removal rate of Cr(VI) was recorded for at least three cycles, while an air pump that was located at the base of the filter provided air (D.O. > 5 mg L$^{-1}$).

According to the results (data not presented here), the Cr-SCW-filter and Cr-WE-filter were not able to achieve complete Cr(VI) reduction in all of the cycles performed under batch operation and did not reach a steady-state condition. Additionally, the d-COD removal rate recorded in these filters ranged between 5 and 35%. This was due to rapid biomass growth along the Cr-filter, which led to spatial heterogeneity and insufficient exploitation of the filter. To avoid these problems, the filter was operated under SBR operating mode with recirculation.

3.1.2. Sequencing Batch Mode Operation with Recirculation

The Cr-SCW-filter or Cr-WE-filter was then loaded with SCW or WE and Cr(VI) at the final concentrations referred to in Section 2.3, while the recirculation rates of 0.5, 1.0, and 2.0 L min$^{-1}$ were tested. The Cr-SCW-filter operated successfully for the recirculation rate of 0.5 L min$^{-1}$ as complete Cr(VI) reduction was achieved for all of the initial concentrations of Cr(VI) (5–110 mg L$^{-1}$) and d-COD (1000–25,000 mg L$^{-1}$) tested. Table 1 presents Cr(VI) reduction rates (mg L$^{-1}$ h$^{-1}$ and g m$^{-2}$ d$^{-1}$) as well as percentage d-COD removal (%) achieved for all series of experiments that were performed in the Cr-SCW-filter for the recirculation rate of 0.5 L min$^{-1}$. Table 1 shows that, as initial d-COD concentration increased, higher Cr(VI) reduction rates were achieved. This was observed only for initial d-COD concentrations of up to 21,000 mg L$^{-1}$. For higher initial d-COD concentrations (25,000 mg d-COD L$^{-1}$), the Cr(VI) reduction rates were lower when compared to those that were recorded for 21,000 mg d-COD L$^{-1}$, but higher than those obtained for concentrations of 1000, 5000, and 13,000 mg d-COD L$^{-1}$. It is known in bioreactors that metals can be removed via weak adsorption onto organic substrates. In such a case, increase of organic load leads to higher Cr(VI) removal rates. However, in this study no loss of Cr(VI) was found through adsorption during the operation of the non-biomass reactors. Besides biological reduction, biosorption was reported to be significant for the efficient removal of Cr(VI) biologically in the Cr-SCW-filter. Biological reduction is the reduction by biological means of metallic ions that are normally multivalent. The reduction agent (reductant) can be organic compounds from living or dead biomass, natural organic matter, and industrial organic waste. Microorganisms may also act as a catalyst or medium for metal reduction. The reduction of Cr(VI) in this study can be achieved both by the presence of microorganisms and by just the presence of high concentrations of organic compounds in the examined wastewaters, acting also as reductant agents. However, the specific contribution between these two mechanisms is difficult to be assessed in such a complex system, as the studied one. Also, the combined effect of metals and organic carbon pollutants on microbial activity and community composition is not so clear, since few studies have addressed this point [41]. According to Nakatsu et al., [41] if species richness is reduced in sites contaminated with complex mixtures with high organic and metal concentrations, the communities may be less resilient. Also, according to Hawley et al., [10] various factors affect the rate of microbial reduction, including carbon concentrations. Therefore, the high organic load of 25,000 mg d-COD L$^{-1}$ of SCW may have affected the structure of the microbial community, and therefore the bioreduction rate.

The maximum Cr(VI) removal rate recorded was 35.99 mg L$^{-1}$ h$^{-1}$ for initial Cr(VI) and d-COD concentrations of 60 and 21,000 mg L$^{-1}$, respectively (Figure 2a). For the filter of 9 cm diameter, or cross section area of 63.62 cm$^2$, the mass of Cr(VI) removed per square meter of the filter's surface per day was 815.1 g m$^{-2}$ d$^{-1}$. Based on the results of the statistical analysis, the initial d-COD concentration was found to have significant effects on Cr(VI) removal for many of the paired data of each initial Cr(VI) concentration that was tested. For each initial Cr(VI) concentration the pairs in which significant differences were observed are: 5 mg Cr(VI) L$^{-1}$: (1000–5000; 5000–13,000; 13,000–21,000), 10 mg Cr(VI) L$^{-1}$: (1000–5000; 21,000–25,000), 30 mg Cr(VI) L$^{-1}$: (5000–13,000; 13,000–21,000; 21,000–25,000), 60 mg Cr(VI) L$^{-1}$: (5000–13,000; 13,000–21,000; 21,000–25,000), and 100 mg Cr(VI) L$^{-1}$: (13,000–21,000).

**Table 1.** Hexavalent chromium (Cr(VI)) reduction and dissolved chemical oxygen demand (d-COD) removal rates for the different initial Cr(VI) and d-COD concentrations tested in the Cr-SCW-filter operating under sequencing batch reactor (SBR) with a recirculation rate of 0.5 L min$^{-1}$.

| Initial Cr(VI) (mg L$^{-1}$) | Initial d-COD (mg L$^{-1}$) | Cr(VI) Reduction Rate (mg L$^{-1}$ h$^{-1}$) | Cr(VI) Reduction Rate (g m$^{-2}$ d$^{-1}$) | d-COD Removal (Cr-SCW-Filter) (%) | Total d-COD Removal (SCW-Filter) (%) |
|---|---|---|---|---|---|
| 5.02 ± 0.08 | 1076 ± 22 | 2.23 ± 0.04 | 50.53 ± 0.9 | 25.28 ± 1.25 | - |
| 10.01 ± 0.14 | 1053 ± 33 | 1.48 ± 0.03 | 33.58 ± 0.68 | 22.74 ± 1.04 | - |
| 5.00 ± 0.04 | 5098 ± 87 | 5.00 ± 0.18 | 150.99 ± 5.43 | 17.21 ± 0.64 | - |
| 10.05 ± 0.13 | 5185 ± 36 | 10.05 ± 0.21 | 227.66 ± 4.75 | 26.65 ± 1.09 | - |
| 30.26 ± 0.02 | 5035 ± 121 | 10.12 ± 0.11 | 161.25 ± 1.75 | 29.69 ± 0.81 | 97.57 ± 1.12 |
| 60.10 ± 0.69 | 5018 ± 141 | 6.67 ± 0.19 | 142.13 ± 4.05 | 27.35 ± 1.19 | - |
| 100.94 ± 1.46 | 5010 ± 69 | 4.16 ± 0.08 | 94.27 ± 1.81 | 34.33 ± 1.21 | - |
| 5.01 ± 0.08 | 13155 ± 362 | 6.68 ± 0.24 | 151.19 ± 5.43 | 27.78 ± 1.03 | - |
| 10.03 ± 0.12 | 13200 ± 121 | 12.53 ± 0.29 | 283.59 ± 6.56 | 33.13 ± 0.66 | - |
| 30.02 ± 1.14 | 13295 ± 305 | 24.01 ± 0.34 | 543.81 ± 7.70 | 37.19 ± 0.99 | 99.06 ± 0.56 |
| 60.76 ± 0.18 | 13425 ± 251 | 12.79 ± 0.22 | 289.67 ± 4.98 | 33.89 ± 1.14 | - |
| 101.66 ± 0.29 | 13075 ± 315 | 4.59 ± 0.09 | 104.03 ± 2.03 | 42.49 ± 2.12 | - |
| 5.03 ± 0.04 | 21610 ± 805 | 8.68 ± 0.32 | 196.51 ± 7.24 | 12.26 ± 0.24 | - |
| 10.07 ± 0.03 | 21600 ± 925 | 14.28 ± 0.33 | 323.29 ± 7.47 | 15.25 ± 0.76 | - |
| 29.68 ± 0.11 | 22190 ± 358 | 29.68 ± 0.63 | 672.13 ± 14.27 | 19.69 ± 0.99 | - |
| 59.75 ± 0.29 | 21080 ± 287 | 35.99 ± 0.45 | 815.10 ± 10.19 | 18.22 ± 0.98 | 99.40 ± 2.04 |
| 102.00 ± 2.44 | 21710 ± 1002 | 5.36 ± 0.09 | 120.00 ± 2.02 | 24.74 ± 1.24 | - |
| 5.03 ± 0.01 | 25175 ± 1205 | 8.98 ± 0.35 | 202.57 ± 7.89 | 16.12 ± 0.86 | - |
| 10.06 ± 0.02 | 25200 ± 925 | 14.42 ± 0.40 | 326.46 ± 9.05 | 17.20 ± 1.09 | - |
| 30.05 ± 0.10 | 25413 ± 1102 | 14.45 ± 0.24 | 327.23 ± 5.43 | 22.09 ± 1.12 | - |
| 60.08 ± 0.25 | 25013 ± 852 | 23.29 ± 0.64 | 527.40 ± 14.49 | 21.64 ± 1.08 | 98.70 ± 1.22 |
| 101.87 ± 2.02 | 25425 ± 912 | 5.20 ± 0.07 | 151.29 ± 2.04 | 26.65 ± 1.34 | - |

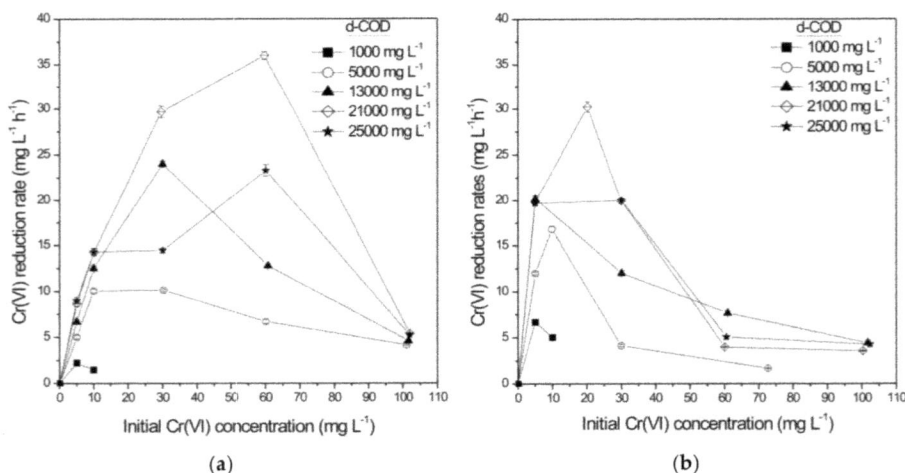

**Figure 2.** Cr(VI) reduction rates in the Cr-SCW-filter operating under SBR with recirculation mode for various initial Cr(VI) and d-COD concentrations and two different recirculation rates: (**a**) 0.5 L min$^{-1}$ and (**b**) 1.0 L min$^{-1}$.

The Cr-WE-filter also operated successfully for the recirculation rate of 0.5 L min$^{-1}$ as 100% Cr(VI) reduction was achieved for all the initial concentrations of Cr(VI) (5–110 mg L$^{-1}$) and d-COD (1000–25,000 mg L$^{-1}$) tested. Table 2 presents Cr(VI) reduction rates (mg L$^{-1}$ h$^{-1}$ and g m$^{-2}$ d$^{-1}$) as well as percentage d-COD removal (%) in the Cr-WE-filter for the recirculation rate of 0.5 L min$^{-1}$. Increase of initial d-COD concentration led to higher Cr(VI) reduction rates only for initial d-COD concentrations of up to 13,000 mg L$^{-1}$. It is probable that, as mentioned above, the high concentration of 21,000 mg d-COD L$^{-1}$ of WE affected the attached microbial community structure, making it less effective at Cr(VI) removal. However, for initial d-COD concentrations of 21,000 and 25,000 mg d-COD L$^{-1}$ the Cr(VI) reduction rates were higher than those obtained for concentrations of 1000 and 5000 mg d-COD L$^{-1}$. The maximum Cr(VI) removal rate achieved using WE was 43 mg L$^{-1}$ h$^{-1}$ for initial Cr(VI) of 60 mg L$^{-1}$ (a higher removal rate compared to SCW, 35.9943 mg L$^{-1}$ h$^{-1}$), but was recorded for lower initial d-COD concentration, 13,000 mg L$^{-1}$ (21,000 mg d-COD L$^{-1}$ for SCW) (Figure 3a).

**Table 2.** Cr(VI) reduction and d-COD removal rates for the different initial Cr(VI) and d-COD concentrations tested in the Cr-WE-filter operating under SBR with a recirculation rate of 0.5 L min$^{-1}$.

| Initial Cr(VI) (mg L$^{-1}$) | Initial d-COD (mg L$^{-1}$) | Cr(VI) Reduction Rate (mg L$^{-1}$ h$^{-1}$) | Cr(VI) Reduction Rate (g m$^{-2}$ d$^{-1}$) | d-COD Removal (Cr-WE-Filter) (%) | Total d-COD Removal (%) |
|---|---|---|---|---|---|
| 5.12 ± 0.04 | 1151 ± 38 | 2.59 ± 0.04 | 58.62 ± 0.9 | · 5.10 ± 0.06 | - |
| 10.22 ± 0.11 | 1109 ± 27 | 1.58 ± 0.02 | 35.76 ± 0.45 | 8.44 ± 0.12 | - |
| 4.56 ± 0.03 | 5108 ± 99 | 5.72 ± 0.09 | 151.05 ± 2.38 | 13.28 ± 0.24 | - |
| 10.01 ± 0.14 | 5005 ± 105 | 12.54 ± 0.27 | 284.28 ± 6.12 | 17.15 ± 0.39 | - |
| 30.82 ± 0.74 | 5010 ± 89 | 12.56 ± 0.31 | 279.08 ± 6.89 | 19.22 ± 0.52 | 91.10 ± 2.04 |
| 59.20 ± 0.33 | 5118 ± 152 | 7.89 ± 0.18 | 178.58 ± 4.07 | 25.15 ± 0.43 | - |
| 101.04 ± 1.03 | 5110 ± 122 | 4.36 ± 0.06 | 98.68 ± 1.36 | 27.33 ± 0.91 | - |
| 4.91 ± 0.05 | 13065 ± 208 | 7.55 ± 0.24 | 170.89 ± 5.43 | 17.08 ± 0.41 | - |
| 10.53 ± 0.18 | 13120 ± 305 | 17.55 ± 0.47 | 397.23 ± 10.63 | 18.23 ± 0.51 | - |
| 31.12 ± 0.22 | 13030 ± 287 | 30.12 ± 0.75 | 681.74 ± 16.97 | 21.85 ± 0.63 | - |
| 60.06 ± 0.49 | 13025 ± 299 | 43.01 ± 1.12 | 973.50 ± 25.35 | 29.25 ± 0.61 | 92.01 ± 1.89 |
| 98.7 ± 0.89 | 13295 ± 327 | 6.32 ± 0.25 | 143.05 ± 5.66 | 30.39 ± 0.78 | - |
| 5.03 ± 0.02 | 21800 ± 758 | 7.06 ± 0.31 | 227.70 ± 10.00 | 15.06 ± 0.44 | - |
| 10.27 ± 0.18 | 21250 ± 993 | 13.69 ± 0.47 | 309.86 ± 10.64 | 16.32 ± 0.57 | - |
| 29.55 ± 0.59 | 22000 ± 875 | 24.73 ± 0.61 | 636.20 ± 15.69 | 18.08 ± 0.61 | - |
| 60.77 ± 1.51 | 21380 ± 741 | 25.01 ± 0.41 | 566.08 ± 9.28 | 21.99 ± 0.49 | 91.80 ± 1.04 |
| 101.00 ± 2.01 | 21010 ± 974 | 5.06 ± 0.05 | 114.53 ± 1.13 | 25.89 ± 0.86 | - |
| 5.13 ± 0.03 | 25250 ± 1108 | 6.26 ± 0.19 | 232.22 ± 7.05 | 4.22 ± 0.04 | - |
| 10.10 ± 0.09 | 25450 ± 899 | 12.46 ± 0.22 | 304.66 ± 5.38 | 10.12 ± 0.21 | - |
| 29.05 ± 0.48 | 25150 ± 948 | 12.20 ± 0.34 | 298.77 ± 8.32 | 13.49 ± 0.38 | - |
| 59.88 ± 0.97 | 25300 ± 832 | 17.29 ± 0.52 | 536.61 ± 16.14 | 17.89 ± 0.47 | 90.50 ± 2.84 |
| 102.10 ± 1.98 | 25400 ± 1032 | 4.52 ± 0.05 | 102.23 ± 1.13 | 19.32 ± 0.64 | - |

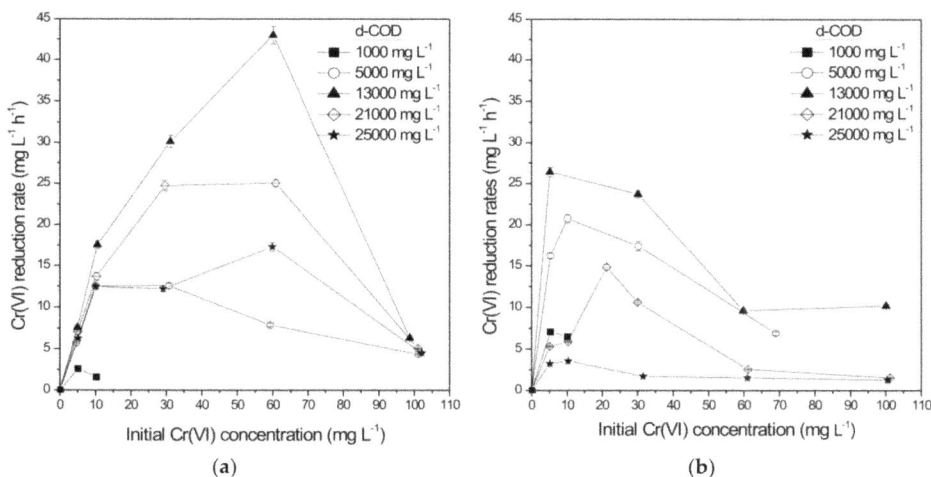

**Figure 3.** Cr(VI) reduction rates in the Cr-WE-filter operating under SBR with recirculation mode for various initial Cr(VI) and d-COD concentrations and two different recirculation rates: (**a**) 0.5 L min$^{-1}$ and (**b**) 1.0 L min$^{-1}$.

Similar to SCW, initial d-COD concentration of WE was found to have significant effects on Cr(VI) removal for many of the paired data of each initial Cr(VI) concentration tested. For each initial Cr(VI) concentration, the pairs in which significant differences were observed are: 5 mg Cr(VI) L$^{-1}$: (1000–5000; 5000–13,000), 10 mg Cr(VI) L$^{-1}$: (1000–5000; 5000–13,000; 13,000–21,000),

30 mg Cr(VI) L$^{-1}$: (5000–13,000; 13,000–21,000; 21,000–25,000), 60 mg Cr(VI) L$^{-1}$: (5000–13,000; 13,000–21,000; 21,000–25,000), and 100 mg Cr(VI) L$^{-1}$: (5000–13,000; 13,000–21,000; 21,000–25,000).

When considering the results of Tables 1 and 2, it is clear that WE led to significantly higher Cr(VI) removal rates than SCW for initial d-COD concentrations of up to 13,000 mg d-COD L$^{-1}$. The same was not observed for initial d-COD concentrations of 21,000 and 25,000 mg d-COD L$^{-1}$, since, when the d-COD concentration was 21,000 mg d-COD L$^{-1}$, the use of WE led to a significantly lower Cr(VI) reduction rate when compared to SCW, while when d-COD concentration was 25,000 mg d-COD L$^{-1}$, Cr(VI) reduction rates had no significant differences between SCW and WE. It is probable that, in these experiments with the less diluted WE, the high concentration of d-COD and/or the (poly)phenolic component of WE (that emanates from the seeds, skins, and piths of grapes), which is the least degradable fraction of WE and associated with phytotoxicity and microbial toxicity, had a negative impact on Cr(VI) reduction [42,43]. According to Ramond at al., [43] microbial communities are significantly impacted by winery wastewaters depending on their spatial locations, probably due to the toxicity of the influent stream (high COD, phenolic content, and low pH). Specifically, they mention that winery wastewaters (with concentrations of about 100 mg L$^{-1}$) have been shown to induce structural changes in environmental microbial communities [43]. It is worth mentioning that in this study the percentage contribution of total phenolics to the COD was small, ranging from 0.5 to 1.5%. Also, minor phenol removal ranging between 0 and 20% was observed during all of the experiments.

Direct comparison with previous studies using cheese whey [16,27–29] is not currently possible, since all were performed in suspended not attached growth systems. Panousi et al. [16] investigated the efficiency of biological groundwater treatment for Cr(VI) removal under anoxic conditions in suspended growth reactors using cheese whey as the carbon source. Since Cr(VI) content in groundwater can be as high as 300 µg L$^{-1}$, they studied low Cr(VI) concentration (up to 200 µg L$^{-1}$). Low substrate concentrations varying between 100 and 200 mgCOD L$^{-1}$ were also tested. Panousi et al. [16] found that, when added to groundwater at a concentration of 200 µg L$^{-1}$, complex fermentable substrates, such as cheese whey (at organic loadings above 120 mg COD L$^{-1}$), can support complete microbial Cr(VI) removal (over 99%). Orozco et al. [27] also revealed (with an initial Cr(VI) concentration of 25 mg L$^{-1}$) that significant specific Cr(VI) removal rates can be attained with fermentable substrates, such as cheese whey and lactose (0.12 mgCr(VI) gTSS$^{-1}$ h$^{-1}$, TSS: Total suspended solids, for initial biomass concentration 3000 ± 200 mgTSS$^{-1}$ L$^{-1}$). Conversely, with non-fermentable substrates, such as citrate and acetate, lower specific Cr(VI) removal rates were obtained (below 0.08 mgCr(VI) gTSS$^{-1}$ h$^{-1}$, for initial biomass concentration 3000 ± 200 mgTSS$^{-1}$ L$^{-1}$). In their experiments, the concentration of the carbonaceous substrate was just 5000 mg L$^{-1}$. Contreras et al. [28,29] tested higher initial Cr(VI) concentrations (up to 300 mg L$^{-1}$) using dehydrated cheese whey of 5000 mg d-COD L$^{-1}$. With regard to the effect of initial Cr(VI), a threshold inhibitory concentration of Cr(VI) within the range of 100 to 300 mg L$^{-1}$ was obtained, while for initial Cr(VI) concentrations that are below 100 mg L$^{-1}$ percentage reduction rates did not exceed 70% (after about 100 h). The toxic effect of high initial Cr(VI) concentration was also illustrated in the present study. Specifically, for the two highest initial d-COD concentrations of 21,000 and 25,000 mg L$^{-1}$, Cr(VI) reduction rates increased as the initial Cr(VI) concentration rose to 60 mg Cr(VI) L$^{-1}$, achieving maximum Cr(VI) reduction rates of 35.99 and 23.29 mg L$^{-1}$ h$^{-1}$, respectively (Figure 2a). However, higher initial Cr(VI) concentrations presented the toxic effect of chromium and lowered Cr(VI) reduction rates considerably (~5 mg L$^{-1}$ h$^{-1}$).

To our knowledge, winery effluent has not been used as a carbon source for hexavalent chromium in either suspended or attached growth systems. An attempt to compare the Cr(VI) reduction rates achieved in this study with similar research works indicates that the Cr(VI) reduction rates observed in this study are among the highest reported in the literature compared to those obtained in attached mixed growth systems using other carbon sources, such as sugar (2 mg L$^{-1}$ h$^{-1}$) for initial Cr(VI) of 100 mg L$^{-1}$ [44], glucose (6.95 mg L$^{-1}$ h$^{-1}$ for 15 mg Cr(VI) L$^{-1}$) [45], sodium acetate (13 mg L$^{-1}$ h$^{-1}$ for 10 mg Cr(VI) L$^{-1}$) [46], and molasses (6.6 mg L$^{-1}$ h$^{-1}$ for 10 mg Cr(VI) L$^{-1}$) [47]. The results of various studies have been summarized in a Table in a previous paper (Table 2) by our research group [25], and show that the biological system used in this study is a very effective method of treating Cr(VI) with SCW, as it can achieve very high Cr(VI) reduction rates (up to 35.99 and 43.01 mg $L^{-1}$ $h^{-1}$, using SCW and WE, respectively) even with high initial Cr(VI) concentrations.

Higher recirculation rates of 1.0 L $min^{-1}$ and 2.0 L $min^{-1}$ were also tested. However, when the recirculation rate was increased to 1.0 L $min^{-1}$, the Cr-SCW-filter and Cr-WE-filter (Tables 3/Figure 2b and 4/Figure 3b, respectively) exhibited lower Cr(VI) removal rates as well as incomplete Cr(VI) removal in some combinations of initial Cr(VI) and d-COD concentrations. With these recirculation rates, the Cr-SCW-filter and Cr-WE-filter could fully treat only significantly lower feed concentrations of Cr(VI). Specifically, only the initial d-COD concentration of 13,000 mg $L^{-1}$ was able to safely remove Cr(VI) concentrations up to 100 mg $L^{-1}$, leading to final Cr(VI) concentrations below the maximum permitted limit of 0.05 mg $L^{-1}$ [7,8]. On the contrary, all of the other initial d-COD concentrations (21,000 and 25,000 mg d-COD $L^{-1}$) were able to completely remove Cr(VI) up to 30 mg $L^{-1}$ and 10 mg $L^{-1}$, for Cr-SCW-filter and Cr-WE-filter, respectively.

**Table 3.** Cr(VI) reduction and d-COD removal rates for the different initial Cr(VI) and d-COD concentrations tested in the Cr-SCW-filter operating under SBR with a recirculation rate of 1.0 L $min^{-1}$.

| Initial Cr(VI) (mg $L^{-1}$) | Initial d-COD (mg $L^{-1}$) | Cr(VI) Reduction Rate (mg $L^{-1}$ $h^{-1}$) | Cr(VI) Reduction Rate (g $m^{-2}$ $d^{-1}$) | d-COD Removal (Cr-SCW-Filter) (%) |
|---|---|---|---|---|
| 5.01 ± 0.03 | 987 ± 10.21 | 6.68 ± 0.29 | 151.19 ± 6.56 | 24.52 ± 0.78 |
| 10.04 ± 0.25 | 996 ± 21.99 | 5.02 ± 0.14 | 113.68 ± 3.17 | 26.27 ± 0.99 |
| 5.01 ± 0.06 | 5123 ± 132 | 12.02 ± 0.14 | 272.12 ± 3.17 | 28.06 ± 1.12 |
| 9.79 ± 0.58 | 5058 ± 99 | 16.79 ± 0.31 | 380.24 ± 7.02 | 32.28 ± 1.07 |
| 29.96 ± 0.22 | 4988 ± 138 | 4.13 ± 0.11 | 93.60 ± 2.49 | 36.54 ± 0.93 |
| 72.59 ± 0.66 | 5029 ± 103 | 1.68 ± 0.03 | 38.03 ± 0.68 | 38.35 ± 1.11 |
| 5.05 ± 0.05 | 13120 ± 322 | 20.18 ± 0.34 | 457.08 ± 7.07 | 33.88 ± 1.27 |
| 30.09 ± 0.04 | 12995 ± 205 | 12.04 ± 0.29 | 272.61 ± 6.57 | 43.75 ± 1.38 |
| 60.94 ± 0.44 | 12800 ± 259 | 7.62 ± 0.25 | 172.51 ± 5.66 | 47.34 ± 1.94 |
| 101.69 ± 1.50 | 12715 ± 223 | 4.42 ± 0.09 | 100.13 ± 2.04 | 49.67 ± 2.02 |
| 4.96 ± 0.14 | 20760 ± 823 | 19.85 ± 0.28 | 449.46 ± 6.34 | 34.06 ± 1.14 |
| 20.09 ± 0.29 | 21400 ± 742 | 30.21 ± 0.58 | 684.26 ± 10.87 | 38.08 ± 0.96 |
| 29.94 ± 0.62 | 21200 ± 906 | 19.96 ± 0.31 | 452.00 ± 7.02 | 40.09 ± 1.21 |
| * 60.09 ± 0.12 | 21420 ± 759 | 3.98 ± 0.03 | 90.08 ± 0.68 | 42.44 ± 1.04 |
| * 100.27 ± 1.68 | 21400 ± 1014 | 3.57 ± 0.06 | 80.80 ± 1.36 | 43.78 ± 1.19 |
| 4.92 ± 0.09 | 25488 ± 1017 | 19.69 ± 0.19 | 445.94 ± 4.30 | 16.28 ± 0.34 |
| 30.00 ± 0.06 | 25475 ± 898 | 20.00 ± 0.24 | 452.88 ± 4.43 | 29.32 ± 0.59 |
| * 60.61 ± 0.74 | 24760 ± 988 | 5.04 ± 0.08 | 114.07 ± 1.81 | 31.07 ± 0.81 |
| * 102.41 ± 1.79 | 25123 ± 1003 | 4.26 ± 0.11 | 96.42 ± 2.48 | 32.06 ± 0.74 |

* Concentrations not able to completely remove Cr(VI).

**Table 4.** Cr(VI) reduction and d-COD removal rates for the different initial Cr(VI) and d-COD concentrations tested in the Cr-WE-filter operating under SBR with a recirculation rate of 1.0 L $min^{-1}$.

| Initial Cr(VI) (mg $L^{-1}$) | Initial d-COD (mg $L^{-1}$) | Cr(VI) Reduction Rate (mg $L^{-1}$ $h^{-1}$) | Cr(VI) Reduction Rate (g $m^{-2}$ $d^{-1}$) | d-COD Removal (Cr-WE-Filter) (%) |
|---|---|---|---|---|
| 5.32 ± 0.03 | 1120 ± 19 | 7.09 ± 0.19 | 160.47 ± 4.3 | 6.50 ± 0.12 |
| 10.12 ± 0.18 | 1080 ± 28 | 6.50 ± 0.11 | 147.12 ± 2.48 | 9.50 ± 0.24 |
| 5.27 ± 0.04 | 4980 ± 95 | 16.23 ± 0.32 | 367.36 ± 7.24 | 14.16 ± 0.37 |
| 10.13 ± 0.11 | 5132 ± 156 | 20.76 ± 0.52 | 469.89 ± 11.77 | 17.26 ± 0.51 |
| 30.25 ± 0.22 | 4990 ± 87 | 17.41 ± 0.59 | 394.06 ± 6.56 | 19.05 ± 0.43 |
| 68.96 ± 0.51 | 5110 ± 190 | 6.90 ± 0.19 | 156.17 ± 4.3 | 21.52 ± 0.57 |
| 5.15 ± 0.02 | 13030 ± 197 | 26.40 ± 0.55 | 597.54 ± 12.45 | 15.44 ± 0.38 |
| 30.15 ± 0.31 | 13100 ± 209 | 23.71 ± 0.41 | 536.66 ± 9.28 | 23.55 ± 0.58 |
| 59.84 ± 0.47 | 12900 ± 274 | 9.62 ± 0.21 | 341.77 ± 7.46 | 24.44 ± 0.67 |
| 100.09 ± 1.21 | 12855 ± 328 | 10.20 ± 0.19 | 230.87 ± 4.3 | 29.67 ± 0.76 |
| 5.06 ± 0.04 | 20520 ± 487 | 5.32 ± 0.09 | 120.41 ± 2.04 | 8.55 ± 0.23 |
| 10.22 ± 0.31 | 21100 ± 854 | 5.89 ± 0.12 | 133.31 ± 2.72 | 9.87 ± 0.19 |
| * 21.19 ± 0.38 | 21150 ± 902 | 14.83 ± 0.34 | 335.66 ± 7.70 | 12.81 ± 0.47 |
| * 30.04 ± 0.29 | 20800 ± 855 | 10.60 ± 0.22 | 239.92 ± 4.98 | 13.09 ± 0.32 |
| * 61.19 ± 0.61 | 20950 ± 932 | 2.56 ± 0.03 | 57.94 ± 0.68 | 14.49 ± 0.42 |
| * 101.17 ± 1.32 | 21100 ± 844 | 1.53 ± 0.02 | 34.63 ± 0.45 | 21.78 ± 0.62 |
| 5.13 ± 0.03 | 25400 ± 1009 | 3.22 ± 0.06 | 72.88 ± 1.35 | 5.32 ± 0.07 |
| 10.32 ± 0.27 | 24900 ± 954 | 3.54 ± 0.05 | 80.13 ± 1.13 | 8.05 ± 0.17 |
| * 31.55 ± 0.37 | 25150 ± 784 | 1.71 ± 0.04 | 15.66 ± 0.09 | 12.02 ± 0.27 |
| * 61.01 ± 0.54 | 25600 ± 1109 | 1.52 ± 0.02 | 34.40 ± 0.45 | 12.88 ± 0.35 |
| * 100.55 ± 0.98 | 24980 ± 1005 | 1.26 ± 0.01 | 28.51 ± 0.22 | 15.09 ± 0.37 |

* Concentrations not able to completely remove Cr(VI).

Additionally, for all initial d-COD concentrations tested, Cr(VI) reduction rates dropped significantly when the initial Cr(VI) concentrations exceeded the value of 30 mg $L^{-1}$ (Table 3). This was due to the high flow velocities that were exerted across the filter with the recirculation rate of 1.0 L $min^{-1}$ that resulted in biomass detachment or in environmental conditions with less effective removal of Cr(VI). The maximum Cr(VI) removal rate achieved for the recirculation rate of 1.0 L $min^{-1}$ in the Cr-SCW-filter was 30.21 mg $L^{-1}$ $h^{-1}$ for initial Cr(VI) and d-COD concentrations of 20 and 21,000 mg $L^{-1}$, respectively (Figure 2b and Table 3), while the maximum rate for the Cr-WE-filter was 26.4 mg $L^{-1}$ $h^{-1}$ for initial Cr(VI) and d-COD concentrations of 5 and 13,000 mg $L^{-1}$, respectively (Figure 3b and Table 4). It must be mentioned that for both the recirculation rates of 0.5 and 1.0 L $min^{-1}$, the feed SCW concentration of d-COD 1000 mg $L^{-1}$ was inadequate (for Cr(VI) above 10 mg $L^{-1}$), and caused microbe growth limitation by carbon throughout the process.

For volumetric flow rate of 1.0 L $min^{-1}$, the initial d-COD concentration of SCW was found to have significant effects on Cr(VI) removal for many of the paired data of each initial Cr(VI) concentration tested. For each initial Cr(VI) concentration, the pairs in which significant differences were observed are: 5 mg Cr(VI) $L^{-1}$: (1000–5000; 5000–13,000), 10 mg Cr(VI) $L^{-1}$: (1000–5000; 5000–13,000; 13,000–21,000), 30 mg Cr(VI) $L^{-1}$: (5000–13,000; 13,000–21,000; 21,000–25,000), 60 mg Cr(VI) $L^{-1}$: (5000–13,000; 13,000–21,000; 21,000–25,000), and 100 mg Cr(VI) $L^{-1}$: (13,000–21,000; 21,000–25,000).

For WE, the initial d-COD concentration was also found to have significant effects on Cr(VI) removal for many of the paired data of each initial Cr(VI) concentration tested. For each initial Cr(VI), concentration the pairs in which significant differences were observed are: 5 mg Cr(VI) $L^{-1}$: (1000–5000; 5000–13,000; 13,000–21,000; 21,000–25,000), 10 mg Cr(VI) $L^{-1}$: (1000–5000; 21,000–25,000), 30 mg Cr(VI) $L^{-1}$: (5000–13,000; 13,000–21,000; 21,000–25,000), 60 mg Cr(VI) $L^{-1}$: (5000–13,000; 13,000–21,000; 21,000–25,000, and 100 mg Cr(VI) $L^{-1}$: (13,000–21,000; 21,000–25,000).

Finally, with a recirculation rate of 2 L $min^{-1}$, the filter was not able to operate sufficiently (for both type of wastewaters used, SCW and WE) due to the detachment of biofilm from the support material. Experiments under continuous operating mode with and without recirculation were also performed. Various Cr(VI) and d-COD feed concentrations were examined with different recirculation rates, however, very low Cr(VI) bioreduction rates were achieved in all the cases (data not shown). Similar observations have also been reported in the literature where the use of continuous biological methods for Cr(VI) treatment presented difficulties due to the eventual loss of active biomass [25,48].

Taking all of the above results into account, it is concluded that SBR with a recirculation rate of 0.5 L $min^{-1}$ proved to be a very effective operating mode since it ensured high Cr(VI) removal rates (35.99 and 43.01 mg $L^{-1}$ $h^{-1}$, for SCW and WE, respectively). Despite the high removal levels of Cr(VI) achieved in this work, the final effluent was still not suitable for discharge since the residual d-COD concentration of the treated wastewater was very high (Figure 4). For the recirculation rate of 0.5 L $min^{-1}$, the d-COD percentage removal ranged between 14 and 42.5% for SCW, and from 4 to 29% for WE, depending on initial Cr(VI) and d-COD feed concentrations (Figures 4a/Table 1 and 5a/Table 2, respectively). This can be explained as the process of Cr(VI) reduction took place over a very short time period (from 0.6 to 25 h and from 0.4 to 23 h for SCW and WE, respectively), thus not allowing for the further removal of d-COD. Contreras et al. have reported higher d-COD percentage removal (in suspended growth systems), but over longer periods of time (up to 100 h) [28,29].

For the recirculation rates of 0.5 and 1.0 L $min^{-1}$, it is noteworthy that the percentage of d-COD removal slightly increased with increasing initial Cr(VI) concentration (Figures 4 and 5). This is rather expected, as high Cr(VI) concentrations require greater amounts of carbon source. It was also observed that the percentages of d-COD removal increased to the maximum when the initial d-COD concentration was 13,000 mg $L^{-1}$. For the concentration of 13,000 mg d-COD $L^{-1}$, the microbial community structure and function appeared to tolerant to both metal and organic load contaminants [41].

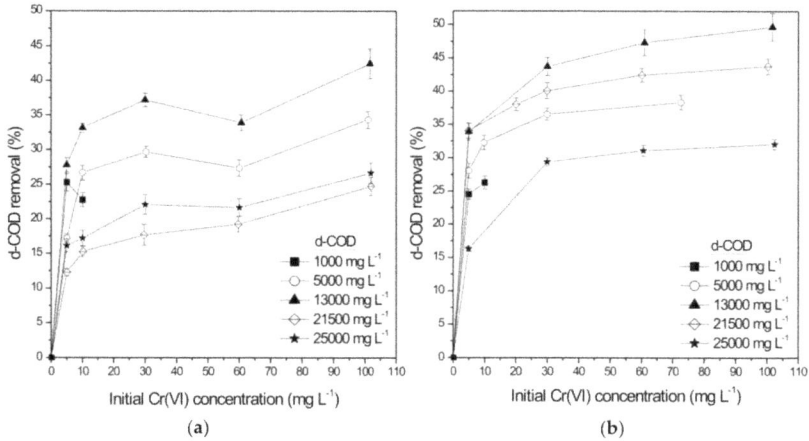

**Figure 4.** Percentage removal of d-COD in the Cr-SCW-filter for two different recirculation rates: (**a**) 0.5 L min$^{-1}$ and (**b**) 1.0 L min$^{-1}$.

For SCW and volumetric flow rate of 0.5 L min$^{-1}$, the following pairs were found to have significant differences for each initial d-COD concentration: 5000 mg d-COD L$^{-1}$: (5–10), 13,000 mg d–COD L$^{-1}$: (5–10), and 21,000 mg d-COD L$^{-1}$: (5–10; 10–30; 60–100). For the same wastewater and a volumetric flow rate of 1.0 L min$^{-1}$, significant differences were found only for the initial d-COD concentration of 13,000 mg d-COD L$^{-1}$: (5–10; 10–30). For WE and volumetric flow rate of 0.5 L min$^{-1}$, significant differences were found in the following pairs for each initial d-COD concentration: 1000 mg d-COD L$^{-1}$: (5–10), 5000 mg d-COD L$^{-1}$: (5–10; 30–60), 13,000 mg d-COD L$^{-1}$: (30–60), while for 1.0 L min$^{-1}$ volumetric flow rate, the pairs: 1000 mg d-COD L$^{-1}$: (5–10), 13,000 mg d-COD L$^{-1}$: (5–30; 60–100) and 21,000 mg d-COD L$^{-1}$: (60–100) had significant differences.

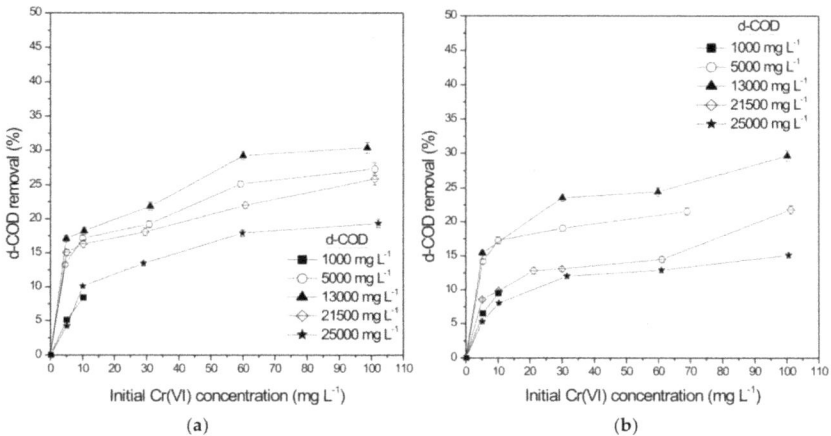

**Figure 5.** Percentage removal of d-COD in the Cr-WE-filter for two different recirculation rates: (**a**) 0.5 L min$^{-1}$ and (**b**) 1.0 L min$^{-1}$.

Figure 6a,b and Figure 7a,b help determine the best operating conditions for the highest viable Cr(VI) and d-COD removal rates in the Cr-SCW-filter and Cr-WE-filter, respectively. The Figure 6a,b and Figure 7a,b show that with low d-COD/Cr(VI) ratios, higher removal rates of both Cr(VI) and d-COD are achieved.

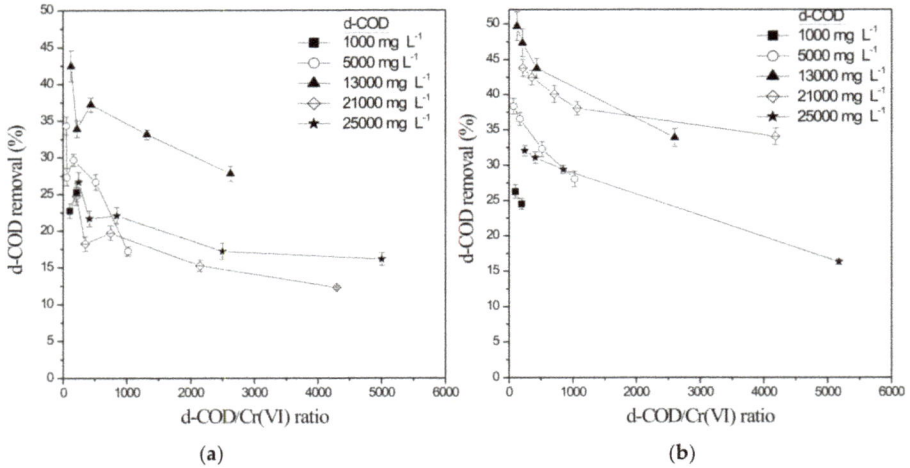

(a)

(b)

**Figure 6.** Effect of (initial d-COD)/(initial Cr(VI)) ratio on d-COD removal in the Cr-SCW-filter for various initial d-COD concentrations under SBR mode with recirculation rates of: (**a**) 0.5 L min$^{-1}$ and (**b**) 1.0 L min$^{-1}$.

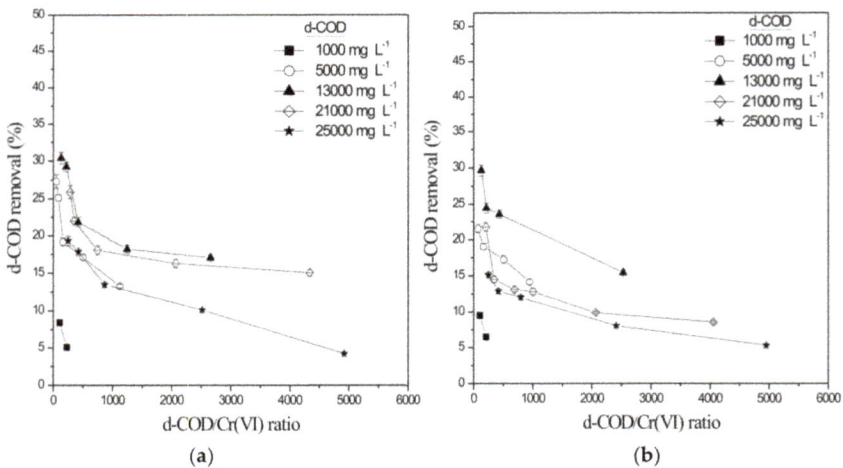

(a)

(b)

**Figure 7.** Effect of (initial d-COD)/(initial Cr(VI)) ratio on d-COD removal in the Cr-WE-filter for various initial d-COD concentrations under SBR mode with recirculation rates of: (**a**) 0.5 L min$^{-1}$ and (**b**) 1.0 L min$^{-1}$.

### 3.2. Packed Bed Reactor Experiments in SCW-Filter and WE-Filter

In the previous section, the successful removal of Cr(VI) in attached growth systems was demonstrated with the use of SCW and WE. This is the first time that SCW and WE have been used as a carbon source for Cr(VI) reduction in biofilters (trickling filters). However, the residual organic load in the Cr-filters' effluent was found to be rather high, and is therefore requiring further treatment before being discharged into receiving waters.

According to Greek legislation, the permissible limit of d-COD for municipal and industrial effluents is 125 mg L$^{-1}$ [49]. Therefore, a second filter (SCW-filter, WE-filter) was constructed, added to the system in series, and then operated as a post-biotreatment step using indigenous microorganisms originating from SCW or WE, respectively (Sections 2.2 and 2.3). The SCW-filter or WE-filter was loaded with the effluent of the Cr-SCW-filter or Cr-WE-filter, respectively (without the presence of Cr(VI)) and operated only under SBR with a recirculation rate of 0.5 L min$^{-1}$.

Kinetic experiments were performed in the SCW-filter or WE-filter, under SBR mode and with recirculation of 0.5 L min$^{-1}$, only for Cr-filter effluents that presented the highest Cr(VI) reduction rate for each initial d-COD concentration tested (5000, 13,000, 21,000, and 25,000 L min$^{-1}$). Tables 1 and 2 show that the total percentage d-COD removal (from both Cr-SCW and SCW-filters or Cr-WE and WE-filters) exceeds the value of 97.57% or 90.5% for SCW and WE, respectively, which is very high for aerobic biological systems. However, it must be pointed out that with high percentage removal rates for the high initial d-COD concentration of 25,000 L min$^{-1}$, the concentration of SCW-filter effluent was 325 mg L$^{-1}$, which was not below the maximum permitted limit of 125 mg L$^{-1}$ (Figure 8a). Additionally, for all of initial d-COD concentrations that was examined in the WE-filter ((5000, 13,000, 21,000, and 25,000 L min$^{-1}$), the concentration of WE-filter effluent was always above the maximum permitted limit of 125 mg L$^{-1}$ (445, 1028, 1753, and 2403 mg L$^{-1}$, respectively) (Figure 8b). Therefore, in these cases a suitable post-treatment step (e.g., a constructed wetland) should be applied to improve the quality of the final outflow [35]. Future research should include experiments in this direction in order to achieve an integrated wastewater treatment system even for high initial d-COD concentrations. SCW treatment has also been examined by our research group in a previous work using pilot-scale horizontal subsurface flow constructed wetlands [50]. Constructed wetlands could successfully treat SCW and provide COD effluent concentrations below the EU legislation limits when hydraulic residence time is greater than two days and COD influent concentration ranges from 1200 to 3500 mg L$^{-1}$.

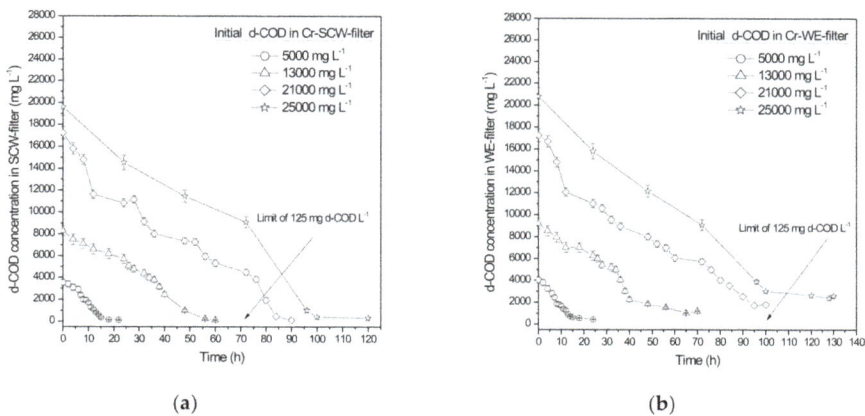

(a)                                                        (b)

**Figure 8.** Percentage removal of d-COD in the: (**a**) SCW-filter or (**b**) WE-filter for various initial d-COD concentrations (depending on the effluent of the Cr-SCE-filter or Cr-WE-filter, respectively).

Experiments on SCW treatment using trickling filters have also been performed by our research team in a previous study without the presence of Cr(VI) [34]. Comparison of the results of that study with the results of the present work (in SCW-filter) indicates that the process of Cr(VI) removal in the

Cr-filter negatively affects the bio-treatment of SCW. Specifically, in the case of SCW treatment following Cr(VI) removal, the duration of the operating cycles increased to 120 h for initial concentrations of about 21,000 mg L$^{-1}$, whereas in the case of SCW treatment without Cr(VI), the duration was significantly shorter (26 h) [34]. This may be attributed to the different biomass communities developing within the biofilters, as well as to the formation of compounds that prevent organic load degradation during the Cr(VI) reduction process.

## 4. Conclusions

The effect of SCW or WE on Cr(VI) reduction was examined in an attached growth system (Cr-SCW-biofilter or Cr-WE-biofilter) under various operating conditions. The post-treatment of the treated wastewater was also examined in a second biofilter that was connected in series (SCW-biofilter or WE-biofilter) to further reduce the organic load. The main conclusions derived from this research work are:

- High percentage biological Cr(VI) reduction can be achieved in an attached growth reactor (99.2–100%) by using an indigenous mixed population and SCW or WE (both very low-cost carbon sources) as the sole electron donor.
- Complete Cr(VI) reduction can be achieved in attached growth reactors operated in batch operation with recirculation of 0.5 L min$^{-1}$ for all initial Cr(VI) (5–110 mg L$^{-1}$) and d-COD (1000–25,000 mg L$^{-1}$) concentrations tested for both agro-industrial effluents (SCW or WE). The reduction rates that are accomplished (35.99 and 43.0 mg L$^{-1}$ h$^{-1}$ for SCW and WE, respectively) are the highest reported in the literature to date. With higher recirculation rates (1.0 L min$^{-1}$) the Cr-SCW-filter or Cr-WE-filter were unable to achieve complete Cr(VI) reduction for initial Cr(VI) concentrations above 30 or 10 mg L$^{-1}$, respectively, while for 2.0 L min$^{-1}$, detachment of biofilm led to inadequate operation of the filter. Continuous operating mode with or without recirculation resulted in very low Cr(VI) bioreduction rates.
- Winery effluents presented slightly higher Cr(VI) reduction rates for initial d-COD concentrations up to 13,000 mg L$^{-1}$. The same was not observed for higher initial concentrations probably due to the presence of higher quantities of phenolic compounds that are associated with microbial toxicity.
- Initial d-COD concentration was found to effect Cr(VI) reduction rate. The feed SCW or WE concentration of 1000 mg d-COD L$^{-1}$ was limiting and caused microbial growth limitation by carbon during the process.
- Due to the high residual d-COD concentration of the treated wastewater, a post-treatment stage was required. The use of mixed indigenous microorganisms originating from SCW or WE provides high degradation rates (total d-COD removal above 97% and 90.5% for SCW and WE, respectively) and durability under various operating conditions. In cases where final d-COD concentrations of the second biofilters are still above the maximum permitted limit of 125 mg L$^{-1}$, a suitable post-treatment step (e.g., a constructed wetland) should be applied to improve the quality of the final outflow.

The combined treatment method presented here uses two biological treatment steps (biofilters) and proved to be very effective as high reduction rates of Cr(VI) and high d-COD removal rates were achieved. These results combined with the low construction and operating costs, suggest the feasibility of using these systems for the biological co-treatment of Cr(VI) (in high concentrations in the range of mg L$^{-1}$, present mainly in industrial wastewaters and not in contaminated groundwaters that usually contain much lower concentrations, in the range of few μg L$^{-1}$) and SCW or WE.

**Author Contributions:** D.V.V., A.G.T. and C.S.A. conceive and design the experiments. T.I.T. and M.K.M. carried out the experiments. A.G.T., C.S.A., T.I.T. and S.P. analyzed the data. All authors discussed the results. A.G.T. wrote the manuscript with support from S.P.

**Conflicts of Interest:** The authors declare no conflict of interest.

## References

1.  Pradhan, D.; Sukla, L.B.; Sawyer, M.; Rahman, P.K.S.M. Recent bioreduction of hexavalent chromium in wastewater treatment: A review. *J. Ind. Eng. Chem.* **2017**, *55*, 1–20. [CrossRef]
2.  Gröhlich, A.; Langer, M.; Mitrakas, M.; Zouboulis, A.; Katsoyiannis, I.; Ernst, M. Effect of Organic Matter on Cr(VI) Removal from Groundwaters by Fe(II) Reductive Precipitation for Groundwater Treatment. *Water* **2017**, *9*, 389. [CrossRef]
3.  Costa, M. Potential hazards of hexavalent chromate in our drinking water. *Toxicol. Appl. Pharmacol.* **2003**, *188*, 1–5. [CrossRef]
4.  Stylianou, S.; Simeonidis, K.; Mitrakas, M.; Zouboulis, A.; Ernst, M.; Katsoyiannis, I.A. Reductive precipitation and removal of Cr(VI) from groundwaters by pipe flocculation-microfiltration. *Environ. Sci. Pollut. Res.* **2017**, 1–7, in press. [CrossRef] [PubMed]
5.  Hossini, H.; Makhdoumi, P.; Mohammadi-Moghadam, F.; Ghaffari, H.R.; Mirzaei, N.; Ahmadpour, M. A review of toxicological, environmental and health effects of chromium from aqueous medium; available removal techniques. *Acta Medica Mediterr.* **2016**, *32*, 1463–1469.
6.  DeFlora, S.; Bagnasco, M.; Serra, D.; Zanacchi, P. Genotoxicity of chromium compounds: A review. *Mutat. Res.* **1990**, *238*, 99–172. [CrossRef]
7.  Anonymous. Directive on the Quality of Water Intended for Human Consumption. In *EC-Official Journal of the European Communities*; 12 December 1998, L330/32; Commission of the European Communities: Brussels, Belgium.
8.  EPA (United States Environmental Protection Agency). *Parameters of Water Quality, Interpretation and Standards*; United States Environmental Protection Agency: Washington, DC, USA, 2001.
9.  Owlad, M.; Aroua, M.; Daud, W.; Baroutian, S. Removal of Hexavalent Chromium-Contaminated Water and Wastewater: A Review. *Water Air Soil Pollut.* **2008**, *200*, 59–77. [CrossRef]
10. Hawley, E.L.; Deeb, R.; Kavanaugh, M.C.; Jacobs, J.R.G. *8-Treatment Technologies for Chromium(VI)*; Chemical Rubber Company CRC Press LLC: Boca Raton, FL, USA, 2005; p. 273.
11. Kaprara, E.; Simeonidis, K.; Zouboulis, A.; Mitrakas, M. Evaluation of current treatment technologies for Cr(VI) removal from water sources at sub-ppb levels. In Proceedings of the 13th International Conference on Environmental Science and Technology, Athens, Greece, 5–7 September 2013.
12. Najm, I.; Brown, N.P.; Blute, N.; Kader, S. Impact of Water Quality on Cr(VI) treatment efficiency and cost results of WaterRF project 4450. In Proceedings of the ACE 2013–2013 AWWA Annual Conference and Exposition, Denver, CO, USA, 9–13 June 2013. Code 101497.
13. Joutey, N.T.; Sayel, H.; Bahafid, W.; El Ghachtouli, N. Mechanisms of hexavalent chromium resistance and removal by microorganisms. *Rev. Environ. Contam. Toxicol.* **2015**, *233*, 45–69. [CrossRef] [PubMed]
14. Wu, X.; Zhu, X.; Song, T.; Zhang, L.; Jia, H.; Wei, P. Effect of acclimatization on hexavalent chromium reduction in a biocathode microbial fuel cell. *Bioresour. Technol.* **2015**, *180*, 185–191. [CrossRef] [PubMed]
15. Molokwane, P.E.; Meli, K.C.; Nkhalambayausi-Chirwa, E.M. Chromium (VI) reduction in activated sludge bacteria exposed to high chromium loading: Brits culture (South Africa). *Water Res.* **2008**, *42*, 4538–4548. [CrossRef] [PubMed]
16. Panousi, E.; Mamais, D.; Noutsopoulos, C.; Mpertoli, K.; Kantzavelou, C.; Nyktari, E.; Kavallari, I.; Nasioka, M.; Kaldis, A. Biological groundwater treatment for hexavalent chromium removal at low chromium concentrations under anoxic conditions. *Environ. Technol. (UK)* **2017**, 1–9. [CrossRef] [PubMed]
17. Panousi, E.; Mamais, D.; Noutsopoulos, C.; Antoniou, K.; Koutoula, K.; Mastrantoni, S.; Koutsogiannis, C.; Gkioni, A. Biological treatment of groundwater with a high hexavalent chromium content under anaerobic and anoxic conditions. *J. Chem. Technol. Biot.* **2016**, *91*, 1681–1687. [CrossRef]
18. Carlos, F.S.; Giovanella, P.; Bavaresco, J.; Borges, C.S.; Camargo, F.A.O. A Comparison of Microbial Bioaugmentation and Biostimulation for Hexavalent Chromium Removal from Wastewater. *Water Air Soil Pollut.* **2016**, *227*, 175. [CrossRef]
19. Singh, R.; Kumar, A.; Kirrolia, A.; Kumar, R.; Yadav, N.; Bishnoi, N.R.; Lohchab, R.K. Removal of sulphate, COD and Cr(VI) in simulated and real wastewater by sulphate reducing bacteria enrichment in small bioreactor and FTIR study. *Bioresour. Technol.* **2011**, *102*, 677–682. [CrossRef] [PubMed]

20. Tekerlekopoulou, A.; Tsiamis, G.; Dermou, E.; Siozios, S.; Bourtzis, K.; Vayenas, D.V. The effect of carbon source on microbial community structure and Cr(VI) reduction rate. *Biotechnol. Bioeng.* **2010**, *107*, 478–487. [CrossRef] [PubMed]

21. Smith, W.A.; Apel, W.A.; Petersen, J.N.; Peyton, B.M. Effect of carbon and energy source on bacterial chromate reduction. *Bioremediat. J.* **2002**, *6*, 205–215. [CrossRef]

22. Rynk, R. Bioremediation with cheese whey. *BioCycle* **2004**, *45*, 26–28.

23. Elangovan, R.; Philip, L. Performance evaluation of various bioreactors for the removal of Cr(VI) and organic matter from industrial effluent. *Biochem. Eng. J.* **2009**, *44*, 174–186. [CrossRef]

24. Chen, Z.F.; Zhao, Y.S.; Zhang, J.W.; Bai, J. Mechanism and Kinetics of Hexavalent Chromium Chemical Reduction with Sugarcane Molasses. *Water Air Soil Pollut.* **2015**, *226*, 363. [CrossRef]

25. Michailides, M.K.; Tekerlekopoulou, A.G.; Akratos, C.S.; Coles, S.; Pavlou, S.; Vayenas, D.V. Molasses as an efficient low-cost carbon source for biological Cr(VI) removal. *J. Hazard. Mater.* **2015**, *281*, 95–105. [CrossRef] [PubMed]

26. Zakaria, Z.A.; Ahmad, W.A.; Zakaria, Z.; Razali, F.; Karim, N.A.; Sum, M.M.; Sidek, M.S.M. Bacterial reduction of Cr(VI) at technical scale - The Malaysian experience. *Appl. Biochem. Biotechnol.* **2012**, *167*, 1641–1652. [CrossRef] [PubMed]

27. Orozco, A.M.F.; Contreras, E.M.; Zaritzky, N.E. Cr(Vi) reduction capacity of activated sludge as affected by nitrogen and carbon sources, microbial acclimation and cell multiplication. *J. Hazard. Mater.* **2010**, *176*, 657–665. [CrossRef] [PubMed]

28. Contreras, E.M.; Orozco, A.M.F.; Zaritzky, N.E. Factors affecting the biological removal of hexavalent chromium using activated sludges. In *Management of Hazardous Residues Containing Cr(VI)*; Balart Murria, M.J., Ed.; Nova Science Publishers: New York, NY, USA, 2011; pp. 109–134.

29. Contreras, E.M.; Ferro Orozco, A.M.; Zaritzky, N.E. Biological Cr(VI) removal coupled with biomass growth, biomass decay, and multiple substrate limitation. *Water Res.* **2011**, *45*, 3034–3046. [CrossRef] [PubMed]

30. Lattanzi, P.; Aquilanti, G.; Bardelli, F.; Iadecola, A.; Rosellini, I.; Tassi, E.; Pezzarossa, B.; Petruzzelli, G. Spectroscopic evidence of Cr(VI) reduction in a contaminated soil by in situ treatment with whey. *Agrochimica* **2015**, *59*, 218–230. [CrossRef]

31. Němeček, J.; Pokorný, P.; Lacinová, L.; Černík, M.; Masopustová, Z.; Lhotský, O.; Filipová, A.; Cajthaml, T. Combined abiotic and biotic in-situ reduction of hexavalent chromium in groundwater using nZVI and whey: A remedial pilot test. *J. Hazard. Mater.* **2015**, *300*, 670–679. [CrossRef] [PubMed]

32. Němeček, J.; Pokorný, P.; Lhotský, O.; Knytl, V.; Najmanová, P.; Steinová, J.; Černík, M.; Filipová, A.; Filip, J.; Cajthaml, T. Combined nano-biotechnology for in-situ remediation of mixed contamination of groundwater by hexavalent chromium and chlorinated solvents. *Sci. Total Environ.* **2015**, *563–564*, 822–834. [CrossRef] [PubMed]

33. Mamais, D.; Noutsopoulos, C.; Kavallari, I.; Nyktari, E.; Kaldis, A.; Panousi, E.; Nikitopoulos, G.; Antoniou, K.; Nasioka, M. Biological groundwater treatment for chromium removal at low hexavalent chromium concentrations. *Chemosphere* **2016**, *152*, 238–244. [CrossRef] [PubMed]

34. Tatoulis, T.I.; Tekerlekopoulou, A.G.; Akratos, C.S.; Pavlou, S.; Vayenas, D.V. Aerobic biological treatment of second cheese whey in suspended and attached growth reactors. *J. Chem. Technol. Biot.* **2015**, *90*, 2040–2049. [CrossRef]

35. Sultana, M.-Y.; Tatoulis, T.I.; Akratos, C.S.; Tekerlekopoulou, A.G.; Vayenas, D.V. Effect of operational parameters on the performance of a horizontal subsurface flow constructed wetland treating secondary cheese whey and Cr(VI) wastewater. *Int. J. Civ. Struct. Eng.* **2015**, *2*, 286–289. [CrossRef]

36. Kyzas, G.Z.; Symeonidou, M.P.; Matis, K.A. Technologies of winery wastewater treatment: A critical approach. *Desalin. Water Treat.* **2016**, *57*, 3372–3386. [CrossRef]

37. Mosteo, R.; Sarasa, J.; Ormad, M.P.; Ovelleiro, J.L. Sequential solar photo-fenton-biological system for the treatment of winery wastewaters. *J. Agric. Food Chem.* **2008**, *56*, 7333–7338. [CrossRef] [PubMed]

38. Andreottola, G.; Foladori, P.; Ziglio, G. Biological treatment of winery wastewater: An overview. *Water Sci. Technol.* **2009**, *60*, 1117–1125. [CrossRef] [PubMed]

39. APHA (American Public Health Association); AWWA (American Water Works Association); WPCF (Water Pollution Control Federation). *Standard Methods for the Examination of Water and Wastewater*, 17th ed.; American Public Health Association, American Water Works Association and Water Pollution Control Federation: Washington, DC, USA, 1989; ISBN 087553161X.

40. Waterman, P.G.; Mole, S. Analysis of phenolic plant metabolites. In *Methods in Ecology*; Lawton, J.H., Likens, G.E., Eds.; Blackwell Scientific: Oxford, UK, 1994; pp. 83–85. ISBN 0632029692.
41. Nakatsu, C.H.; Carmosini, N.; Baldwin, B.; Beasley, F.; Kourtev, P.; Konopka, A. Soil Microbial Community Responses to Additions of Organic Carbon Substrates and Heavy Metals (Pb and Cr). *Appl. Environ. Microbiol.* **2005**, *71*, 7679–7689. [CrossRef] [PubMed]
42. Welz, P.J.; Holtman, G.; Haldenwang, R.; le Roes-Hill, M. Characterisation of winery wastewater from continuous flow settling basins and waste stabilisation ponds over the course of 1 year: Implications for biological wastewater treatment and land application. *Water Sci. Technol.* **2016**, *74*, 2036–2050. [CrossRef] [PubMed]
43. Ramond, J.B.; Welz, P.J.; Tuffin, M.I.; Burton, S.G.; Cowan, D.A. Assessment of temporal and spatial evolution of bacterial communities in a biological sand filter mesocosm treating winery wastewater. *J. Appl. Microbiol.* **2013**, *115*, 91–101. [CrossRef] [PubMed]
44. Tekerlekopoulou, A.G.; Tsiflikiotou, M.; Akritidou, L.; Viennas, A.; Tsiamis, G.; Pavlou, S.; Bourtzis, K.; Vayenas, D.V. Modelling of biological Cr(VI) removal in draw-fill reactors using microorganisms in suspended and attached growth systems. *Water Res.* **2013**, *47*, 623–636. [CrossRef] [PubMed]
45. Zakaria, Z.A.; Zakaria, Z.; Surif, S.; Ahmad, W.A. Biological detoxification of Cr(VI) using wood-husk immobilized Acinetobacter haemolyticus. *J. Hazard. Mater.* **2007**, *148*, 164–171. [CrossRef] [PubMed]
46. Dermou, E.; Velissariou, A.; Xenos, D.; Vayenas, D.V. Biological chromium(VI) reduction using a trickling filter. *J. Hazard. Mater.* **2005**, *126*, 78–85. [CrossRef] [PubMed]
47. Krishna, K.R.; Philip, L. Bioremediation of Cr(VI) in contaminated soils. *J. Hazard. Mater.* **2005**, *121*, 109–117. [CrossRef] [PubMed]
48. Chirwa, N.E.M.; Wang, Y.T. Modeling hexavalent chromium removal in a Bacillus sp. fixed-film bioreactor. *Biotechnol. Bioeng.* **2004**, *87*, 874–883. [CrossRef] [PubMed]
49. Anonymous. Establishment of measures, conditions and procedures for the reuse of treated wastewater and other provisions. *Gazette of the Government (GR)*, 8 March 2011; 2011/354B.
50. Sultana, M.-Y.; Mourti, C.; Tatoulis, T.; Akratos, C.S.; Tekerlekopoulou, A.G.; Vayenas, D.V. Effect of hydraulic retention time, temperature, and organic load on a horizontal subsurface flow constructed wetland treating cheese whey wastewater. *J. Chem. Technol. Biotechnol.* **2016**, *91*, 726–732. [CrossRef]

MDPI

St. Alban-Anlage 66

4052 Basel

Switzerland

Tel. +41 61 683 77 34

Fax +41 61 302 89 18

www.mdpi.com

*Water* Editorial Office

E-mail: water@mdpi.com

www.mdpi.com/journal/water

www.ingramcontent.com/pod-product-compliance
Lightning Source LLC
Chambersburg PA
CBHW051839210326
41597CB00033B/5712